Guide to Cytochromes P450

Structure and Function

David F. V. Lewis

Taylor & Francis
Founded 1798
London and New York

First published 2001
by Taylor & Francis
11 New Fetter Lane, London EC4P 4EE

Simultaneously published in the USA and Canada
by Taylor & Francis Inc.
29 West 35th Street, New York, NY 10001

Taylor & Francis is an imprint of the Taylor & Francis Group

Typeset in Times by
HWA Text and Data Management, Tunbridge Wells
Printed and bound in Great Britain by
Biddles Ltd, Guildford and King's Lynn

British Library Cataloguing in Publication Data
A catalogue record for this book is available from the British Library

Library of Congress Cataloging in Publication Data
Lewis, D.F.V.
Guide to cyctochromes P450 structure and function / David F.V. Lewis
p. cm.
Includes bibliographical references and index.
1. Cytochrome P-450.

QP671.C83 L49 2001
572´.791–dc21 2001023962

ISBN 0 748 40897 5

Guide to Cytochromes P450

This book is dedicated to my darling wife, Hanne

Contents

Colour plates between pages 152 and 153
List of tables ix
List of figures xii
Acknowledgements xiv
Abbreviations xv
Chemical abbreviations xvii
Symbols xviii
Preface xix
Foreword xxi

1 Introduction **1**

1.1 Discovery of an enzyme superfamily 1
1.2 Occurrence and distribution 4
1.3 Isolation and purification 9
1.4 Characterization and detection 9
1.5 Genetic polymorphisms and allelic variants: protein sequence alignments 15
1.6 Biotechnological applications 16
1.7 Nomenclature note: cytochromes, peroxidases and P450 17

2 Evolution of the P450 superfamily **20**

2.1 Introduction 20
2.2 P450 phylogeny 22
2.3 P450 sequences 31
2.4 Nomenclature of P450 genes and enzymes 38

3 The P450 catalytic cycle **51**

3.1 Introduction 51

3.2 Stages in the reaction 54
3.3 Thermodynamics of the P450 reaction, rates of electron transfer and kinetics 60
3.4 Coupling of spin and redox equilibria 62
3.5 Factors influencing the rate of substrate turnover 65
3.6 Appendix – Factors influencing rates and clearance in P450 reactions: QSARs 70

4 Substrate selectivity and metabolism 76

4.1 Introduction 76
4.2 Endogenous substrates 76
4.3 Exogenous substrates 85
4.4 Criteria governing substrate selectivity for human hepatic P450s 102
4.5 Toxicity mediated by P450 enzymes 109
4.6 Quantitative structure–activity relationships (QSARs) in P450 substrates 112

5 Regulation of P450 enzymes 118

5.1 Introduction 118
5.2 Mechanisms of P450 regulation and gene expression 123
5.3 Toxic consequences of induction: predicting toxicity via the COMPACT method 126
5.4 QSARs in P450 inducers and nuclear receptor ligand binding 129
5.5 Inhibition of P450 enzymes 132
5.6 QSARs for P450 inhibitors 138

6 The structures of cytochromes P450 140

6.1 Introduction 140
6.2 X-ray crystallography 141
6.3 Spectroscopy 145
6.4 Site-directed mutagenesis and residue modification 147
6.5 Molecular modelling studies 150
6.6 Molecular orbital calculations 158
6.7 Conclusions 159

Glossary of terms 160
Bibliography 162
References 164
Index 211

Tables

1.1	Structural roles of P450 enzyme apoprotein	5
1.2	Comparison between the I helices of several P450s	6
1.3	Human P450s involved in drug metabolism	7
1.4	Some characteristics of mammalian P450s	10
1.5	Selected type I substrates	13
1.6	Selected type II substrates	13
1.7	Selected reverse type I substrates	13
1.8	UV spectral characteristics of P450	14
1.9	P450 content in human liver: a comparison between two methods	14
1.10	The P450 signature motif in bacterial forms	18
1.11	Summary of P450 superfamily nomenclature	19
2.1	Plant chemicals and drug-metabolizing P450s	30
2.2	P450 sequences and characteristics	31
2.3	Amino acids: their properties and abbreviations	34
2.4	Signature motifs of 18 prokaryotic P450 sequences	35
2.5	Putative I helical segments (distal to the heme) of various P450s	36
2.6	Chromosomal locations of human P450 genes and regulatory factors	40
2.7	The CYP1 family	41
2.8	The CYP2 family	41
2.9	The CYP3 family	42
2.10	The CYP4 family	43
2.11	Families CYP5, CYP6, CYP7, CYP8, CYP9 and CYP10	44
2.12	Steroidogenic and related families CYP11, CYP17,CYP19, CYP24, CYP26 and CYP27	45
2.13	Fungal P450 families CYP51 to CYP66	46
2.14	Plant P450 families CYP71–CYP79	47
2.15	Plant P450 families CYP80–CYP99	48
2.16	Bacterial CYP families CYP101–CYP120	48
2.17	Nematode P450s from *Caenorhabditis elegans*	49
2.18	Insect P450s	49
2.19	Mollusc P450s	50
2.20	New and old names of selected mammalian P450s	50

3.1	Dioxygen species bond data	60
3.2	Molecular orbital data on P450cam ligands	61
3.3	Redox potentials (mV) of components in various P450 systems	64
3.4	Redox partners and heme exposure	64
3.5	Physicochemical data for P450 catalysis	66
3.6	Kinetics and thermodynamics of stages in the catalytic cycle	67
3.7	Reduction rate constant and % high-spin content of phenobarbital-induced microsomal P450	69
3.8	Extent of tyrosine exposure and % high spin CYP101	69
4.1	Reactions catalysed by P450 enzymes	77
4.2	Endogenous roles of xenobiotic-metabolizing P450s	78
4.3	Classes of plant P450 substrates	81
4.4	Plant P450 functionalities	82
4.5	Fungal and bacterial P450 substrates	82
4.6	Rat hepatic P450s (families CYP1–3) and endogenous steroid metabolism	83
4.7	A summary of endogenous substrate metabolism by P450 enzymes	84
4.8	Substrates, inhibitors and inducers of P450 families CYP1, CYP2, CYP3 and CYP4	87
4.9	Human drug-metabolizing P450s	88
4.10	CYP1A substrates	89
4.11	CYP2A substrates	91
4.12	CYP2B substrates	91
4.13	CYP2C substrates	94
4.14	CYP2D substrates	98
4.15	CYP2E substrates	100
4.16	CYP3A substrates	102
4.17	Structural descriptors for 48 substrates of 8 human P450s	105
4.18	Characteristics of human P450 substrates	107
4.19	QSARs for human P450 substrates	108
4.20	Human P450 enzymes involved in activation of carcinogens and toxins	111
4.21	Structural data for selected CYP1A2 substrates	113
4.22	Structural data for selected CYP2C9 substrates	115
4.23	Structural data for selected CYP2D6 substrates	115
4.24	Substrates, inhibitors and inducers of the major human cytochrome P450s involved in xenobiotic metabolism	116
5.1	Average levels of P450 enzymes (%) in human liver	119
5.2	Human P450s: genetics and variation in levels	119
5.3	Receptors associated with P450 expression and induction	120
5.4	Receptor and other factors associated with P450 induction of endogenous substrate metabolism	120
5.5	Regulation of P450 genes	124
5.6	Nuclear receptors associated with P450 gene regulation	126

5.7 Examples of toxic or carcinogenic P450 inducers 129
5.8 QSARs for P450 inducers and ligand-binding affinity 130
5.9 Ah receptor binding affinity and planarity of PAHs 131
5.10 Selective P450 inhibitors for different isoforms 135
5.11 Inhibition of P450 by divalent metal ions 137
5.12 QSARs for P450 inhibitors 139
6.1 Structurally characterized P450s and their redox partners 142
6.2 Experimental timescales (seconds) 148
6.4 A summary of site-directed mutagenesis experiments in the CYP2 family SRS regions 150
6.3 A summary of site-directed mutagenesis and residue modification in the CYP1A subfamily 151
6.5 A summary of CYP3A4 site-directed mutagenesis and interactions 152
6.6 Homology models of P450s produced from CYP102 154
6.7 Comparison between heme geometries in CYP102 156

Figures

1.1	Ultra-violet absorption spectrum of a P450 carbon monoxide complex showing the characteristic Soret peak at around 450 nm	3
1.2	P450 research areas and application fields	4
1.3	A multiple sequence alignment between CYP102 and CYP2 family proteins	see colour section between p152 and p153
1.4	Heterologously-expressed human P450s: current applications	17
2.1	Phylogenetic tree of cytochrome P450 protein sequences inferred using the UPGMA tree drawing algorithm with Kimura calculated distances	23
2.2	A summarized and abbreviated form of the P450 phylogenetic tree mapped against the generally accepted development of terrestrial biota	24
2.3	Universal phylogenetic tree based on sequence comparisons	32
3.1	The reaction cycle and enzymatic intermediates known in P450 reactions	52
3.2	The P450 catalytic cycle in schematic form showing the heme iron spin-state changes and oxygen species	55
4.1	Steroid hormone biosynthetic pathways in the adrenal cortex showing reactions catalyzed by specific P450s	79
4.2	A structural overlay between lauric acid and cholesterol molecules showing the similarity between part of their carbon skeleton.	81
4.3	A selection of human CYP1A2 substrates showing the preferred sited of metabolism	90
4.4	A selection of known CYP2A6 substrates showing the preferred sites of metabolism	92
4.5	A selection of known CYP2B6 substrates showing the preferred sites of metabolism	93
4.6	A selection of known CYP2C9 substrates showing the preferred sites of metabolism	95
4.7	A selection of known CYP2C19 substrates showing the preferred sites of metabolism	96

4.8 A selection of known CYP2C8 substrates showing the preferred sites of metabolism 97
4.9 A selection of known CYP2D6 substrates showing the preferred sites of metabolism 99
4.10 A selection of human CYP2E1 substrates showing the preferred sites of metabolism 101
4.11 A selection of known CYP3A4 substrates showing the preferred sites of metabolism 103
4.12 A decision tree scheme for differentiating between human P450 substrates from the CYP1, CYP2 and CYP3 families on the basis of their physicochemical and structural characteristics 104
5.1 Graph showing the linear correlation (R = 0.98) between binding to the Ah receptor (pEC_{50}) and molecular planarity (area/depth^2) for 7 polyaromatic hydrocarbons 131
5.2 Graph showing a linear relationship (R = 0.987) between P450 inhibition (percentage loss of P450) and ionic radius (Å) for 5 divalent metal ions 137
6.1 Structure of the CYP102 hemoprotein domain showing tertiary and secondary structural elements including α-helices and β-sheets see colour plates between p152 and p153
6.2 Diagrammatic representation of the CYP102 hemoprotein domain showing the location of various structural features 153
6.3 The putative active site region of human CYP1A2 containing the substrate caffeine orientated for N-3 demethylation see colour plates between p152 and p153
6.4 Putative active site region of CYP2A6 showing the substrate, coumarin, docked close to the heme moiety for 7-hydroxylation see colour plates between p152 and p153
6.5 Putative active site region of CYP2C9 showing the substrate, tolbutamide, docked close to the heme iron for 4-methyl hydroxylation see colour plates between p152 and p153
6.6 Putative active site region of CYP2C19 showing the substrate, omeprazole, docked close to the heme group for 5-methyl hydroxylation see colour plates between p152 and p153
6.7 Putative active site region of CYP2D6 showing the substrate, metoprolol, docked close to the heme group for O-demethylation see colour plates between p152 and p153
6.8 Putative active site region of human CYP2E1 showing the substrate, *p*-nitrophenol, docked close to the heme iron for 2-hydroxylation see colour plates between p152 and p153

Acknowledgements

I am indebted to my sponsors, GlaxoWellcome, Merck Sharp & Dohme, Exxon, the University of Surrey and the European Union, for substantial and generous funding of my research over the last five years. In particular, I would like to thank Maurice Dickins, Sandeep Modi, Alan Beresford, Peter Eddershaw and Mike Tarbit (GW), Howard Broughton and David Evans (MSD), Michael Bird (Exxon) and the many others with whom I have collaborated over the years, especially Dennis Parke, Brian Lake, Costas Ioannides, Peter Goldfarb, Gordon Gibson, Miloslav Dobrota, Alan Wiseman and Angela Namor (University of Surrey), John Gorrod (University of Essex), Peter Lee-Robichaud (University of Southampton) and Peter Hlavica (University of München). I would also like to acknowledge the help of my students, past and present, particularly Miriam Jacobs and Estelle Watson. Additionally, I very much appreciate the excellent typing of the manuscript by Sheila Evans and Lisa Jacobs, without whom this book would not have been completed on schedule. Finally, I wish to thank the publishers, Taylor & Francis, for inviting me to produce this work and I also extend my gratitude to Professor Dennis Parke for writing the Foreword.

Abbreviations

Adx	Adrenodoxin
AhR	Aryl hydrocarbon receptor
ARNT	Ah receptor nuclear transporter
CAR	Constitutive androstane receptor
CD	Circular dichroism
cDNA	Complementary DNA
CNDO	Complete neglect of differential overlap
CO	Carbon monoxide
CRE	Cyclic AMP response element
CYP	Cytochrome P450
DNA	Deoxyribonucleic acid
EH	Epoxide hydrase
ER	Estrogen receptor
ESR	Electron spin resonance
EXAFS	Extended X-ray absorption fine edge spectroscopy
FAD	Flavin adenine dinucleotide
FMN	Flavin mononucleotide
FXR	Farnesoid X receptor
GI	Gastro-intestinal
GPCR	G-protein coupled receptor
GR	Glucocorticoid receptor
HNF	Hepatocyte nuclear factor
HOMO	Highest occupied molecular orbital
HRE	Hormone response element
HS	High-spin
HSP	Heat shock protein
IE	Ionization energy
INDO	Intermediate neglect of differential overlap
IR	Infrared
LS	Low-spin
LUMO	Lowest unoccupied molecular orbital
LXR	Liver X receptor

MB	Mössbauer spectroscopy
MCD	Magnetic circular dichroism
MD	Molecular dynamics
MO	Molecular orbital
mRNA	Messenger RNA
NADH	Nicotinamide adenine dinucleotide
NADPH	Nicotinamide adenine dinucleotide phosphate
NCE	New chemical entity
NMR	Nuclear magnetic resonance
ORD	Optical rotatory dispersion
PAGE	Polyacrylamide gel electrophoresis
Pdx	Putidaredoxin
PPAR	Peroxisome proliferator-activated receptor
PXR	Pregnane X receptor
QSAR	Quantitative structure–activity relationship
RAR	Retinoic acid receptor
RNA	Ribonucleic acid
RR	Resonance Raman
RXR	Retinoid X receptor
SDS	Sodium dodecyl sulphate
UV	Ultraviolet
VDR	Vitamin D receptor
XRE	Xenobiotic response element

Chemical abbreviations

AAF	2-Acetylaminofluorene
AFB_1	Aflatoxin B_1
AIA	Allylisopropylacetamide
BNF	β-Naphthoflavone
DDT	*p,p*-Dichlorodiphenyltrichloroethane
DEN	Diethylnitrosamine
DMN	Dimethylnitrosamine
Glu-P-1	2-amino-6-methyldipyrido[1,2-*a*;3′,2′-*d*]imidazole
Glu-P-2	2-aminodipyrido[1,2-*a*;3′,2′-*d*]imidazole
IQ	2-amino-3-methylimidazo[4,5-*f*]quinoline
LTB_4	Leukotriene B_4
MEHP	Mono(2-ethylhexyl)phthalate
MeIQ	2-amino-3,8-dimethylimidazo[4,5-*f*]quinoline
MeIQx	2-amino-3,8-dimethylimidazo[4,5-*f*]quinoxaline
NNK	4-(methylnitrosamino)-1-(3-pyridyl)-1-butanone
PAH	Polyaromatic hydrocarbon
PB	Phenobarbital
PCB	Polychlorobiphenyl
PCN	Pregnenolone 16α-carbonitrile
PhIP	2-amino-1-methyl-6-phenylimidazo[4,5-*b*]pyridine
TAO	Troleandomycin
TCDD	2,3,7,8-Tetrachlorodibenzo-*p*-dioxin
Trp-P-1	3-amino-1,4-dimethyl-5*H*-pyrido[4,3-*b*]indole
Trp-P-2	3-amino-1-methyl-5*H*-pyrido[4,3-*b*]indole
3MC	3-Methylcholanthrene

Symbols

Å	Ångstrom unit (10^{-8} cm)
c	centi (10^{-2})
D	debye (3.33564×10^{-30} Cm)
Da	Dalton (1.66×10^{-24}g)
e	electronic charge (1.60219×10^{-19} C)
E^0	redox potential (V or mV)
eV	electron volt (96.485 kJ mol^{-1})
F	Faraday constant (96484.6 C)
J	joule (1 calorie = 4.184 joules)
K	equilibrium constant
k	kilo (10^3)
k	rate constant
k_{cat}	catalytic rate constant
K_D	dissociation constant
K_m	Michaelis constant
m	milli (10^{-3})
M	molar
n	nano (10^{-9})
P	partition coefficient
R	gas constant (8.31441 J K^{-1} mol^{-1})
T	absolute temperature
ΔE	energy change
ΔG	Gibbs free energy change
ΔH	enthalpy change
ΔS	entropy change
μ	micro (10^{-6}) and dipole moment

Note: In statistical analyses, the following symbols are employed as standard:

n = number of observations
s = standard error
R = correlation coefficient
F = variance ratio (F-test)

Preface

> Were I to await perfection my book would never be finished.
>
> (Lao Tzu)

In this volume, I am utilizing the material provided in the previous book (*Cytochrome P450: Structure, Function and Mechanism*, Taylor & Francis, London 1996) and updating where necessary. It has not been my intention, however, to repeat material already presented in the former book, but rather to build upon that foundation by extending the ground which has been covered since 1996, and the extensive updating of the text includes the citing of many year 2000 references. The intention is, however, for this Guide text to be somewhat simplifying in nature for those just coming into the field; although it should be recognized that the full picture is often likely to be more complicated.

In writing this book, I decided that it was not just going to be a second edition of the previous one. However, I wanted to use the first book as a template for this one, especially in an organizational sense, such that the chapter topics are more or less the same and appear in a similar sequence. In this text, each chapter is intended to stand on its own, although there is some degree of cross-referencing between chapters and, consequently, occasional repetition is inevitable, although this has been kept to a minimum. The Introduction is much shorter than in the previous book, and some material has either been moved to other chapters or drastically reduced in size. Although there are inevitably some similarities, this book represents a complete rewrite of the first one, which has itself only been used as a rough guide to the general structure and content of this work.

Although the subject matter on P450 is vast by any standards, single-volume works serve a useful purpose for directing readers to source material and can also be easily referred to, despite the fact that they can never be entirely comprehensive. However, being as up-to-date as any book on this subject can, this will be helpful for those who may wish to obtain a 'fast-track' entry into the P450 field, where recent review articles in particular have been cited as frequently as possible. The main focus in most chapters is on human P450s and drug metabolism in Man due to current interest, as has been highlighted recently in an article published in the Washington Post (Brown, 2000).

We now know a substantial amount about this enzyme system: how it works, what its structure looks like and why it evolved during the course of time. In the various chapters presented (Introduction, Evolution, Activation, Metabolism, Regulation and Structure) these areas are explored, hopefully in a readily understandable way such that an overview of the field can be obtained, and also serve as a resource to where the reader may be directed for regions of more detailed study.

DFVL, August 2000

Foreword

Cytochrome P450 – undoubtedly the most popular research topic in biochemistry and molecular biology over the past half century, due largely to the world-wide development of the pharmaceutical industry – is itself a misnomer, a paradox, for it is neither a cytochrome, nor a single entity, but a vast family of haem-thiolate enzyme proteins with redox properties, known as 'mono-oxygenases' because of their ability to activate molecular dioxygen into highly reactive oxygen species (ROS), and then to insert the oxygen into a wide variety of substrates. These cytochrome P450 (CYP) enzymes are ubiquitous in all five biological kingdoms, with the exception of some primitive micro-organisms that evolved more than 3.5 billion years ago. Although human, and other mammalian, P450 enzymes are the most extensively investigated, the past decade has seen a wide interest in the P450s of birds, fish, insects, plants, bacteria, and other biological species. The substrates and reactions of these enzymes are equally ubiquitous, ranging from drug metabolism to the biosynthesis and metabolism of fatty acids, steroids, prostanoids, vitamins D and bile acids, alkaloids, terpenes, and other phytoalexins, the detoxication of carcinogens, pesticides, and other xenobiotics, metabolic activation of procarcinogens, and the generation of ROS which damage DNA and other vital biological entities.

Iron, in any chemical form, is capable of producing lethal levels of ROS in biological systems, and hence requires tight regulation. With P450 this is achieved through substrate binding to the enzyme, which changes the spin state of the molecule, thereby enabling stabilization of bound ROS, and oxygenation of the substrate. So in most cases, it is only when a suitable chemical substrate is bound to the P450 that O_2 also becomes bound, is reduced to ROS, and so oxygenates the substrate. This decreases the likelihood of the leakage of ROS in the absence of substrate.

Over 200,000 chemicals are metabolized by the P450 family of enzymes, catalysing many different types of reactions including oxidations, reductions, and dehalogenations, catalysed by >120 different P450 families. Although initially characterized by the UV spectra of their reduced CO-complexes – from which their name 'cytochrome P450' is derived – more precise methods, such as immunoblotting with specific antibodies, and full amino acid/gene sequence analysis, are now used. With these more refined techniques, some 1200 P450 enzymes have been isolated and sequenced; the major families now recognized are CYP1A, 2A, 2B, 2C, 2D, 2E, 3, 4, 5, 7, 11 (animals), 51–66 (fungi), 71–100 (plants), 101–120 (bacteria).

Individual chapters of this monumental new work on P450 are devoted to its evolution (Chapter 2), the catalytic cycle (Chapter 3), substrate selection and metabolism (Chapter 4), regulation (Chapter 5), and structure (Chapter 6). The CYP superfamily is ancient and probably dates back 3500 million years, and the development of the P450 phylogenetic tree (Chapter 2) agrees closely with known pathways of evolution of the plant and animal kingdoms, with the separation of the mitochondrial and microsomal P450s, and the development of sexual reproduction, at around 1000 million years ago. Some 400 million years ago, the major role of the P450 system changed from metabolism of endogenous substrates (cholesterol) to encompass the detoxication of dietary phytoalexins and environmental chemicals. In the catalytic cycle (Chapter 3), the organic substrate (RH) combines with O_2 in the presence of two reducing equivalents – supplied by NADH or NADPH and transferred in two consecutive stages via the mediation of one or two redox partners (redoxin or cytochrome b_5), depending on the P450 enzyme involved. Whereas mitochondrial and most bacterial P450s exhibit high selectivity in their choice of substrates, and specificity in their reactions, the mammalian microsomal P450s – associated with the oxidative metabolism of drugs and environmental chemicals – exhibit a wide diversity of reactions and of substrates (Chapter 4). Many P450 enzymes catalyse pathways of endogenous metabolism, including steroid biogenesis, oestradiol and testosterone metabolism, and metabolism of retinol and retinoic acid. There is a strong association between CYP1 and the activation of procarcinogens – CYP1A1 substrates including the carcinogenic planar, polycyclic, aromatic hydrocarbons, and the CYP1A2 substrates comprising the carcinogenic heterocyclic amines and cooked-food mutagens. Other CYP2 families are concerned with the metabolism of environmental chemicals (coumarin) and drugs (phenobarbital); CYP2E is concerned with the metabolism of small molecules (acetone), is induced by ethanol, and is associated with the generation of ROS – probably due to poorly coupled redox cycles. CYP3 substrates are large compounds of high molecular weight (erythromycin). CYP4 substrates are long-chain fatty acids, metabolized by (ω-)- and (ω-1)-hydroxylation, induced by phthalate esters, benoxaprofen, etc., and associated with the proliferation of peroxisomes, induction of peroxide formation, and ROS generation, and is considered to be a major mechanism of epigenetic carcinogenesis in small rodents. Chapter 5 deals with the regulation of P450 enzymes, including the role of nuclear receptors, mRNA stabilization, and post-translational modification. Regulation of the CYP1 family is mediated by the Ah receptor, which has a high affinity for the environmental pollutant, TCDD (2,3,7,8–tetrachlorodibenzo-*p*-dioxin), and for other planar polyaromatic hydrocarbons. The mechanisms of regulation of different CYP families are reviewed; CYP1 and CYP2E enzymes are major activators of procarcinogens, and a technique

(COMPACT) is described which identifies potential risks from activation of chemicals to carcinogens by either of these two enzyme families. Chapter 6 concerns the structures of the P450s, and reviews the methodology available to determine their nature, including molecular modelling, X-ray crystallography, spectroscopy, and molecular biology. Molecular orbital, and related quantum-mechanical calculations of electronic structure have been applied to many P450 systems, their substrates, inhibitors, and active sites, leading to a greater understanding of drug metabolism, drug toxicity, environmental pollution, and chemical carcinogenicity.

First published in 1996, this authoritative monograph on cytochrome P450 has been completely revised and rewritten, because of the many recent developments in the field. The author of this excellent work, Dr David Lewis, was the first to apply advanced computer graphics technology, and molecular orbital studies, to elucidation of the structures of cytochrome P450, and the oxidative metabolism of drugs and environmental chemicals. He has continued for the past fifteen years in the development of these and other mathematical techniques to determine the structures and enzymic activities of the P450s. Among the most valuable features of this new book is the abundance of illustrations, notably some 30 figures and 80 tables, many of which are invaluable compilations of QSARs. The book is a dedicated work of scholarship, and shows a novel way forward for biochemists and molecular biologists involved in the elucidation of structure/function of similar complex families of enzymes and other biological entities.

Professor Dennis V. Parke

Chapter 1

Introduction

It is by logic that we prove, but it is by intuition that we discover.

(Henri Poincaré)

1.1 Discovery of an enzyme superfamily

The development of P450 research has largely paralleled that of drug metabolism over the past forty years or so, and there are strong connections between the two areas (Gibson and Skett, 1994; George and Farrell, 1991; Forrester et al., 1992; Cholerton et al., 1992; Alvares and Pratt, 1990; Lu, 1998a and b; Spatzenegger and Jaeger, 1995; Watkins, 1990; White, 1998; Wrighton and Stevens, 1992). It is unlikely that any other enzyme system has been studied so extensively and with such an array of physicochemical and biochemical techniques, some of which will be reviewed herein although these have been described in more detail previously (Lewis, 1996a). Understanding P450 structure and function, including its mechanism of action, has represented a major scientific challenge during the last forty years or more since its discovery in the late 1950s, although there are still several areas of controversy where questions remain to be fully answered, together with others that are generally understood but require further elucidation (Estabrook, 1996). Some of these aspects have been reviewed (Ruckpaul, 1978; Ruckpaul and Rein, 1984) and the reader is also referred to an interesting account of $P450_{cam}$ research covering a twenty-five year period (Mueller et al., 1995).

It is now forty years since Garfinkel (1958) and Klingenberg (1958) first reported the existence of an unusual cytochrome, which later became known as P450 due to its characteristic Soret absorption maximum at 450 nm in the UV spectrum (see Figure 1.1) of its carbon monoxide adduct (Omura and Sato, 1962, 1964). Much has changed since those early days of P450 detection, and interest in the enzyme has expanded enormously over the four decades since its discovery. At the current time of writing in 2000, there are over 1200 individual P450 genes known and, doubtless, many more will be reported (Nelson, 1998, 1999; Anzenbacher and Anzenbacherova, 2000) in the next nomenclature update which is due to be published in the near future. Furthermore, the corpus of knowledge in this area is both extensive and profound. The number of publications in the P450 field has regularly exceeded 1000 research papers per annum for the last decade, and has also shown a steady annual increase during this period (Estabrook, 1996; Koymans et al., 1993). Figure 1.2 indicates the broad nature of P450 applications and research areas of current interest.

Initially, it was thought that P450 was a single unique cytochrome but it soon became clear that the enzyme exists as multiple forms, each with different properties in respect of their substrate selectivity and certain physicochemical characteristics (Omura et al., 1993). (See the nomenclature note in Section 1.7 for further information.) For example, the absorption maxima in the UV spectrum of carbon monoxide adducts show a distinct variation around the 450 nm mark for enzymes isolated from different sources, or from purified microsomal P450 preparations following initial pre-treatment with various chemicals such as phenobarbital and 3-methylcholanthrene (Ruckpaul and Rein, 1984). This multiplicity of forms led to the enzymes being termed 'mixed-function' oxidases and, more recently, 'mono-oxygenases' to describe their ability to insert oxygen into a large variety of substrates and wide range of structural classes of compounds (Porter and Coon, 1991; Okita and Masters, 1992; Ortiz de Montellano, 1986, 1995). In fact, although it is still widely used, the name 'cytochrome' is not now regarded as the most appropriate description for these hemoproteins as P450s differ significantly from other cytochromes in several ways and also because, strictly, they are termed heme-thiolate enzymes rather than redox proteins (such as the cytochromes a, b and c for example) although they do, of course, possess redox properties as will be described later.

As mentioned previously, the appearance of a vivid orange-yellow cellular pigment following complexation of microsomal preparations with carbon monoxide, and by virtue of its intense Soret absorption maximum at around 450 nm, was responsible for the discovery (Garfinkel, 1958; Klingenberg, 1958) of the enzyme system itself. Over a period of about 3.5 billion years, the P450 system has been steadily evolving by utilizing the chemical potential of molecular oxygen for undergoing organic reactions via the controlled formation of a reactive oxygen species (e.g. superoxide, peroxide, etc.) and with the mediation of an iron-porphyrin complex as the heme prosthetic group because porphyrin is an ideal ligand for regulating the iron spin-state equilibrium (Frausto da Silva and Williams, 1991), which is an important aspect of P450 catalysis. This type of reaction is facilitated by the modulating influence of the proximal cysteine ligand (thiolate) whereas the rest of the hemoprotein is employed for channelling both water molecules and the reducing equivalents (i.e. protons/electrons) necessary for the reaction to proceed (reviewed in Lewis and Pratt, 1998), together with enabling organic substrates to bind. Clearly, the P450 apoprotein had evolved to bind the heme moiety (as, incidentally, do other cytochromes and naturally hemoproteins in general) and to 'steer' selected substrates towards oxidative metabolism at particular positions, while also allowing for the reaction to be controlled by the consecutive two-stage reduction of bound oxygen (reviewed in Lewis and Pratt, 1998). There is, moreover, additional provision for electron transfer pathways from P450 redox

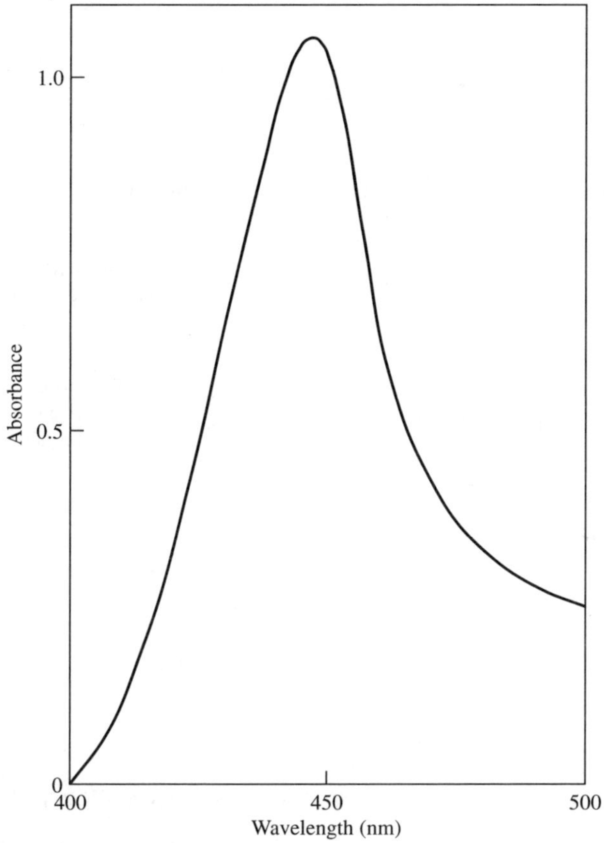

Figure 1.1 Ultra-violet absorption spectrum of a P450 carbon monoxide complex showing the characteristic Soret peak at around 450 nm (data obtained from Estabrook, et al., 1963).

partner(s) which may also involve the entire system being embedded in a phospholipid bilayer, such as the endoplasmic reticulum membrane in the case of hepatic microsomal P450s, for example (reviewed in Bernhardt, 1995).

It does not seem unreasonable, therefore, to propose that the P450 tertiary structure evolved in such a way as to bring together the important elements (see Table 1.1) required for its catalytic function, namely, binding of substrate, heme, oxygen and redox partner(s). These, by and large, seem to have remained essentially the same from one P450 to another despite the changes in local cellular environment between the bacterial, mitochondrial and microsomal systems (Degtyarenko and Archakov, 1993) although it is now apparent that some exceptions to this conservation exist, such as in those P450s involved in unusual functions, e.g. allene oxide synthase (AOS), prostacyclin synthase (PGIS) and thromboxane synthase (TXAS). In fact, the different functionalities displayed by these enzymes can be rationalized

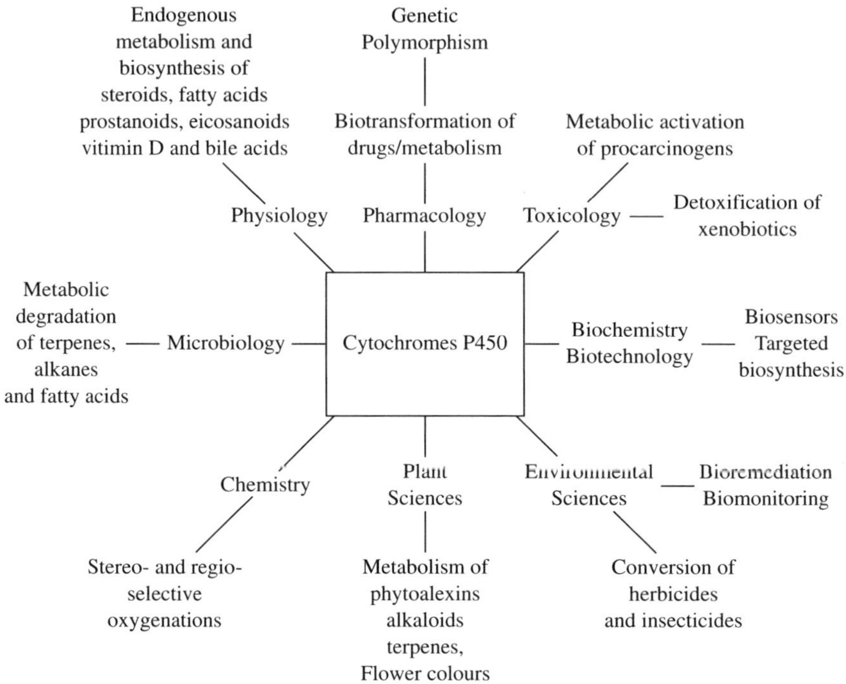

Figure 1.2 P450 research areas and application fields. (Adapted from Berhardt, 1995.)

in terms of relatively small but crucial changes (see Table 1.2) within the catalytic centre of the P450 itself (Mansuy and Renaud, 1995). Consequently, what was once regarded as a single enzyme is now known to exhibit a multiplicity of forms (> 1000) with a variety of functionalities and a vast number of potential substrates (Porter and Coon, 1991; Guengerich, 1991a and b, 1997; Nelson, 1999; Stegeman and Livingstone, 1998; Mansuy, 1998).

1.2 Occurrence and distribution

Cytochrome P450 enzymes are present in all five biological kingdoms (Nebert et al., 1989a and b) although it would appear that certain primitive species of bacteria do not contain any forms of the enzyme, possibly indicating that the ancestral P450 gene developed around 3.5 billion years ago (Nelson et al., 1993, 1996). In mammalian species, P450s are present in most tissues; the greater proportion being found in the liver, where P450 enzymes are located in the smooth endoplasmic reticulum (ER) membrane of hepatocytes (Ruckpaul and Rein, 1984; Stier, 1976). The endoplasmic reticular membrane is a structure of surprisingly large surface area (comprising 7–11 m^2 gm^{-1} of liver weight) relative to its volume

Table 1.1 Structural roles of P450 enzyme apoprotein

Function of P450	*Structural features involved*
1. Binding of heme	I and L helices, invariant cysteine and usually two basic amino acid residues for ion-pairing with heme propionates
2. Binding and activation of oxygen	I Helix region distal to the heme moiety
3. Binding redox partners	Basic amino acid residues for ion-pairing
4. Enabling proton transfer to oxygen	Internal ion-pairs within the heme environment
5. Binding different substrates	B′, F and I helices, β_1 and β_4 sheets (SRS regions)
6. Regulating these activities within a membrane	N-terminal peptide of 30–40 residues in length

SRS = Substrate recognition site (Gotoh, 1992)

and, consequently, the ER is ideal as an efficient catalytic surface for supporting an array of xenobiotic-metabolizing enzymes including those involved in Phase 1 metabolism such as the P450s (Lewis and Pratt, 1998; Gibson and Skett, 1994).

As between 12 and 15% of the endoplasmic reticulum is composed of cytochromes P450, it can be estimated that P450 enzymes account for about 1–2% of an individual hepatocyte by weight (Ruckpaul and Rein, 1984). This extraordinarily well-adapted system for catalysing the oxidation of xenobiotics (primarily) has been the paradigm for the design of enzyme mimics (Wiseman et al., 2000) and it appears, for example, that relatively simple iron porphyrin complexes embedded in a zeolite support are able to mediate some of the mono-oxygenase reactions normally carried out by P450s (Parton et al., 1994) albeit lacking the exquisite selectivity of many of these enzymes.

In addition to the major site of P450 enzymes in the liver (see Table 1.3), there are also various forms of the cytochromes P450 expressed in other organs and tissues such as the kidney, breast, prostate (Williams et al., 2000a) skin and nasal epithelium, gonads, placenta, brain, lung, spleen, pancreas and gastro-intestinal tract (Waterman, 1992; Schenkman and Griem, 1993; Hellmold et al., 1998; Hakkola et al., 1998, 1996). P450s have also been characterized in species other than mammalia, including birds, fish, reptiles, insects, molluscs, arthropods, crustacea, fungi, plants and bacteria (Schenkman and Griem, 1993; Pinot et al., 1999; Gorman et al., 1998; Chapple, 1998; Scott et al., 1998; Walker, 1998; Hallahan et al., 1993; Hallahan and West, 1995; Nelson, 1998). (A special issue of *Comparative Biochemistry and Physiology* (Volume 121C, Nos. 1-3, 1998) contains much information on non-mammalian P450s.) However, much interest currently centres around mammalian P450s (Funae and Imaoka, 1993) and, in particular, the human forms (see Table 1.3) of the enzyme (Guengerich, 1989a and b, 1992a–d, 1994, 1995a; Guengerich et al., 1998, 1992; Rendic and DiCarlo, 1997). Consequently, human P450s have been extensively investigated, and heterologous expression systems are now becoming a reliable means of studying enzyme–substrate interactions of isolated P450 enzymes (Estabrook et al., 1991; Gonzalez and Korzekwa, 1995).

Table 1.2 Comparison between the I helices of several P450s

Species	*CYP*	*Abbreviation*	*Sequence*										*Enzyme reaction type*
						↓	↓		↓	↓			
Human	3A4	nif	F	I	F	A	G	Y	E	T	T	S	nifedipine *N*-oxidase
Human	5A1	TXS	F	L	I	A	G	Y	E	I	I	T	thromboxane synthase
Rat	5A1	TXS	F	L	I	A	G	H	E	I	T	T	thromboxane synthase
Mouse	5A1	TXS	F	L	I	A	G	H	E	V	I	T	thromboxane synthase
Pig	5A1	TXS	F	L	I	A	G	Y	E	I	I	T	thromboxane synthase
Flax	74	AOS	N	S	W	G	G	F	K	I	L	L	allene oxide synthase
Guayule	74	AOS	N	T	F	G	G	V	K	I	L	F	allene oxide synthase
Human	19	arom	M	L	I	A	A	P	D	T	M	S	androgen aromatase
Cow	19	arom	M	L	I	A	A	P	D	T	M	S	androgen aromatase
Rat	19	arom	M	L	I	A	A	P	D	T	M	S	androgen aromatase
Mouse	19	arom	M	L	I	A	A	P	D	T	M	S	androgen aromatase
Chicken	19	arom	M	L	I	A	A	P	D	T	L	S	androgen aromatase
Trout	19	arom	M	V	I	A	A	P	D	T	L	S	androgen aromatase
Yeast	51	14DM	V	L	M	G	G	Q	Q	T	S	A	lanosterol 14α-demethylase
Rat	51	14DM	L	L	L	A	G	Q	Q	T	S	S	lanosterol 14α-demethylase
Human	51	14DM	L	L	L	A	G	Q	Q	T	S	S	lanosterol 14α-demethylase
Candida	51	14DM	V	L	M	G	G	Q	Q	T	S	A	lanosterol 14α-demethylase

Notes:
1 The usual distal helix sequence AG *x* ET is shown by CYP3A4 where *x* denotes any residue
2 CYP5A1 does not possess the conserved distal threonine
3 CYP74 does not possess either the distal treonine or a preceeding acidic residue
4 CYP19 possesses a proline before the acidic residue, aspartate
5 CYP51 does not possess an acidic residue before the distal threonine
↓ denotes normally conserved residues

References: Nelson et al., 1996; Mansuy and Renaud, 1995

Table 1.3 Human P450s involved in drug metabolism

CYP	*% Hepatic P450*	*Variation*	*% Involvement in drug metabolism*	*Extrahepatic location that is associated with xenobiotic metabolism*
1A1	<1		2.5	Lung, kidney, GI tract, skin, placenta and lymphocytes
1A2	~13	40 fold	8.2	?
1B1	<1		?	Skin, kidney, mammary, prostate, uterus, fetus
2A6	~4	30–100 fold	2.5	Lung, nasal membrane
2B6	<1	50 fold	3.4	GI tract, lung
2C8, 2C9	~18	25–100 fold	15.8	GI tract, larynx, lung
2C18, 2C19	~1	25–100 fold	8.3	GI tract, larynx, lung
2D6	≤2.5	1000 fold	18.8	GI tract
2E1	≤7	20 fold	4.1	Lung, placenta
2F1	<1		~1.3	Lung, placenta
3A4, 3A5	≤28	20 fold	34.1	GI tract, placenta, fetus
4A11	?		?	Kidney
4B1	?		?	Lung, placenta

? = unknown

Notes:
The % hepatic P450 complement data represent average values although the extent of variability is given.

The % of drug oxidations carried out by these forms has been derived from drugs in clinical use at the time and is, therefore, subject to change.

With the exception of the polymorphic forms CYP2D6 and CYP2C19, there is a good correlation ($R = 0.95$) between % hepatic P450 and % involvement in drug metabolism (Lewis, 1998a) based on the data presented above.

Reference: Rendic and DiCarlo, 1997

It appears that there exists a close coupling between P450s and the enzymes of Phase 2 metabolism such as the conjugases, together with that of epoxide hydrase (EH) and P450 (Mackenzie, 1990; Thomas et al., 1990). Furthermore, the fluid nature of the ER membrane facilitates the transverse movement of embedded P450 enzymes, which are thus able to 'trawl' the phospholipid bilayer structure such that potential substrates may be readily oxygenated prior to further metabolism (Hollebone, 1986). Although a full description of the characteristics of the ER is beyond the scope of this book, the interested reader is referred to some excellent accounts in the literature (see, for example, Gibson and Skett, 1994; Ruckpaul and Rein, 1984) and there is also evidence for the role of the ER in the folding and degradation of glycoproteins (Parodi, 2000). However, it is important to note that extensive studies have revealed the oligomeric nature of the protein quaternary structures within the ER membrane (Finch and Stier, 1991; Schwarz, 1991) indicating that about 6 to 8 P450 enzymes tend to cluster around a central reductase. These hexameric (or octameric) units appear to be able to undergo transverse diffusion within the ER membrane phospholipid, which itself can show lateral diffusion (via what is termed a 'flip-flop' mechanism), such that relatively lipophilic substrates can readily become actively transported to their sites of metabolism at the P450 reactive centre (reviewed in Archakov and Bachmanova, 1990). Consequently, there is interest in both the membrane topology of microsomal P450s (Brown and Black, 1989; Black, 1992) and in the structural features of the oxido-reductase redox partner (Black and Coon, 1982; Porter and Kasper, 1986) which has now been resolved crystallographically (Wang et al., 1997).

With the numbers of P450s in any one individual mammalian species currently thought to be around 50 or more (Estabrook, 1998) and the likelihood of more P450s being discovered in many of the previously less well-studied species (Nelson, 1998, 1999), it is important to recognize that the classification of these enzymes within families and subfamilies, sometimes irrespective of species, leads to an appreciation that there are often significant species differences between P450s and also in the substrates which they metabolize (Guengerich, 1997). Moreover, the complement of P450 enzymes may differ considerably from one mammalian species to another, together with their inducibilities by various chemicals and substrate selectivities (Lewis et al., 1998a; Soucek and Gut, 1992; Nedelcheva and Gut, 1994; Smith, 1991; Guengerich, 1997). Consequently, it can sometimes be difficult to extrapolate the results obtained from experiments carried out in laboratory-bred rodents, for example, to the likely outcome of compound metabolism in *Homo sapiens* (Lewis et al., 1998a). It is also known that there are sex differences in mammalian P450s (Schenkman et al., 1967) and the levels of individual P450s are affected by both diet and nutritional status (Ioannides, 1999; Parke and Ioannides, 1994). Furthermore, there are various cellular factors controlling the expression of different P450s in tissues which lead to an alteration in the levels of these enzymes in different tissues (reviewed in Gibson and Skett, 1994; Wolff and Strecker, 1992). In humans, there are marked variations in the hepatic P450 complement, for example, due to the genetic differences in ethnogeographical groups and phenotypic changes (Price-Evans, 1993; Bachmann, 1996), some of which have been associated with cancer susceptibility (Crespi et al., 1991; Nebert et al., 1996; Crofts et al., 1993; Puga et al., 1997). However, it is possible to provide average values (Rendic and DiCarlo, 1997) for the P450 levels in human liver based on cohort studies in Caucasians, for example, and these are presented in Table 1.3. In the adrenal gland, however, there are P450s which are important for the biosynthesis of steroid hormones from cholesterol, and these are present in the adrenal cortex mitochondria (Takemori and Kominami, 1984; Hanukoglu, 1992) as well as the endoplasmic reticulum, although there are differences between these steroidogenic P450s and the biosynthetic pathways they mediate depending on the cellular organelles involved (Lewis and Lee-Robichaud, 1998).

1.3 Isolation and purification

Following the identification and discovery of multiple forms of the enzymes, several groups of co-workers embarked on the isolation and purification of various P450s, mainly from mammalian sources (see, for example, Guengerich, 1983). However, in recent years there has been an increasing interest in the isolating and gene sequencing of P450s from non-mammalian species (see Nelson et al., 1996 and a special issue of *Comparative Biochemistry and Physiology*, Volume 121C, Nos. 1–3, 1998) although bacterial forms of the enzyme have been readily isolatable and easily purified due to their cytosolic location (see, for example, Sato and Omura, 1978). The most exhaustive treatment of the methodology for purifying mammalian microsomal P450s appears in an excellent review by Ryan and Levin (1990) and more recent accounts (Gibson and Skett, 1994; Phillips and Shephard, 1998) together with those of others (Ruckpaul and Rein, 1984; Archakov and Bachmanova, 1990; Guengerich, 1987; Schenkman and Kupfer, 1982) are also recommended to the interested reader.

Essentially, the procedures involved in purifying hepatic P450s from microsomal sources include the following stages:

1 Homogenization of liver tissue
2 Centrifugation of microsomal suspensions
3 Chromatographic separation
4 Electrophoretic separation using SDS-PAGE

Further purification and final isolation of the P450 proteins using isoelectric focusing and HPLC has also been described (Ruckpaul and Rein, 1984) as being necessary to produce immunochemically homogeneous P450s and these may then be characterized in terms of their spectral and catalytic properties. It is found that the apparent molecular weights of P450s separated by SDS-PAGE are similar to the actual values determined by other experimental methods (Black and Coon, 1986). In fact, there is a ~92% correlation between the true molecular weights of P450s and those observed using SDS-PAGE (see Lewis, 1996a) and it is possible that some discrepancies may arise as a result of loss of the N-terminal membrane binding sequence of between 30 and 40 amino acid residues. Comparison between P450 molecular weights and SDS-PAGE data shows that there is approximately 4–5 kDa difference between the two, thus corresponding to 30–40 amino acids which would be consistent with removal of the N-terminal membrane 'anchor' (Lewis, 1996a).

1.4 Characterization and detection

Cytochromes P450 can be characterized using a number of physical methods and also by molecular biological techniques, such as cDNA sequencing, together with biochemical and immunological procedures (Phillips and Shephard, 1998). The molecular weights of P450s appear to be a characteristic property to a certain

Table 1.4 Some characteristics of mammalian P450s

CYP	*Species*	λ_{max} *(nm.) CO adduct*	*MWt. (kDa)*	*Number of residues*
1A1	rat	447	56	524
1A1	mouse	449	55	524
1A1	rabbit	448	56	518
1A2	rat	447	54	513
1A2	mouse	448	55	513
2A1	rat	451	48	492
2A2	rat	452	48	492
2A4	mouse	451	48	494
2A5	mouse	451	48	494
2B1	rat	450	52	491
2B2	rat	451	52	491
2B4	rabbit	451	49	491
2B5	rabbit	451	49	491
2C2	rabbit	451	50	490
2C3	rabbit	450	51	490
2C5	rabbit	450	48	487
2C6	rat	450	48	490
2C7	rat	448	51	490
2C11	rat	451	50	500
2C12	rat	448	51	490
2C13	rat	449	50	490
2C14	rabbit	451	50	490
2D1	rat	448	52	504
2D2	rat	448	52	500
2D9	mouse	449	50	504
2E1	rat	452	51	493
2E1	rabbit	452	51	493
2E2	rabbit	452	52	493
3A2	rat	449	51	504
3A6	rabbit	449	52	501
4A1	rat	452	52	511
4A4	rabbit	450	52	506
4B1	rabbit	449	56	506

References: Omura et al., 1993; Nelson, et al., 1993

extent and these seem to depend primarily on the number of amino acid residues in the protein which, incidentally, comprises a single polypeptide chain spanning around 400–500 residues (see Table 1.4). For example, it can be demonstrated that there is a ~ 99% correlation between apoprotein molecular weight and amino acid residue number for a variety of P450s (Lewis, 1996a).

Some characteristics of P450s isolated from mammalian sources (namely rat, mouse and rabbit) are presented in Table 1.4. These include their molecular weights and residue numbers, together with the Soret band absorption maximum in the UV spectrum of their CO adducts (Omura et al., 1993), although the actual electronic absorption spectra extend into the visible region. UV/visible spectrophotometry has been used extensively in the detection and characterization of P450s, where the influence of substrate binding has a marked effect on the appearance of the overall spectrum, especially with respect to the positions and intensities or the major absorption bands (Schenkman et al., 1981; Gibson and Skett, 1994). The reasons for such characteristics in P450 electronic spectra are due to energy level transitions within the heme locus of the enzyme itself: these are profoundly influenced by the nature of the heme ligands, the environment of the heme moiety and the presence of a bound substrate or inhibitor (reviewed in Lewis, 1996a). Three specific types of substrate binding spectra (Schenkman, 1970; Schenkman et al., 1981) have been identified, namely Type I, Type II and Reverse Type I (the latter being sometimes referred to as Modified Type II, however).

Type I binding spectra are characterized by a decrease in intensity of the Soret absorption peak at 420 nm coupled with a concomitant increase in the band intensity of the 390 nm absorption (reviewed in Gibson and Skett, 1994). This type of UV absorption spectrum is indicative of the substrate's influence on the P450 heme iron spin-state equilibrium, because low-spin P450 absorbs at around 418 nm (actually within the range 416–420 nm) whereas the high-spin form gives rise to the 390 nm absorption (with a range of 385–394 nm). Consequently, a Type I spectrum shows that the binding of substrate within the P450 heme locus brings about a shift in the iron(III) spin equilibrium from low-spin to high-spin (reviewed in Lewis, 1996a). It is possible to calculate the percentages of the two spin-states from an analysis of the Type I binding spectrum, which may then be used to estimate the relevant equilibrium constant for the process, together with the associated thermodynamic parameters (ΔG, ΔH and ΔS) following construction of the van't Hoff plot which assesses the effect of temperature on the spin equilibria (Schenkman and Kupfer, 1982). Most P450 substrates exhibit Type I binding spectra, and it is often the case that the magnitude of the Type I spectral change indicates both the extent and rate of metabolism, especially as there can be tight coupling between spin and redox equilibria in both the microsomal and bacterial P450 systems (Sligar and Gunsalus, 1979; Sligar et al., 1979). Clearly, a lowering of the P450 redox potential (i.e. becoming less negative) facilitates reduction by the appropriate redox partner in the system concerned and this has an effect on the overall rate of substrate turnover, as has been shown from experimental studies (Blanck et al., 1983; Schwarze et al., 1985; Petzold et al., 1985) although there is apparently some degree of 'leakiness' in the microsomal system (Blanck et al., 1991).

The Type II spectral change is exhibited by compounds which tend to act as P450 inhibitors via ligation of the heme iron (reviewed in Lewis, 1996a). The structures of such molecules usually possess atoms containing non-bonded electrons, such as a nitrogen lone pair, and where there is also accessibility in the structure of the ligand so that it may bind freely at the heme locus. It appears that the pK_a of the nitrogen is a factor in heme binding affinity of amines (Byfield et al., 1993) but the bond angle between the amine nitrogen and heme plane is also an important quantity (Lewis, 1996a) in those cases where steric hindrance can occur. Evidence for the Type II binding process involves a UV spectrum which is characterized by a decrease in absorption at around 390–405 nm with a concomitant increase at about 425–435 nm (Schenkman et al., 1981). Consequently, there is a shift in the absorption maxima corresponding to low- and high-spin iron(III) towards longer wavelengths coupled with a shift in spin-state equilibrium from high- to low-spin iron. Furthermore, there is crystallographic evidence to suggest that typical Type II ligands cause the iron atom to move into the plane of the heme, and this would be consistent with a favouring of the low-spin state (reviewed in Lewis, 1996a) as high-spin iron possesses a greater ionic radius than the low-spin form (Shannon and Prewitt, 1970).

As the Reverse Type I change is characteristically a 'mirror-image' of the Type I, where there is an increase in the 420 nm absorption coupled with a decrease in the 390 nm peak, this type of spectral change was originally referred to as Modified Type II because of its resemblance to the Type II binding spectrum. However, it is not thought that heme ligation occurs in this case but that, instead, displacement of the distal ligand (likely to be a water molecule or, possibly, a hydroxide ion) combined with substrate binding to a hydrophobic region of the heme pocket may take place (Schenkman et al., 1981). Tables 1.5, 1.6 and 1.7 list several Type I, Type II and Reverse Type I substrates, respectively. It can be appreciated that, from inspection of these tables, Reverse Type I substrates tend to constitute relatively hydrophobic molecules, such that displacement of bound water via some form of lipophilic interaction within the heme environment could occur on substrate binding. Methodology for evaluating the binding of hydrophobic substrates has been reported (Backes and Canady, 1981) and correlations between the three different types of substrate binding spectra and spin-state changes have also been established (Kumaki et al., 1978; Schenkman et al., 1981).

In addition to the substrate binding spectra for iron(III)P450, there is also a characteristic UV absorption spectrum exhibited by the reduced form (i.e. iron(II)P450) especially when the carbon monoxide (CO) complex is investigated (Sato and Omura, 1978). Table 1.8 indicates that the wavelength of the Soret absorption maximum of the reduced P450 CO adduct tends to be a characteristic of the particular P450 enzyme involved. Although close to the average value of about 450 nm, this peak varies slightly depending on the P450 and Table 1.4 gives a number of these values for different P450 enzymes (Omura et al., 1993). However, the wavelength of the Soret peak in the UV spectrum of the reduced P450 CO complex is not now regarded as being ideal for characterizing P450s due to the fact that, when the number of known P450 enzymes started to increase significantly, it was found that several individual forms could give rise to identical Soret maxima thus causing ambiguity in the analysis. Consequently, more precise methods were later developed to characterize P450s such as immunoblotting with specific antibodies, although full protein (or gene) sequence analysis provides an unambiguous procedure for identifying unique P450 enzymes (Black and Coon, 1986).

Immunoquantification using monoclonal antibodies raised against purified P450s represents a standard procedure (Leeder et al., 1996; Stresser and Kupfer,

Table 1.5 Selected type I substrates

Aldrin	Halothane	Testosterone
Aminopyrine	Hexane	Toluene
Arachidonic acid	Hexobarbital	Vinyl chloride
Benzpyrene	Lauric acid	R-Warfarin
Benzphetamine	Methoxychlor	Zoxazolamine
Biphenyl	Naphthalene	
Camphor	Pentobarbital	
Chlordane	Phenacetin	
DDT	Phenobarbital	
Dieldrin	Propranolol	
Endrin	Secobarbital	

Table 1.6 Selected type II substrates

Aniline
Imidazole
Pyridine
Cyanide
Ethyl isocyanide
Nitric oxide
Pyrrolidone

Table 1.7 Selected reverse type I substrates

Caffeine
Theophylline
Butanol
Ethanol
Methanol
Propan-2-ol
Isoamyl alcohol
Isobutyl alcohol

Reference: Schenkman, et al., 1981

1999) for determining the presence of individual P450 enzymes in given samples, and a recent article describes the development of a panel of such antibodies for immunoblotting human hepatic P450s (Edwards et al., 1998). Table 1.9 compares the relative amounts of P450 isoforms in human liver in terms of percentages based on immunological quantification against those obtained from monooxygenase activities toward key marker substrates of the relevant P450 enzymes. By and large, it can be appreciated from an inspection of Table 1.9 that there is a good agreement between the two methods of determination, demonstrating that the procedures are essentially complementary despite the fact that some selective substrates may not be entirely specific for a particular P450 enzyme. Nevertheless, the use of marker substrates (Clarke, 1998; D.A. Smith et al., 1998) and the

Table 1.8 UV spectral characteristics of P450

	Absorption bands (nm)				
Iron state	*α*	*β*	*Soret*	*Near UV*	*P450 enzyme*
Fe (III) low-spin	569	535	417	460	CYP101
	568	535	418	460	CYP2B
	568	535	418	457	CYP1A
Fe (III) high-spin	646	540	391	–	CYP101
	644	547	394	–	CYP1A
	645	540	394	–	CY11A1
Fe (II) high-spin	–	542	408	–	CYP101
	–	544	413	–	CYP2B
	–	542	411	–	CYP1A
Fe (II) low-spin	–	550	447	363	CYP101.CO
	–	552	451	370	CYP2B.CO
	–	550	448	–	CYP1A.CO

References: Ruckpaul and Rein, 1984; Lewis, 1986, 1996a; White and Coon, 1980.

Note:
Figure 1.1 shows the characteristic Soret absorption band for the CO adduct of CYP21 (Estabrook et al., 1963).

Table 1.9 P450 content in human liver: a comparison between two methods

CYP	*Percentage*[a]	*Percentage*[b]	*Marker substrate*	*Reaction*
1A2	9	12.5	Phenacetin	O-deethylation
2A6	13	16.9	Coumarin	7-Hydroxylation
2C8	15	ND	ND	ND
2C9	14	15.7	Tolbutamide	4-hydroxylation
2C19	11	16.7	S-Mephenytoin	4-hydroxylation
2D6	4	3.2	Debrisoquine	4′-hydroxylation
2E1	3	2.7	Chlorzoxazone	6-hydroxylation
3A4, 3A5	21	18.4	Midazolam	1′-hydroxylation
4A11	10	14.0	Lauric acid	12-hydroxylation

Reference: Edwards et al., 1998.

Notes:
a = immunoquantification
b = based on enzyme activity
ND = not determined

There is a reasonably good agreement ($R = 0.89$) between the two methods (a and b) used, with the possible exception of CYP2C19.

evaluation of P450-mediated metabolism remains an important technique for the estimation of P450 content which will, however, vary significantly in human populations depending on ethnicity, age and gender differences.

1.5 Genetic polymorphisms and allelic variants: protein sequence alignments

From the use of selective substrates for various human P450s, it has been established that there are substantial genetic variations which exist between both individuals and within certain ethnogeographical groups (Price-Evans, 1993; Gonzalez and Gelboin, 1991, Nebert, 1997a and b; Nebert and Carvan, 1997). Polymorphisms in drug metabolism, in particular, are known for several human P450 enzymes, such as CYP2D6 (Eichelbaum and Gross, 1990; Heim and Meyer, 1991) and CYP2C19, although there is also evidence for variations in CYP2A6, CYP2C9 and CYP1A2, for example (Smith et al., 1998a; Miners and Birkett, 1998).

These genetic polymorphisms have been shown to be the result of allelic variants in the particular P450 genes concerned (Sachse et al., 1997; Ingelman-Sunberg et al., 1995; Tucker et al., 1998) and this has resulted in several investigations of pharmacogentic phenotyping and genotyping in human cohorts (Gonzalez and Idle, 1994; Rodrigues, 1999; Aithal et al., 1999). Many (~ 30) allelic variants of CYP2D6 are known (Daly et al., 1996) and a significant number of these correspond with specific mutations (single or multiple) in the coding sequence of the P450 gene. Somewhat fewer allelic variants have been reported for other P450s and, for example, the situation in the human CYP2C subfamily (notably CYP2C19) has been reviewed (Goldstein and De Morais, 1994). It is possible that genetic polymorphisms in drug metabolism which relate to allelic variation in human P450s are linked with susceptibility towards certain disease states. For example, a variant in CYP1A1 has been reported which appears to cosegregate with lung cancer susceptibility in smokers (Kawajiri et al., 1992; Kawajiri and Hayashi, 1996). Also, there has been an association reported (Aithal et al., 1999) between CYP2C9 polymorphisms and risk of bleeding complications from the use of the anticoagulant drug, warfarin.

Mutants in human and other mammalian P450s are extremely useful for determining the likely sites of interaction with either substrates or redox partners of the P450 enzyme itself (Lewis, 1998a). Substrate recognition sites (SRSs) in the CYP2 family have been defined by Gotoh (1992) based on protein sequence alignment of various CYP2 family enzymes. Figure 1.3 shows an example of protein sequence alignment of selected CYP2 family P450 enzymes (Lewis, 1998b) which have had their individual amino acid labels (for single letter codes see Table 2.3) coloured according to type of residue (see legend to Figure 1.3). Those amino acids which differ in allelic variants of the same P450 are emboldened, as are those which have been the subject of site-directed mutagenesis experiments (reviewed in Lewis, 1998a). The locations of individual SRSs are also indicated, and these show that mutants which give rise to alteration in enzyme activity lie within five of the six SRSs. This alignment provides a framework for homology modelling of P450s from crystal structure templates, together with enabling a phylogenetic analysis of P450 families and subfamilies (see Chapters 6 and 2, respectively, for more details).

1.6 Biotechnological applications

The use of heterologously expressed human P450s for studying drug metabolism *in vitro* is now commonplace (Parkinson, 1996; Eddershaw and Dickins, 1999). For example, human CYP1A2 expressed in a yeast vector has been employed in the quantification of drug–drug interactions between tacrine and fluvoxamine (Becquemont et al., 1998), thus leading to a capability for the early prediction of potentially adverse drug reactions in man. A number of the potential uses of expressed P450s are shown in Figure 1.4, and a detailed account of expression systems has been produced by Doehmer and Griem (1993) and by Gonzalez and Korzekwa (1995), whereas the employment of these recombinant DNA approaches in carcinogenesis mediated by P450s has also been reviewed (Miles and Wolf, 1991; Gonzalez et al., 1991).

For reports of individual P450s being used in expression systems, the interested reader is referred to publications relating to CYP1A2 (Sandhu et al., 1994), CYP2B (Kedzie et al., 1991), CYP2D6 (Rowland et al., 1993), CYP2E1 (Gillam et al., 1994) and CYP4A1 (Faulkner et al., 1995) where the latter utilized electrocatalysis procedures to mimic the endogenous reaction of fatty acid ω-hydroxylation. In addition, heterologous expression systems and recombinant DNA approaches in the P450 field have been reviewed (Estabrook et al., 1991; Langenbach et al., 1992; Gonzalez and Gelboin, 1992). It would appear that yeast represents the preferred vector for heterologous gene expression of P450s (Wiseman, 1993, 1996a and b; Winkler and Wiseman, 1992; Imaoka et al., 1996) and genetic engineering of P450s in both bacterial and yeast systems has also been reviewed (Yabusaki and Ohkawa, 1991). An overview of P450 biotechnology, which covers many of the areas shown in Figure 1.4 has been provided by Munro (1994), whereas the use of engineered P450s for biodegradation and bioremediation has also been reviewed (Guengerich, 1995b; Kellner et al., 1997). Examples of mutant P450s being employed in bioremediation include reports of naphthalene (Kulisch and Vilker, 1991), diphenylmethane (Fowler et al., 1994), and pyrene degradation via genetically engineered CYP101 (England et al., 1998) together with the use of mutant CYP101 in the oxidative metabolism of polyhalogenated hydrocarbons (Kikuchi et al., 1994; Wackett et al., 1994). Moreover, engineered P450s have been shown to perform both biosynthetic reactions (Duport et al., 1998) and mutant strains of CYP101 are known to act as peroxidases instead of showing the usual mono-oxygenase reaction (Joo et al., 1999). The use of P450 mimics as biocatalysts has also been reviewed (Wiseman, 1994, 1996c and d; Wiseman et al., 2000) and it has been reported that iron porphyrin complexes in a zeolite cage support can

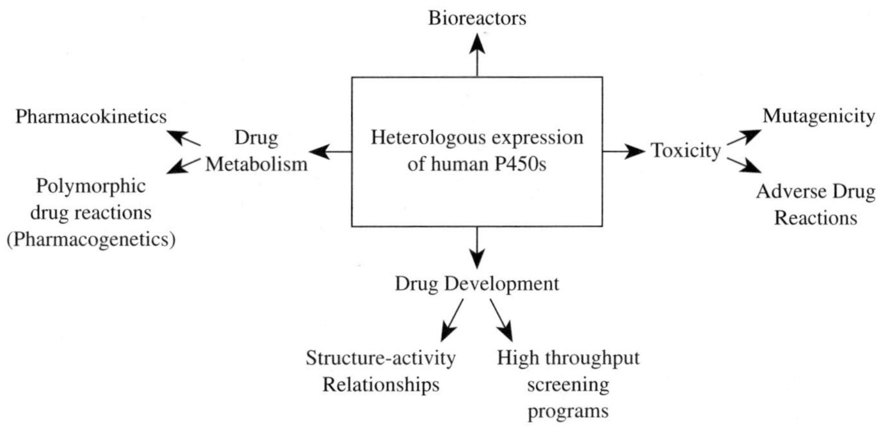

Figure 1.4 Heterologously-expressed human P450s: current applications.

reproduce certain P450-like mono-oxygenase reactions (Parton et al., 1994), thus indicating the considerable future potential for P450 mimics in this area.

Finally, site-directed mutagenesis of P450s is able to give rise to altered catalytic properties (Li et al., 2000; Johnson, 1992; Johnson et al., 1992) and substrate regio-selectivity, whereas the potential regions of redox partner interaction can be probed using this technique, as has been shown for CYP17 binding with cytochrome b_5 (Lee-Robichaud et al., 1995, 1998, 1999) and for reductase binding sites in CYP2B4 (Schulze et al., 2000; Lehnerer et al., 1999, 2000). This area is extensive, however, and some of the work carried out in P450s from families CYP1, CYP2 and CYP3 will be summarized later. However, the interested reader is referred to a review of CYP2 family mutations (Lewis, 1998a) which also includes information on allelic variants.

1.7 Nomenclature note: cytochromes, peroxidases and P450

There are several distinct types of redox proteins and enzymes which, historically, have been termed cytochromes. It would appear that the only common factor in all of these enzymes and proteins is the possession of a heme group, but their primary, secondary and tertiary structures are quite different. A number of cytochromes, e.g. cytochromes a, b and c, are electron transport proteins that tend to be tightly coupled physiologically (e.g. in mitochondria) for electron transfer down a redox potential gradient (Lehninger et al., 1993). However, the use of the term 'cytochrome' as applied to P450 (EC 1.14.14.1) is a misnomer, and these are now regarded as heme-thiolate proteins or, better, heme-thiolate enzymes. The other cytochromes tend to possess at least one histidine residue which ligates the heme iron whereas, in P450, the iron ligand is a cysteinate residue (see Table 1.10

Table 1.10 The P450 signature motif in bacterial forms

CYP	*Abbreviation*	*Sequence*										*Species*
									↓			
101	cam	F	G	H	G	S	H	L	C	L	G	Pseudomonas putida
102	BM-3	F	G	N	G	Q	R	A	C	I	G	Bacillus megaterium
103	PinF1	F	G	A	G	V	H	R	C	L	G	Agrobacterium tumefaciens
104	PinF2	F	G	S	G	P	H	H	C	P	G	Agrobacterium tumefaciens
105A1	SU1	F	G	F	G	V	H	Q	C	L	G	Streptomyces griseolus
105B1	SU2	F	G	F	G	V	H	Q	C	L	G	Streptomyces griseolus
105C1	Chop	F	G	H	G	M	H	Q	C	L	G	Streptomyces spp.
105D1	SoyC	F	G	F	G	V	H	Q	C	L	G	Streptomyces griseolus
106	meg	F	G	N	G	P	H	F	C	L	G	Bacillus megaterium
107A1	eryF	F	G	Q	G	I	H	F	C	M	G	Saccharopolyspora erythraea
107B1	orf	F	G	H	G	I	H	F	C	V	G	Saccharopolyspora erythraea
108	terp	F	G	W	G	A	H	M	C	L	G	Pseudmonas spp.
109	ORF405	F	G	F	G	I	H	F	C	L	G	Bacillus subtilis
110	ana	F	G	G	G	S	R	R	C	L	G	Anaboena cyanobacter
111	lin	F	G	S	G	Q	H	V	C	V	G	Pseudamonas incognita
112	BJ-1	F	G	Y	G	P	H	A	C	I	G	Bradyrhizobium japonicum
113	eryK	F	G	H	G	V	H	F	C	L	A	Saccharopolyspora erythraea
114	BJ-4	F	G	H	G	I	H	F	C	L	G	Bradyrhizobium japonicum

References: Nelson, 1995; Nelson et al., 1996.

Notes:
Consensus prokaryote P450 signature motif: F G *x* G *x* H/R *x* C L G/A where *x* is any amino acid residue (although there are certain preferences) as can be seen from the above list.

↓ denotes heme-ligating invariant cysteine

Table 1.11 Summary of P450 superfamily nomenclature

CYP	*Functionality or species*
1	Metabolism of foreign compounds
2	Metabolism of foreign compounds
3	Metabolism of foreign compounds
4	Metabolism of long-chain fatty acids
5	Thromoboxane biosynthesis
6	Insect forms
7	Steroid metabolism
8	Prostacyclin biosynthesis
9	Insect forms
10	Mollusc forms
11	Steroid biosynthesis
17	Steroid biosynthesis
19	Steroid biosynthesis
21	Steroid biosynthesis
24	Vitamin D_3 metabolism
26	Retinoid metabolism
27	Bile acid biosynthesis
51–70	Fungal forms
71–100	Plant forms
101–140	Bacterial forms

Reference: Nelson et al., 1996.

for the conserved cysteine signature motif in bacterial P450s). This partially explains the fact that the major function of P450s lies in the activation of a dioxygen molecule, and subsequent single oxygen insertion into an organic substrate (reviewed in Lewis, 1996a). The peroxidases, however, are in an entirely different class of enzymes and their prime function is in the breakdown of hydrogen peroxide to water (Marnett and Kennedy, 1995). Peroxidases also contain a heme moiety but the ligand is usually histidine, although the enzyme chloroperoxidase possesses a cysteine heme ligand and, consequently, displays some spectral characteristics in common with P450s. However, the crystal structures of these enzymes show that they are quite distinctly different (Li and Poulos, 1994; Marnett and Kennedy, 1995). Furthermore, peroxidases usually only have sufficient space in their active sites to admit a single molecule of hydrogen peroxide, whereas P450s generally possess larger active sites to allow for the occupancy of both organic substrates and oxygen. Table 1.11 provides a summary of P450 nomenclature based on the functionality of the specific enzyme family or the particular species involved and the reader is referred to the following P450 websites for further information: http://www.icgeb.trieste.it/p450 and http://drnelson.utmen.edu/CytochromeP450.html. These also provide links with other individual websites of several workers in this area, and contain much useful source material in the P450 field.

Chapter 2

Evolution of the P450 superfamily

We are the products of editing, rather than authorship.

(George Wald)

2.1 Introduction

This planet teems with life. Over one million separate species compete for survival in a multitude of ecological niches from the tropics to the polar ice-caps. The vast majority of these life-forms utilize oxygen as a primary source for metabolism and biosynthesis: it was not always so (Lovelock, 1988). The estimated age of the Earth is about 4550 million years, whereas it is thought that the crust may not have formed until around 3800 million years ago (Margulis and Sagan, 1995). It is generally accepted that, at the time when earliest life began, i.e. with the archaebacteria developing *circa* 3800 million years ago at least (Mojzsis et al., 1996), the Earth's atmosphere was essentially reducing in nature with very little oxygen present. Consequently, these primitive prokaryotic organisms would have been anerobic in nature and probably utilized the thermal energy produced by underwater vulcanism prevalent around 3500 million years ago. As at least one P450 is known to be present in a species of thermophile bacteria, it would appear that the enzyme superfamily is extremely ancient and, presumably, dating back as far as 3500 million years before present. However, P450s have not been isolated from *E. coli* or various other archaebacterial anerobes and, consequently, it seems likely that P450 would not have arisen prior to the advent of the thermophiles where, presumably, the rich supplies of both iron and sulphur from volcanic outgassing of the planet's core could have represented a primary source of the constituents required for a heme-thiolate enzyme system utilizing iron-sulphur redoxin as a redox partner.

It is thought that the rise in atmospheric oxygen levels, brought about by the emergence of cyanobacteria utilizing photosynthesis, triggered the development of early life in the shallow seas from the primitive stromatolites to more complex eukaryotic organisms (Margulis and Sagan, 1995). A general estimate for the changeover from a reducing to an oxidizing atmosphere would appear to be around 2000 million years ago, as recorded for the occurrence of banded-iron formations

(BIFs) which were produced when iron comes out of solution due to contact with free gaseous oxygen. The increase in atmospheric oxygen enabled life to develop from single-celled organisms to the progressively more complex forms because, although molecular oxygen is normally chemically unreactive with organic compounds, enzymes like P450 can unlock the chemical potential of this element by converting oxygen from its triplet ground state to the active singlet excited state such that reactions with organic substrates are possible.

Consequently, once oxygen levels rose significantly due to the photosynthetic activity of blue-green algae (Knoll, 1992) which started to proliferate around 1800 million years ago, enzyme systems could utilize this oxygen for metabolism such that eukaryotic organisms were able to develop more rapidly – thus leading to the so-called Cambrian 'explosion' of new life forms which occurred about 550 million years before present. In fact, the ancestors of all modern animal kingdoms can be discerned from analysis of the multi-cellular forms fossilized in the Burgess Shale deposits, and it is thought that the Cambrian expansion may have occurred following a major extinction of about 70% of all previous species due to an ice age around 600 million years ago, possibly linked with the supercontinental plate tectonics cycle (Hedges et al., 1996).

By 450 million years ago, colonization of land began, firstly by plants and then by certain animal species during the Devonian period about 400 million years before present (Harland et al., 1989). It is thought that, via a co-evolutionary process, animals elaborated the P450 system from an essentially endogenous substrate-metabolizing role to one of detoxifying metabolism for potentially lethal plant products (Gonzalez and Nebert, 1990). This attractive scenario would explain the extensive branching of the P450 phylogenetic tree, and its concordance with the established timescale for evolutionary development of metazoans over the last 370 million years (Lewis et al., 1998b). It may also be no coincidence that the mid-point on the atmospheric oxygen level growth curve (Cloud, 1976) corresponds almost exactly with the period of land colonization by animal species around 397 million years ago, which could represent the most rapid rate of oxygen increase with time (Lewis, 1996a). However, about 80–90% of life disappeared around 250 million years ago in a major extinction which has been linked with formation of the Pangaea supercontinent. This would have resulted in a loss of ecologically important continental shelf regions, whereas increased vulcanism is likely to represent another contributory factor in the extensive loss of species at the start of the Triassic period. As metazoans developed from reptiles to birds and mammals, there appears to have been an accompanying break-up of Pangaea which may have been responsible for the ordinal diversification of animal species (Hedges et al., 1996).

The most recent large extinction event occurred 65 million years ago, about 145 million years after the previous one and it is thought that there may be a cyclical pattern to the major extinctions over the past 500 million years (Ridley, 1996). The Cretaceous-Tertiary boundary represents the point marking the demise of large reptilian species and this may have enabled the consequent rise of the mammalia, including primates and, eventually, mankind. Most large extinctions seem to occur every 145 million years and appear to be associated with the supercontinental cycle of expansion and contraction of tectonic plate drift (Lewis et al., 1998b). Consequently, it is possible that life evolved due to geological change, and may have actually self-regulated it via homeostatic mechanisms (Lovelock, 1988).

2.2 P450 phylogeny

The next question we shall consider is, how well does the development of life concord with elaboration of the P450 phylogenetic tree?

Ever since the number of P450s which have been sequenced reached a significant amount, there has been an interest in the possible ways in which the enzyme superfamily has evolved over the last 3.5 billion years. It is possible to make comparisons between any two or more protein (or DNA) sequences such that, using appropriate algorithms, the approximate evolutionary distance between them can be calculated (Gotoh and Fujii-Kuriyama, 1989); this is usually based on a specific rate of mutation which seems to be fairly constant for a given protein family (Creighton, 1993). From an evaluation of P450 divergence times, it would appear that the unit evolutionary period (i.e. the time required for a 1% change in amino acid residues) for the superfamily is about 4.5 million years, although a more careful analysis indicates that this rate is not linear with time, and may be linked with the rise in atmospheric oxygen over the last 2.1 billion years (Lewis, 1996a; Lewis et al., 1998b). The mutation rates of proteins have been described mathematically by Tajima and Nei (1984) on the basis of nucleotide sequence changes and divergence times between two related proteins. This has led to the UPGMA (unweighted pair group method of phylogenetic analysis (Nebert et al., 1991a; Nelson & Strobel, 1987)) method for the formulation of phylogenetic trees from analysis of the relevant sequences (see Figure 2.1 as an example). Moreover, evolutionary distances can be estimated from the following equation:

$$2tk = -\ln(1 - d/n)$$

where t is the time required for two proteins n residues long, differing at d sites, to diverge with a rate k of amino acid substitutions per site per annum.

Thus, it is feasible to establish a phylogenetic tree for a given set of proteins that maps out their development over a period of time which, in the case of the P450 superfamily, can be presented as shown in Figure 2.2, which displays a simplified P450 phylogenetic tree that demonstrates how the superfamily has developed over an evolutionary timescale. This figure also gives a comparison between major P450 divergences and currently accepted stages in the evolution of life on Earth (Lewis, 1996a). The concordances between these two elaborations are quite striking, and thus suggest that the two processes are evolutionarily linked,

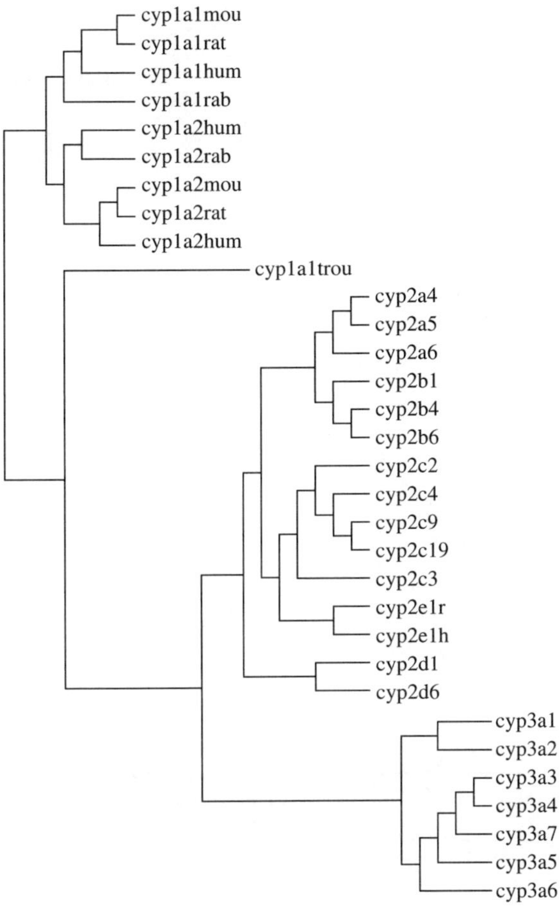

Figure 2.1 Phylogenetic tree of cytochrome P450 protein sequences inferred using the UPGMA tree drawing algorithm with Kimura calculated distances. Protein sequences from the CYP1, CYP2 and CYP3 families were aligned using ClustalW with manual readjustment based on known secondary conformation (Lewis and Sheridan, 2001).

possibly via an increasing concentration of oxygen in the terrestrial atmosphere (Lewis et al., 1998b; Lewis, 1996a). However, the role of an oceanic environment for the protection of early biological forms from solar UV radiation, in the absence of a developed ozone layer, has also been stated (Cleaves and Miller, 1998) and it is likely, therefore, that the earliest organisms to have developed P450 systems were thermophilic archaebacteria occupying abyssal oceanic vents in the vicinity of underwater volcanoes, where there would have been a plentiful supply of both iron and sulphur – key elements for P450s and their Fe_nS_n redoxin redox partners. It has also been advanced that an early role for P450 enzymes may have involved

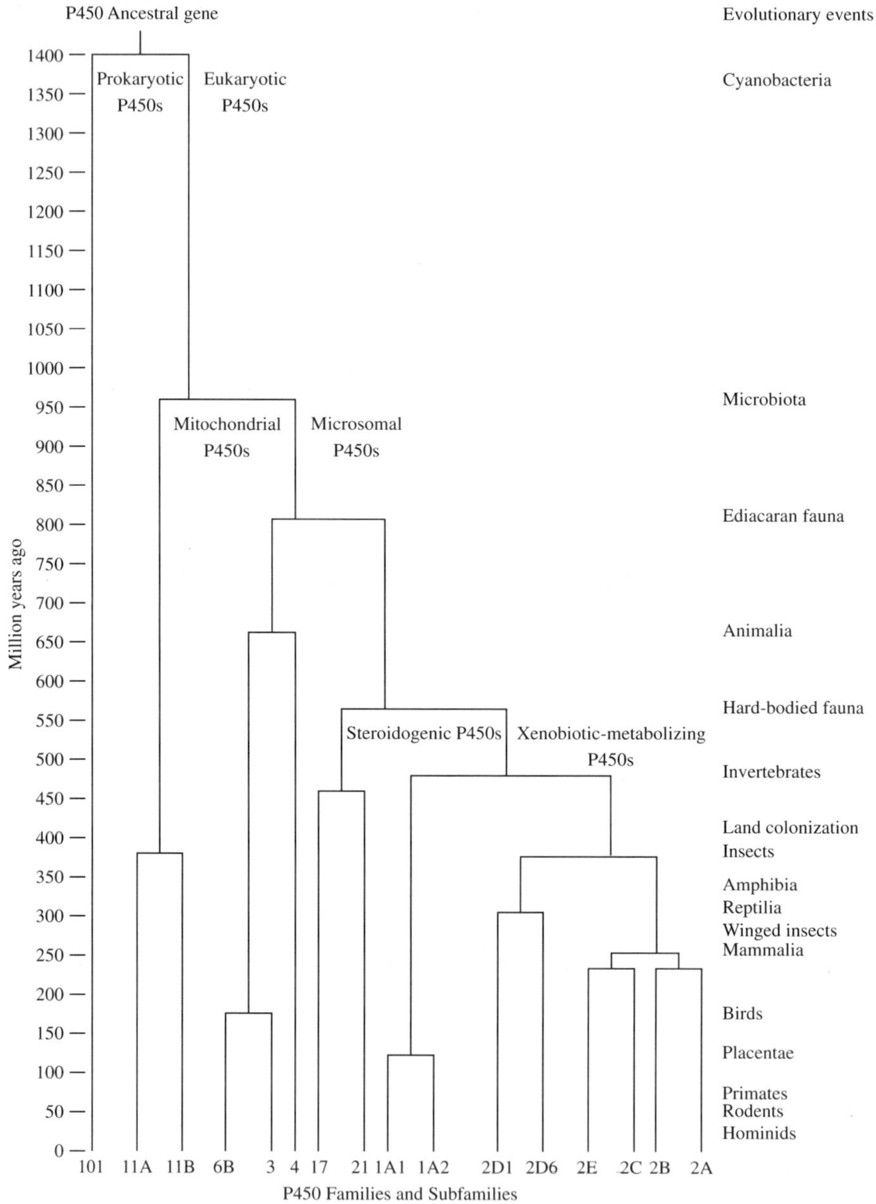

Figure 2.2 A summarized and abbreviated form of the P450 phylogenetic tree mapped against the generally accepted development of terrestrial biota (Lewis, et al., 1998b).

detoxification of reactive oxygen species harmful to anerobes (Wickramasinghe and Villee, 1975) and it is plausible that P450s could have utilized carbon substrates to detoxify oxygen in the Earth's early atmosphere.

The most recent dating for the divergence between eukaryotic and prokaryotic forms at 2.1 billion years ago (Doolittle et al., 1996) corresponds with the established time when the atmosphere changed from being predominantly a reducing one to that of an oxidizing nature (Canfield and Teske, 1996; Hattori et al., 1983). An analysis of the P450 phylogenetic tree also points to a major branching at around this period of time in the late Proterozoic where the main stem of the eukaryotic P450s separated from the bacterial ones. Nelson has shown that metazoan P450 evolution can be conceptualized in terms of higher order groupings, or clans, which enable an understanding of the complexity of animal P450 organization (Nelson, 1998). It will possibly necessitate a similar system for the analysis of bacterial P450 sequences as the published reports of these increase in number, and there is some evidence for a higher order structure even from consideration of a relatively small amount of prokaryote P450s (Degtyarenko and Archakov, 1993). However, it is important to recognize that one bacterial P450, namely CYP102 from *Bacillus megaterium*, clearly belongs to the eukaryotic P450 class and, consequently, there may be other instances where overlaps exist between prokaryotic and eukaryotic P450 types. Nevertheless, there is generally a clear distinction between bacterial P450s and those of the other biological kingdoms, such that one can explore the way in which eukaryotic P450s developed during the course of evolution. With reference to Figure 2.2, one can appreciate that the overall elaboration of metazoan P450s correlates well with the known evolutionary events which occurred over the last 1000 million years (Lewis et al., 1998b). Beginning with the earliest bifurcation of the eukaryotic P450 branch, it can be readily observed from inspection of Figure 2.2 that microsomal and mitochondrial P450s separated at around 950 million years ago. This is close to the accepted divergence time between the plant and animal kingdoms at about 1000 million years ago, which is also thought to represent the development of sexual reproduction. As mitochondrial P450s are crucial to the biosynthesis of steroid hormones from cholesterol which, eventually, gives rise to androgens and estrogens, it is possible that there is concordance between the emergence of mitochondrial P450s of the CYP11 family and the development of microbiota displaying sexual characteristics. Furthermore, microsomal P450s would be a natural consequence of more complex multi-cellular organisms such as plants, fungi and primitive animals. Consequently, one can speculate that microsomal, membrane-spanning, P450 enzymes may have occurred as a direct result of eukaryotic evolution.

At around 800 million years ago, a bifurcation of the microsomal P450 branch corresponding to the separation of the CYP3 and CYP4 families from the main microsomal P450 stem occurs. This period represents the time when the Ediacaran fauna were extant, and it is thought that pseudocoelomes such as *C. elegans* may have diverged from schizocoelomate species at this time (Doolittle et al., 1996).

Presumably, the role of P450 at this stage of biological development was primarily endogenous, being involved in the metabolism of eicosanoids and other fatty acids, together with steroidogenesis and possibly regulation of steroid hormone levels. Divergence of animalia is thought to have started about 650 million years ago, and recently reported fossil evidence (Brasier and McIlroy, 1998) indicates pre-Cambrian metazoan radiation *circa* 600 million years ago. This period of the late Neoproterozoic corresponds to a branching of CYP3 and CYP4 families which suggests that the separate roles of fatty acid and steroid metabolism may have become more well-defined at this stage; whereas the 600 million years ago time-frame is also regarded as the likely point for hemoglobin/myoglobin divergence, and this would be consistent with an increasing importance of oxygen in biological development.

By 550 million years ago the Cambrian 'explosion' of new life-forms produced the ancestors of virtually all animal phyla which radiated between 543 and 533 million years ago, and the emergence of hard-bodied animalia (on the basis of considerable fossil evidence) is paralleled by a major split in the microsomal P450 branch where the xenobiotic-metabolizing P450s (CYP1 and CYP2) diverged from those involved in steroid biosynthesis, namely, CYP17 and CYP21. Although P450s of CYP1 and CYP2 are associated with exogenous compound metabolism, it is possible that they may also possess endogenous roles and, at this stage in terrestrial evolution, it is likely that their involvement in regulation of steroids could have been the driving force which necessitated their demarcation from those P450s specifically designed for steroid hormone biosynthesis (Lewis et al., 1998b). This is further underscored by the divergence of CYP1 and CYP2 at around 470 million years ago when invertebrates developed and the branching of the CYP6 family from CYP3, which also occurred at this time, appears to be consistent with the appearance of invertebrate species since CYP6 seems to be primarily an insect P450 family. Colonization of land by plant species is thought to have started at around 450 million years ago and this point also marks the divergence between gnathostome and lamprey (Doolittle et al., 1996) whereas there is a bifurcation in the steroidogenic P450 stem, corresponding to the splitting of CYP17 and CYP21, which maps onto this timeframe in the Ordovician period, although the possible significance of this remains to be understood unless, of course, this simply represents speciation events.

Land colonization by animal species may have commenced by 400 million years ago, towards the start of the Devonian, when the projected growth curve for atmospheric oxygenation reached about 50% of the present level. At this point, it has been estimated that piscine P450s of the CYP1 family (Morrison et al., 1995) separated from the main stem and this could also correspond with the emergence of lungfish utilizing the increasing oxygen content of the atmosphere resulting from expanding plant life. The change from an aquatic to a largely terrestrial environment may also have coincided with the development of insect wings from ancestral gills (Averof and Cohen, 1997). Amphibia appeared at around 370 million years ago, shortly after insect species may have emerged, and this stage during the late Devonian period concords with initial branching of the CYP2 family which began with divergence of the CYP2D subfamily. Probably, this resulted from the need for detoxification of plant toxins, and the CYP2D enzymes are known to be primarily associated with the oxidation of nitrogenous bases, including many plant alkaloids. This is an example of a co-evolutionary process (Gonzalez and Nebert, 1990) whereby animal species developed detoxifying enzyme systems in order to metabolize poisonous chemicals biosynthesized by plants to deter animal predators which colonized land after plant species had become established.

By 300 million years ago, i.e. the recognized timeframe for bird–mammal–reptile divergence (Doolittle et al., 1996), the P450 tree indicates initial bifurcation of the CYP2D subfamily and it has been estimated that a further branching of the CYP1 stem led to the cleavage of the CYP1B and CYP1A subfamilies. The endogenous role of these enzymes remains to be elucidated but there is evidence that estrogen metabolism could represent one feature, although the detoxification of plant flavonoids also appears to constitute one of their xenobiotic functionalities despite clear indications for induction by and activation of polyaromatic hydrocarbons (Nebert and Gonzalez, 1987; Nebert et al., 1989a and b). An extensive elaboration of the CYP2 family appears to have occurred at around 250 million years ago when there was an initial bifurcation, followed by further divergence about 230 million years ago to form the CYP2A, CYP2B, CYP2C and CYP2E subfamilies, the CYP2D subfamily having already split much earlier. This seems to correspond with, firstly, the rise of reptilian species and, subsequently, that of mammalia; their radiation apparently being linked with continental drift (Hedges et al., 1996). Clearly, the significant rise in metazoan expansion during the Triassic and Jurassic periods would have necessitated the development of an extensive and yet sensitive enzyme 'arsenal' for the detoxification of exobiotics, although it is also thought that these P450s possess key roles in endogenous substrate metabolism and regulation (Lewis, 1996a).

Between 140 and 120 million years ago, speciation of the CYP2D subfamily appears to have occurred, whereas divergence of the CYP6 family is thought to have taken place at around 170 million years ago. This period of time is associated with the radiation of avian species (~ 150 million years ago) and the emergence of angiosperms or flowering plants (~ 125 million years ago). This probably marks the onset of bird–insect–plant co-evolution where both birds and certain insects would have required enzymes for the detoxification of plant chemicals, and an intriguing possibility is that phytoestrogens could have modulated estrus in herbivores whereas some of the developing P450s arising at this time may be able to exert a regulatory effect on hormonal levels via endogenous steroid metabolism.

The appearance of placental mammals at about 120 million years ago coincides with a bifurcation of the CYP1A subfamily into CYP1A1 and CYP1A2. The variations in CYP1A2 levels, which occur during the monthly cycle, could be associated with a possible endogenous role of this enzyme in the metabolism of estradiol (Horn et al., 1995) and, therefore, there may be an explanation for this concordance between P450 elaboration within the CYP1 family and evolutionary events. By 100 million years ago the oxygen content of the atmosphere reached its current level, and 80 million years ago represents the time at which mammalian radiation occurred. At this point, the microsomal P450 tree shows many branches corresponding to speciation events, with primate radiation taking place from about 40 million years ago. Further branching of mammalian P450s occurred at 20 million years ago and 17 million years ago as a result of rodent speciation and rat/mouse divergence, respectively; whereas hominids probably developed around 15 million years ago, although *Homo sapiens* as a distinct species did not emerge until about 1 million years ago.

The well-documented polymorphisms in human populations where certain ethnogeographical groups exhibit 'poor-metabolizer' phenotypes towards drug-metabolizing P450s, such as CYP2D6 and CYP2C19, could represent the results of thousands of years of dietary preferences (Nebert, 1997a). Consequently, the pharmacogenetics of P450 substrates may have been largely determined by dietary differences and specific food sources arising from early human migrations. It would also appear that the P450 complements of other species will have been largely determined by their varying food sources, and the striking differences found between New World monkeys and Old World primates may constitute one example (Lewis et al., 1998b). It is possible to explain that CYP1A2 is present in New World monkeys, whereas Old World primates tend to possess CYP2A instead, on the grounds of continental drift and dietary requirement. Opening of the South Atlantic probably began around 80 million years ago, which corresponds with the emergence of pro-simian primates and also divergence of CYP1A1 and CYP1A2. Consequently, there could have been a geographical separation of primate populations which would have evolved by adapting to their differing habitats and environment. Specifically, New World primates tend to be arboreal omnivores whereas those of the Old World are herbivorous and ground-dwelling. Thus, a likely explanation is that the Old World group would require CYP2A for the metabolism of plant flavonoids, whereas New World monkeys needed CYP1A2 for clearing aromatic amino (and amido) compounds present in their more carnivorous diet. *Homo sapiens*, however, has retained roughly equivalent amounts of both CYP1A2 and CYP2A6 due to an initially vegetarian diet recently supplemented by the practice of eating cooked meat, which is rich in heterocyclic amines that are typical CYP1A2 substrates.

Another example of P450 enzymes being specifically required for the metabolism of potentially toxic plant products is afforded by the black swallowtail butterfly, *Papilio polyxenes*, which possesses a unique P450 (CYP6B1) that is able to metabolize xanthotoxin (Guengerich, 1993). This furocoumarin, produced by plants of the *Apiaceae* and *Rutaceae* families, is highly toxic to most insect species but the swallowtail butterfly has, via a co-evolutionary process, developed a P450 which enables its larvae to feed on such plants with immunity (Guengerich, 1993).

Many plant flavones and phytoalexins are biosynthesized by P450s, whereas the anthocyanins responsible for most flower colours are also produced by plant

P450 enzymes (Holton et al., 1993). Some plant flavones possess weakly estrogenic properties and, moreover, a substantial number of flavonoids also act as P450 inhibitors (Rendic and DiCarlo, 1997). Consequently, it is possible that many plant toxins act upon the P450 systems of potential animal predators in a variety of ways, although these could be regarded as affecting the population size of both plant and animal species, thus achieving a self-regulatory balance over a period of time.

This homeostatic mechanism for regulating population dynamics in co-evolutionary processes could represent an example, at the metabolic level, of the 'Daisyworld' hypothesis expounded by Lovelock (1988). Nebert and co-workers (1990) have elaborated upon the potential role of phytoalexins and other plant flavonoids in plant-animal 'warfare' via their action on the Ah receptor (AhR) and its associated gene battery which regulates the expression of CYP1 family P450s and related enzymes involved in metabolism. Although the endogenous roles of the CYP1 family enzymes remain to be fully elucidated, there is evidence to suggest an involvement in development and growth, whereas relatively high levels of AhR are found in chondrocytes (Stegeman et al., 1996; Iwata and Stegeman, 2000). Moreover, enzymes of the CYP1 family (which are regulated by the AhR) are known to metabolize endogenous estrogens, such as estradiol, where there is some regioselectivity of hydroxylation in that CYP1B1 tends to favour 4-hydroxylation, whereas CYP1A2 primarily mediates 2-hydroxylation (Rendic and DiCarlo, 1997). Consequently, phytoestrogens and other plant flavones may be able to act upon both the AhR and CYP1 enzymes themselves and thus provide some degree of regulatory control over the normal levels of endogenous estrogens such that modulation of the estrus cycle in animals can occur following ingestion of the relevant plant species.

Table 2.1 shows various plant products which are either metabolized by or inhibit animal P450s known to be associated with the Phase 1 metabolism of drugs and other foreign compounds. In fact, the likely reason why there is an appropriate complement of drug-metabolizing enzymes in human liver, for example, stems from an evolutionary origin which required detoxifying systems to eliminate potentially deleterious plant toxins (Gonzalez, 1989; Gonzalez and Nebert, 1990; Gonzalez and Gelboin, 1992).

There are similarities between the binding sites of G-protein coupled receptors (GPCRs), such as the β-adrenoceptor, and those of the P450s which metabolize the same classes of chemicals that act as ligands for the GPCRs (e.g. neurotransmitters and their antagonists). For example, the same types of amino acid residues (namely aspartate, serine and phenylalanine) are present both in the putative active sites of CYP2D subfamily enzymes and also in the ligand-binding sites of tryptaminergic receptors and adrenoceptors, where the disposition of their sidechains is also similar. Moreover, this analogy extends to some of the nuclear receptors which bind steroids with respect to those P450s involved in steroid biosynthesis, for example. In particular, this is found for the estrogen receptor ligand-binding site and the putative active site of aromatase (CYP19) where it

Table 2.1 Plant chemicals and drug-metabolizing P450s

CYP	*Chemical substrates or inhibitors*
1A	Flavones, psoralens, phytoalexins, alfatoxin B_1, tannins
2A	Coumarin, xanthotoxin
2B	Cocaine, nicotine
2C	Betulinic acid
2D	Quinine, quinidine, yohimbine, sparteine
2E	Ethanol, ethylene glycol
3A	Quinine, quinidine, pyrrolizidine alkaloids, aflatoxin B_1

Reference: Rendic and DiCarlo, 1997.

Notes:

1 Most modern pharmaceuticals and agrochemicals have been designed on the basis of analogy with natural plant products and, therefore, it is not surprising that the same P450s are involved in their metabolism.

2 Some natural products metabolized by xenobiotic-metabolizing mammalian P450s are also listed above.

appears that a combination of acidic and basic residues bind the steroidal ligand or substrate in each case. Furthermore, there also seems to be some degree of similarity in tertiary fold between proteins of the steroid hormone receptor superfamily and P450s, where α-helical arrangements appear to dominate the common structural cores of the two types of protein, and is thus at least suggestive of an evolutionary link. It is also possible that the steroidogenic P450s may have developed from those enzymes which metabolize long-chain fatty acids, as the carbon skeletons of these two classes of substrates are superimposable (Lake and Lewis, 1996).

Therefore, the P450 superfamily of Phase 1 enzymes seems to have evolved over a period of about 3.5 billion years, first to detoxify molecular oxygen, and it appears that they subsequently acquired the ability (once eukaryotes developed) to harness its chemical potential for the oxidative metabolism of endogenous compounds such as steroids, prostanoids and eicosanoids. Indeed, long-chain fatty acids may have been the templates for constructing the steroid nucleus and, furthermore, the substrate-binding sites for the P450s which metabolize carboxylic acids and steroids show considerable similarity. P450 enzymes were then utilized in plant–animal co-evolution both to biosynthesize toxins (in plants) and to detoxify xenobiotics (in animals), with this latter role being augmented by that of drug metabolism in the 20th century. Moreover, the human drug-metabolizing enzyme polymorphisms evident at present probably arose through dietary preferences developing in ethnogeographical population groups as a result of their changing habitats and food sources caused by human migrations.

2.3 P450 sequences

Table 2.2 provides information regarding the numbers of P450 sequenced to date. Although this list is increasing all the time, currently about 1200 P450s have had their gene sequences determined, and around 50 human P450s are known, whereas about 60 rat P450s have been identified, for example. The most recent published update is that of Nelson et al. (1996) and the reader is referred to the relevant P450 website for further information (http://drnelson.utmem.edu/Cytochrome P450.html). P450s are classified into gene families on the basis of a greater than 40% sequence identity, whereas there is a less than 40% identity between one P450 family and another. Nelson has shown that metazoan P450s can be classified into higher order groupings, termed clans (Nelson, 1998, 1999) with the nematode, *C. elegans*, forming a separate clan comprised of families CYP14, CYP23, CYP33, CYP34, CYP35 and CYP36, although this organism also possesses P450s which form part of other clans, such as those in clans 3 and 4. The sequences from *C. elegans* are interesting in that they provide a means of studying P450 evolution in the context of metazoan development, including that of mitochondrial origins, as this species appears to lack those P450s normally regarded as mitochondrial forms. However, it is generally accepted (Woese, 1987; Wheelis et al., 1992) that the mitochondria of animal species arose via the incorporation of purple bacteria into eukaryotic cells; and this process of endosymbiosis extends both to other animal cell organelles and to the chloroplasts of plant species, which were thought to have arisen from cyanobacteria by a similar method (Margulis and Sagan, 1995). The fact that one species of amoeba, known as *Pelomyxa*, symbiotically utilizes soil bacteria instead of mitochondria shows that eukaryotic cells probably encaspulated various types of prokaryotes which were then eventually modified into the different organelles such as centrioles, cilia and flagella (thought to have arisen from the spirochaetes) whereas the endoplasmic reticulum could have come from other eubacterial species, possibly of the Gram-positive type. Figure 2.3 presents a simplified universal phylogenetic tree based on rRNA comparisons (Woese, 1987; Wheelis et al., 1992; Knoll, 1992; Nelson et al., 1993), showing these relationships in a graphical format, together with the possible origins of eukaryotic cell organelles (Lewis, 1996a).

Table 2.2 P450 sequences and characteristics

	Known	*Estimated*
P450s sequenced	750	1,000+
P450s per mammalian species	30–50	60–100
Reactions catalysed	40	60
Number of substrates	1,000+	1,000,000+
Number of inducers	200+	1000+
Number of human P450s	50	60
Number of rat P450s	60	70

References: Nelson, 1998; Coon et al., 1996; Estabrook, 1998.

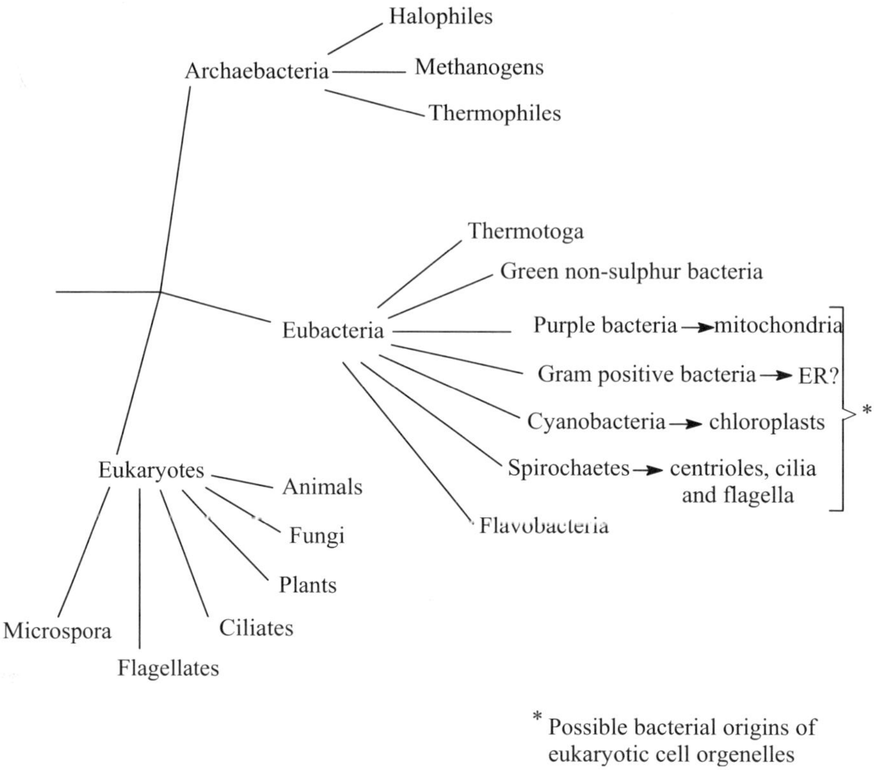

Note: P450s have been found in most species, including Archaea, although not in E.coli (Munro and Lindsay, 1996)

References: Wheelis, et al., 1992; Yang, et al., 1985; Woese, 1987; Knoll, 1992; Nelson, et al., 1993

Figure 2.3 Universal phylogenetic tree based on sequence comparisons (Lewis, 1996a).

Such considerations may point to the evolution of the P450 superfamily in respect of the similarities between the mitochondrial P450 system and that of many bacterial forms, such as $P450_{cam}$ and $P450_{terp}$, where these redox systems all possess the same type of electron transfer chain (Lewis and Hlavica, 2000). As has been reported by Degtyarenko and Archakov (1993), the Class 1 P450s utilize iron-sulphur redoxins to transfer electrons from NADH via an FAD-containing flavoprotein reductase; whereas Class 2 P450s, comprising the microsomal forms and $P450_{BM3}$ from *Bacillus megaterium*, all employ a single reductase flavoprotein containing both FAD and FMN cofactors for transferring electrons from NADPH to the P450 hemoprotein (reviewed in Lewis and Hlavica, 2000). It is possible to consider the evolutionary development of the P450 redox system in terms of the changeover from Fe_nS_n redoxins with their limited range (*circa* –300 mV) of redox potentials (Frausto da Silva and Williams, 1991) to those of the hemoproteins where the FeII/III redox potential can be varied from –400 mV to +400 mV by altering the nature of the heme ligands, and via control of the heme environment (Stellwagen, 1978). In the case of the P450 hemoprotein, a redox potential close to that of either redoxin or flavoprotein has been determined by the use of a proximal cysteine ligand (in its thiolate state) in conjunction with a distal water (or hydroxyl) sixth ligand that is readily displaced by an incoming substrate which can thus modulate the heme iron redox potential, and this is discussed further in the following chapter. The development of a linked FAD and FMN reductase probably arose from the requirement for a more controlled mechanism of electron transfer to the hemoprotein by incorporation of both components into a single polypeptide which enabled membrane binding of the flavoprotein in the endoplasmic reticulum of eukaryotic cells. However, in the case of $P450_{BM3}$ (CYP102) there is a fusion of the flavoprotein and hemoprotein domains into a single polypeptide chain, which appears to be responsible for the high catalytic activity of this enzyme (Lewis and Hlavica, 2000).

Comparisons between the deduced protein sequences of P450s has facilitated the study of how the overall tertiary structure may have evolved, such that the similarities and differences between the various classes and types can be made. For example, it has been established that all P450s contain a 10-residue signature motif which includes the thiolate (cysteine) ligand, and this region (which precedes the L helix) is highly conserved across the entire P450 superfamily of over 1200 genes. This P450 signature can be formulated as follows:

$$\text{F G/S } x \text{ H/R } x \overset{\downarrow}{\text{C}} x \text{ G/A}$$ where ↓ denotes the proximal heme ligand, cysteine

and where *x* is any amino acid (see Table 2.3 for amino acid codes) although there are certain preferences, as shown in Table 2.4 which provides details of 18 bacterial P450 signature motifs. This glycine-rich decapeptide, which forms a region proximal to the heme, contains a basic residue (usually arginine) two positions upstream of the invariant cysteine (arrowed) and begins with a well-conserved phenylalanine which could be involved in both heme binding and electron transfer from a redox partner. This conserved basic residue is also thought to represent a potential electrostatic contact point with oppositely charged acidic residues on either redoxin or reductase, depending on the class of P450, although cytochrome b_5 may also be able to bind with this sidechain in some instances such as is encountered in the hepatic microsomal system (Lewis and Hlavica, 2000).

In addition to the P450 signature, there are other regions of conservation between sequences across the superfamily which relate to key aspects of enzyme structure and function. In particular, the oxygen-binding pocket is likely to be formed from a ‘kink’ in the I helix distal to the heme where a particularly well-conserved sequence has been identified (see Table 2.5). This involves a pentapeptide containing a highly conserved alanine and glycine at its N-terminus, whereas the C-terminus usually includes a threonine preceded by an acidic residue. Essentially, it is thought that the glycine residue at this position is responsible for the unusual

Table 2.3 Amino acids: their properties and abbreviations

Amino Acid	*Codes*		*Type of residue*	*Molecular mass (Da)*	*pI*
Glycine	Gly	G	Nonpolar	57.05	5.97
Alanine	Ala	A	Hydrophobic	71.09	6.00
Valine	Val	V	Hydrophobic	99.14	5.96
Leucine	Leu	L	Hydrophobic	113.16	5.98
Isoleucine	Ile	I	Hydrophobic	113.16	6.02
Phenylalanine	Phe	F	Aromatic	147.18	6.48
Tyrosine	Tyr	Y	Aromatic	163.18	5.66
Tryptophan	Trp	W	Aromatic	186.21	5.89
Serine	Ser	S	Polar	87.08	5.68
Threonine	Thr	T	Polar	101.11	5.60
Cysteine	Cys	C	Hydrophobic	103.15	5.07
Methionine	Met	M	Hydrophobic	131.19	5.74
Asparagine	Asn	N	Amide	114.11	5.41
Glutamine	Gln	Q	Amide	128.14	5.65
Aspartic acid	Asp	D	Acidic	115.09	2.77
Glutamic acid	Glu	E	Acidic	129.12	3.22
Lysine	Lys	K	Basic	128.17	9.74
Arginine	Arg	R	Basic	156.19	10.76
Histidine	His	H	basic	137.14	7.59
Proline	Pro	P	Hydrophobic	97.12	6.30

Reference: Lehninger et al. 1993.

Notes:
pI is the isoelectric pH value and, as such, is a measure of the average pK_a of the amino acid.

deformation in the distal helix which enables binding of an oxygen molecule to the heme iron. The threonine residue appears to be important to ensure a typical mono-oxygenation of substrates by P450, as its mutation tends to cause uncoupling of the reaction. Moreover, it is also clear that the putative I helices of unusual P450s (see Table 2.5) indicate a departure from the normal oxygen-binding and mono-oxygenase activity in that there are key changes to the relevant amino acid positions, such as the threonine, the preceding acidic residue and/or the conserved upstream glycine.

In addition to the I and L helices which are involved in heme binding, several other portions of the general P450 sequence have been associated with substrate interactions. These are referred to as substrate recognition sites (SRSs), and Gotoh (1992) has identified a total of six SRSs from an analysis of CYP2 family P450 sequences. Extensive studies using site-directed mutagenesis experiments have shown that five separate regions of polypeptide tend to constitute the substrate-binding site for various P450s, including those of families CYP1, CYP2 and CYP3. However, only the CYP2 family has been investigated in detail, and work in this area has been reviewed fairly recently (Lewis, 1998a). The SRSs generally show significant variation between several P450 sequences which have been aligned, and this is due to the fact that a major characteristic of hepatic mammalian

Table 2.4 Signature motifs of 18 prokaryotic P450 sequences

CYP	*F*	*G/S*	x	*G*	x	*H/R*	x	*C*	x	*G/A*	*Gen EMBL accession number*	*Species*	*Trivial name*
101	F_{350}	G	H	G	S	H	L	C	L	G_{359}	M12546	*P. putida*	cam
102	F_{393}	G	N	G	Q	R	A	C	I	G_{402}	J04832	*B. megaterium*	BM-3
103	F_{362}	G	A	G	V	H	R	C	L	G_{371}	M19352	*A. tumefaciens*	PinF1
104	F_{349}	G	S	G	P	H	H	C	P	G_{358}	M19352	*A. tumefaciens*	PinF2
105A1	F_{347}	G	F	G	V	H	Q	C	L	G_{356}	M32238	*S. griseolus*	SU1
105B1	F_{354}	G	F	G	V	H	Q	C	L	G_{363}	M32239	*S. griseolus*	SU2
105C1	F_{344}	G	H	G	M	H	Q	C	L	G_{353}	M31939	*Streptomyces*	ChoP
105D1	F_{319}	G	F	G	V	H	Q	C	L	G_{368}	X63601	*S. griseolus*	SoyC
106	F_{349}	G	N	G	P	H	F	C	L	G_{358}	X16610	*B. megaterium*	meg
107A1	F_{345}	G	Q	G	I	H	F	C	M	G_{354}	M54983	*S. erythraea*	eryF
107B1	F_{345}	G	H	G	I	H	F	C	V	G_{354}	M83110	*S. erythraea*	ORF405
108	F_{370}	G	W	G	A	H	M	C	L	G_{379}	M91440	*P. spp.*	terp
109	F_{344}	G	F	G	I	H	F	C	L	G_{353}	M36988	*B. subtilis*	ORF405
110	F_{463}	G	G	G	S	R	R	C	L	G_{372}	M38044	*A. cyanobacter*	ana
111	F_{348}	G	S	G	Q	H	V	C	V	G_{357}	L23310	*P. incognita*	lin
112	F_{338}	S	Y	G	P	H	A	C	I	G_{347}	L02323	*B. japonicum*	BJ-1
113	F_{358}	G	H	G	V	H	F	C	L	A_{367}	L05776	*Sacc. erythraea*	eryK
114	F_{317}	G	H	G	I	H	F	C	L	G_{326}	L12971	*B. japonicum*	BJ-4

↓
invariant
cysteine
(heme ligand)

References: Nelson, D.R. in: Cytochrome P450 (P.R. Ortiz de Montellano, Ed.) Plenum, New York, 1995.

Notes:
x = any amino acid residue
See Table 2.3 for codes to amino acids.

Table 2.5 Putative I helical segments (distal to the heme) of various P450s

CYP	Species	Distal heme motif										Endogenous function
									↓			
3A4	human	F	I	F	A	G	Y	E	T	T	S	steroid 6β-hydroxylase
5A1	human	F	L	I	A	G	Y	E	I	I	T	thromboxane synthase (TXS)
5A1	rat	F	L	I	A	G	H	E	I	T	T	thromboxane synthase (TXS)
5A1	mouse	F	L	I	A	G	H	E	V	I	T	thromboxane synthase (TXS)
5A1	pig	F	L	I	A	G	Y	E	I	I	T	thromboxane synthase (TXS)
74	flax	N	S	W	G	G	F	K	I	L	L	allene oxide synthase (AOS)
74	guayule	N	T	F	G	G	V	K	I	L	F	allene oxide synthase (AOS)
19	human	M	L	I	A	A	P	D	T	M	S	aromatase
19	cow	M	L	I	A	A	P	D	T	M	S	aromatase
19	rat	M	L	I	A	A	P	D	T	M	S	aromatase
19	mouse	M	L	I	A	A	P	D	T	M	S	aromatase
19	chicken	M	L	I	A	A	P	D	T	L	S	aromatase
19	trout	M	V	I	A	A	P	D	T	L	S	aromatase
19	*Oryzias latipes*	M	V	I	A	A	P	D	T	L	S	aromatase
51	yeast*	V	L	M	G	G	Q	H	T	S	A	lanosterol C_{14}-demethylase
51	rat	L	L	L	A	G	Q	H	T	S	S	lanosterol C_{14}-demethylase
51	human	L	L	L	A	G	Q	H	T	S	S	lanosterol C_{14}-demethylase
51	*Issatchenkia*	V	L	M	G	G	Q	H	T	S	S	lanosterol C_{14}-demethylase
51	*S. pombe*	L	L	M	A	G	Q	H	T	S	A	lanosterol C_{14}-demethylase
51	*Ustilago maydis*	L	L	M	A	G	Q	H	T	S	S	lanosterol C_{14}-demethylase
51	pig	L	L	M	A	G	Q	H	T	S	S	lanosterol C_{14}-demethylase
51	*Candida tropicalis*	V	L	M	G	G	Q	H	T	S	A	lanosterol C_{14}-demethylase
51	*Candida albicans*	I	L	M	G	G	Q	H	T	S	A	lanosterol C_{14}-demethylase
51	*Candida krusei*	V	L	M	G	G	Q	H	T	S	A	lanosterol C_{14}-demethylase

Reference to sequences: Nelson et al., 1996.

Notes:
* yeast strain is *Saccharomyces cerevisiae* ↓ denotes distal threonine

1 There is normally a conserved distal threonine (arrowed) associated with oxygen binding and activation.
2 An acidic residue usually precedes the conserved threonine which may constitute a distal charge relay.
3 TXS sequences do not possess a conserved threonine residue.
4 AOS sequences do not possess either an acidic residue or distal threonine.
5 CYP51 sequences do not possess an acidic residue prior to conserved threonine.
6 CYP19 sequences do not possess a conserved glycine three residues upstream of threonine and all contain a proline residue prior to aspartate.
7 CYP3A4 exhibits a common distal I helix relative to most P450s; the others in the above list show altered activities that are partly explained by their unusual I helices.

microsomal P450s is their broad structural diversity in substrates; this aspect will be explored further in some of the following chapters.

However, protein sequence alignment of P450s indicates that, in addition to the heme and oxygen-binding regions, there are other conserved motifs which probably have a structural role, although the binding sites for redox proteins (such as reductase, redoxin and cytochrome b_5) also tend to exhibit some conservation within a given class of P450 enzymes. A number of basic residues appear to have been conserved in certain positions within P450 sequences for the binding of redox partners, and these tend to cluster in a region close to the binding sites of redox partners, but proximal to the heme face, in the case of the redoxin or cytochrome b_5 site (reviewed by Lewis and Hlavica, 2000). There are also two basic residues required for the formation of ion-paired interactions with the heme propionates and, in most microsomal P450s, a tryptophan residue is conserved for heme binding and possibly also for facilitating electron transfer from redox partners. The fact that bacterial P450s which utilize redoxins as redox partners do not possess a tryptophan at the same point, but instead have a C-terminal tryptophan on the redoxin, indicates that this key component for the electron transfer process may have been incorporated into the P450 sequence when eukaryotic cells required P450 functionality in the endoplasmic reticulum. Another addition apparent in microsomal P450 sequences is an N-terminal segment of about 30 amino acids which is generally regarded as representing a membrane anchor peptide, and this region usually contains a relatively large number of hydrophobic residues that are compatible with binding of the protein within an essentially lipophilic component of the phospholipid bilayer. Following the N-terminal membrane anchor peptide there is usually a polyproline motif which is believed to possess an essentially structural role, and other relatively small but highly conserved motifs also contribute to maintaining the P450 tertiary fold such as an internal ion-pair composed from the termini of a tetrapeptide, E*xx*R, although this grouping may also be involved in the catalytic activity of P450 enzymes.

Consideration of available bacterial P450 crystal structures shows that there is a common structural core, and sequence comparisons indicate that there is greater homology within the C-terminal portion of P450s with respect to the N-terminal half. The regions involved in heme and oxygen binding largely correspond to the C-terminal 50% of P450 sequences; whereas most of the variable SRSs are within the N-terminal segment, especially SRS1 which is also the largest section of peptide thought to be in contact with the substrate and, in addition, probably includes some residues that may be involved in substrate recognition and access. It is possible to consider the development of the P450 tertiary structure as being one of elaboration of the N-terminal section about a common C-terminal core, and inspection of known P450 crystal structural motifs such as the globin fold of α-helices suggests that P450s may have a common ancestor with globin proteins, or that one could have developed from the other. However, the Greek key helical bundle and Rossmann domain which have been identified as structural motifs in

P450s are well-known super-secondary structure elements that are also found in other proteins (Branden and Tooze, 1991).

In conclusion, therefore, it is assumed that the P450 structure evolved initially to detoxify oxygen when cyanobacteria began to produce sufficient quantities of free O_2 to render it toxic to anerobic organisms. Then, following the development of eukaryotes and multi-cellular animals, the P450 system utilized the reactive potential of O_2 for oxidative metabolism of chemically stable organic compounds, such as fatty acids, steroids, eicosanoids and prostanoids, in an endogenous role. P450s then played a part in the co-evolution of plants and animals, both in the biosynthesis of plant products and for the detoxification of poisonous chemicals developed as predator deterrents (e.g. phytoalexins and phytoestrogens). There has also been a role of plant P450s in the attraction of insect pollinators as flower colour pigments are also products of P450 reactions (Holton et al., 1993), so the co-evolutionary development of plants, insects and animals all indicate a key involvement for P450 enzymes. More recently, man-made chemicals such as pharmaceuticals and agrochemicals also represent P450 substrates and, of course, many of these were designed from naturally occurring plant products thus explaining why *Homo sapiens* is equipped with drug-metabolizing enzymes for the oxidation of a large number of synthetic compounds. However, as only about fifty P450s possess a known physiological role (Estabrook, 1998) it is possible that further functions may yet be discovered for those P450 enzymes generally regarded as being solely for xenobiotic metabolism. In particular, the regulation of steroid hormone levels and development (or maintenance) of gender characteristics in the neonate, together with cartilage formation, have all been postulated as P450-mediated functions. There are likely to be many more endogenous roles of P450s, e.g. in cutin monomer biosynthesis (Tijet et al., 1998), which presumably developed as a method of plant protection from predatory grazing. Consequently, the next few years will probably see the elucidation of many hitherto poorly understood metabolic pathways in various species, not only from the animal kingdom but in other forms of life, where members of the P450 superfamily are implicated as being physiologically important. Clearly the functional versatility and structural flexibility of P450 enzymes have played an important part in evolution, especially that of mammalian species (Negishi et al., 1996a and b).

In studying the major role of P450 in the evolution of life on Earth, however, the analysis of new gene sequences will be extremely useful for defining many of the key stages in development of the biota; where long-existent P450 families like CYP51 represent a prime 'molecular clock' in establishing an accurate timescale for the P450 phylogenetic tree (Nelson, 1998).

2.4 Nomenclature of P450 genes and enzymes

There is a systematic naming of P450s which has been in place since 1989, and periodically updated since (Nebert et al., 1989a and b, 1991a; Nelson et al., 1993, 1996). The nomenclature system makes use of the symbol CYP as an abbreviation for cytochrome P450, which is italicized when referring to the gene (see Table 2.6 for chromosomal locations of human P450 genes). An alphanumerical designation is employed for naming P450 families, subfamilies and individual proteins; most of these are listed in Tables 2.7–2.19. However, the nomenclature of new P450 sequences is determined by their similarity with others which have been previously assigned. The current systematic method bears the following designations (Tables 2.7 to 2.12), some of which are related to positions on the steroid nucleus (i.e. positions 11, 17, 19, 21, 24 and 27) that are metabolized by that form of P450. Also, certain ranges of CYP numbers are allocated to species within a given biological kingdom, e.g. plant P450s are 71–99, bacterial P450s are 101 and higher, fungal forms are 51–70, whereas some of the lower families tend to be exclusively composed of P450s from, for example, insect species (CYP6 and CYP9) and molluscs (CYP10).

The literature is riddled with mixtures of old and new names for P450s, and this is obviously confusing to many. Table 2.20 gives a listing of some of these instances, although this is by no means exhaustive and the reader is referred to the various nomenclature updates cited above for further information.

However, the general rules governing the assignment of P450 genes requires that there is a $\leq 40\%$ identity between a P450 in one family and that in another family, and that there is a $> 40\%$ identity between subfamilies of a given P450 family. Although this system was initially set up on an arbitrary basis, it appears to have been an appropriate choice as the original demarcation has worked satisfactorily in the majority of cases. Furthermore, the signature motif has enabled one to differentiate P450 proteins from those of others which appear to be related proteins but, in fact, are not P450s at all – such as the nitric oxide synthases (NOS), for example (Masters et al., 1996; Ortiz de Montellano et al., 1998; Knowles and Moncada, 1994; Renaud et al., 1993). The P450 signature is not sufficiently identifiable in NOS despite the fact that this protein is another heme-thiolate enzyme, and the crystal structure of NOS clearly show that this is not of the same class of hemoprotein as the P450 superfamily.

The total number of known P450 genes are subdivided into 15 mammalian families (and 29 subfamilies), 30 plant families, 50 bacterial families, 16 fungal families, 7 insect families, 2 mollusc families and 12 nematode families, thus bringing the total number of P450 families to date to over 120. Individual mammalian species tend to possess a relatively large number of individual P450 genes. For example, 60 P450s have been sequenced in the rat, 45 in the mouse and 32 in the rabbit, whereas humans have at least 36 P450 genes although it is likely that more will be discovered in the future. In contrast, insect species appear to have fewer P450 genes; at present, less than 20 have been reported for *Drosophila melanogaster* and *Musca domestica*. Apart from the exclusively insect families, CYP6, CYP9, CYP12, CYP15 and CYP18, there are several P450 genes from the CYP4 family present in all insect species investigated thus far, although the insect CYP4 subfamilies differ from those found in mammalia (Lake and Lewis, 1996). In addition to the tabulated P450 families, a comparison between some of the

Table 2.6 Chromosomal locations of human P450 genes and regulatory factors

CYP	*Location (Human Genome mapping)*	*Inducers*	*Nuclear receptors involved*
Xenobiotic metabolism			
1A1	15q22–q24	PAHs, TCDD	AhR, ARNT
1A2	15q22–q24		AhR, ARNT
1B1	1		AhR, ARNT
2A6	19q13.1–q13.2	PB	HNF-4
2A7	19q13.1–q13.2		
2B6	19q12–q13.2	PB	CAR
2C8	10q24.1	PB	RAR/RXR?
2C9	10q24.1	PB	RAR/RXR?
2C18	10q24.1	PB	RAR/RXR?
2C19	10q24.1	PB	RAR/RXR?
2D6	22q13.1		
2E1	10q24.3–qter	ACE, EtOH	
2F1	19q13.2		
2J2			
3A4	7q22–qter	DEX, PCN, RIF	GR, PXR
3A53A7	7q22.17q22.1		
Endogenous metabolism			
4A9		CLOF	PPAR
4A11	1	CLOF	PPAR
4B1	1p12–p34		
4F2			
4F3			
7	8q11–q12		
11A1	15q23–q24	ACTH (cAMP)	Ad4BP/SF-1
11B1	8q21–q22	ACTH (cAMP)	Ad48P/SF-1
11B2	8q21–q22		
17	10q24.3	LH	
19	15q21	FSH	
21	6p	ACTH	
27	2q33–qter	PTH	VDR

References: Smith et al., 1998a; Guengerich, 1995a; Gibson and Skett, 1994; Honkakoski and Negishi, 2000.

Notes:

PAH	= polyaromatic hydrocarbon	TCDD	= 2,3,7,8-tetrachlorodibenzo-*p*-dioxin
PB	= phenobarbital	EtOH	= ethanol
DEX	= dexamethasone	PCN	= pregnenolone 16α-carbonitrile
RIF	= rifampicin	CLOF	= clofibrate
ACTH	= aryl hydrocarbon receptor	RXR	= retinoid-X receptor
CAR	= constitutive androstane receptor	PXR	= pregnane-X receptor
RAR	= retinoic acid receptor	PPAR	= peroxisome proliferator-activated receptor
VDR	= vitamin D receptor	HNF-4	= hepatocyte nuclear factor 4

Table 2.7 The CYP1 family

Species	*CYP1A1*	*CYP1A2*	*CYP1A3*	*CYP1B1*
Rat	1A1	1A2		1B1
Human	1A1	1A2		1B1
Rabbit	1A1	1A2		
Dog	1A1	1A2		
Hamster	1A1	1A2		
Macaque	1A1			
Guinea pig	1A1	1A2		
Trout	1A1		1A3	
Plaice	1A1			
Red sea bream	1A1			
Toadfish	1A1			
Scup	1A1			
Butterfly fish	1A1			
Tom cod	1A1			
Sea bass	1A1			
Ice cod	1A1			
Mouse	1a1	1a2		1b1
Chicken		1A2		

Reference: Nelson et al., 1996.

Note:
The fish CYP1 genes are now regarded as belonging to a separate subfamily.

Table 2.8 The CYP2 family

Species	*CYP2A*	*CYP2B*	*CYP2C*	*CYP2D*	*CYP2E*
Rat	2A1, 2A2, 2A3	2B1, 2B2, 2B3, 2B12, 2B14, 2B15, 2B16	2C6, 2C7, 2C11, 2C12, 2C13, 2C22, 2C23, 2C24	2D1, 2D2, 2D3, 2D4, 2D5, 2D18	2E1
Mouse	2a4, 2a5, 2a12	2b9, 2b10, 2b13	2c29	2d9, 2d10, 2d11, 2d12, 2d13	2e1
Human	2A6, 2A7, 2A13	2B6	2C8, 2C9, 2C27, 2C28	2D6, 2D7, 2D8	2E1
Hamster	2A8, 2A9		2C25, 2C26, 2C27, 2C28	2D20	2E1
Rabbit	2A10, 2A11	2B4, 2B5	2C1, 2C2, 2C3, 2C4, 2C5, 2C14, 2C15, 2C16, 2C30		2E1, 2E2
Dog		2B11	2C21	2D15	

Continued...

Table 2.8 (continued)

Species	*CYP2A*	*CYP2B*	*CYP2C*	*CYP2D*	*CYP2E*
Macaque		2B17	2C20, 2C37	2D17	2E1
Guinea pig		2B18		2D16	
Goat			2C31		
Pig			2C32, 2C33, 2C34, 2C35, 2C36	2D21	
Cow				2D14	
Marmoset				2D19	2E1

Species	*CYP2F*	*CYP2G*	*CYP2H*	*CYP2J*	*CYP2K*
Human	2F1			2J2	
Mouse	2f2				
Rat		2G1		2J3	
Rabbit		2G1		2J1	
Chicken			2H1, 2H2		
Trout					2K1
Killifish					2K2

Species	*CYP2L*	*CYP2M*	*CYP2N*	*CYP2P*	*CYP2Q*
Trout		2M1			
Killifish			2N1	2P1, 2P2, 2P3	
Lobster	2L1				
Xenopus					2Q1

Reference: Nelson et al., 1996.

Table 2.9 The CYP3 family

Species	*CYP3A*
Rat	3A1, 3A2, 3A9, 3A18, 3A23
Human	3A4, 3A5, 3A7
Rabbit	3A6
Macaque	3A8
Hamster	3A10
Mouse	3a11, 3a13, 3a16
Dog	3A12
Guinea pig	3A14, 3A15, 3A17, 3A20
Goat	3A19
Marmoset	3A21
Pig	3A22

Reference: Nelson et al., 1996.

Table 2.10 The CYP4 family

Species	*CYP4A*	*CYP4B*	*CYP4C*	*CYP4D*	*CYP4E*	*CYP4F*	*CYP4G*
Rat	4A1, 4A2, 4A3, 4A8	4B1				4F1, 4F4 4F5, 4F6	
Rabbit	4A4, 4A5	4B1					
Human	4A9, 4A11	4B1				4F2, 4F3	
Mouse	4a10, 4a12 4a14	4b1					
Guinea pig	4A13						
Cockroach			4C1				
Mosquito			4C2	4D5, 4D6, 4D7			
Drosophila			4c3	4d1, 4d2, 4d8	4e1, 4e2 4e3, 4e4		4g1
Beetle roach			4C4, 4C5, 4C66				
House fly				4D3, 4D4, 4D9			4G2, 4G3

Species	*CYP4H*	*CYP4J*	*CYP4K*	*CYP4L*	*CYP4M*	*CYP4N*	*CYP4P*
Mosquito	4H1, 4H2 4H3, 4H4 4H5, 4H6 4H7, 4H8 4H9	4J1, 4J2, 4J3	4K1				
Drosophila							4p1
House fly						4N1, 4N2	
Tobacco hornworm				4L1, 4L2	4M1, 4M2 4M3		

Reference: Nelson et al., 1996.

previously used names of mammalian P450s and their current CYP nomenclature is shown in Table 2.20. Moreover, for the bacterial P450s shown in Table 2.16, there is an indication of their non-systematic names together with the reactions catalysed by each one (where known) and some of these names stem from the substrate concerned. Many of the P450s involved in steroid biosynthesis possess CYP numbers which relate to positions on the steroidal nucleus where metabolism occurs, and some of the previously used names also show a degree of similarity

Table 2.11 Families CYP5, CYP6, CYP7, CYP8, CYP9 and CYP10

Species	*CYP5*	*CYP6*	*CYP7*	*CYP8*	*CYP9*	*CYP10*
Human	5		7A1, 7B1	8		
Rat	5		7A1, 7B1			
Pig	5					
Mouse	5		7a1, 7b1			
Rabbit			7A1			
Hamster			7A1			
Cow				8		
House fly		6A1, 6A3, 6A4, 6A5, 6A6, 6A7, 6C1, 6C2, 6D1				
Drosophila		6a2, 6a8, 6a9			9b1, 9b2, 9c1	
Swallowtail butterfly		6B1, 6B3				
Cotton bollworm		6B2				
Papilio glaucus		6B4				
Tobacco budworm					9A1	
Pond snail						10

References: Nelson et al., 1996; Cohen and Feyereisen, 1995.

Notes:
CYP5 is a thromboxane synthase.
CYP7 is a cholesterol 7a-hydroxylase.
CYP8 is a prostacyclin synthase.

with the new system. For example, $P450_{11\beta1}$ is now referred to as CYP11B1 and $P450_{17\alpha}$ is called CYP17. Finally, CYP51 is a P450 family which spans several species from fungal forms to mammalia, including *Homo sapiens*. This enzyme specifically catalyses the C-14 demethylation of lanosterol (formerly named $P450_{14DM}$) and this describes a functionality which has been conserved for hundreds of millions of years (Lewis et al., 1999a). Consequently, this particular P450 family constitutes an extremely useful molecular clock for the detailed analysis of the P450 phylogenetic tree (Nelson, 1998). However, a fuller understanding of the way in which the P450 superfamily elaborated over time will be obtained as more P450 genes are sequenced, particularly from evolutionarily distant species.

Table 2.12 Steroidogenic and related families CYP11, CYP17,CYP19, CYP24, CYP26 and CYP27

Species	*CYP11*	*CYP17*	*CYP19*	*CYP21*	*CYP24*	*CYP26*	*CYP27*
Human	11A1, 11B1, 11B2	17	19	21	24	26	27
Cow	11A1, 11B1, 11B2, 11B4	17	19	21			
Pig	11A1, 11B1, 11B2, 11B4	17		21			
Rat	11A1, 11B1	17	19		24		27
Rabbit	11A1, 11B1, 11B2, 11B3						27
Sheep	11A1	17		21			
Goat	11A1, 11B1						
Chicken	11A1	17	19				
Trout	11A1	17	19				
Hamster	11A1	17					
Mouse	11b1, 11b2	17		21			
Bullfrog							
Guinea pig		17					
Dogfish		17					
Goldfish			19				
Japanese quail			19				
Zebra finch			19				
Channel catfish			19				
Medaka			19				

Reference: Nelson et al., 1996.

Notes:
CYP11A1 is a cholesterol side-chain cleavage.
CYP11B1 is a steroid 11β-hydroxylase.
CYP17 is a steroid 17α-hydroxylase.
CYP19 is an androgen aromatase.
CYP21 is a steroid 21-hydroxylase.
CYP24 is vitamin D_3 24-hydroxylase.
CYP26 is a retnoic acid hydroxylase.
CYP27 is a sterol 27-hydroxylase.

Table 2.13 Fungal P450 families CYP51 to CYP66

Species	*CYP51*	*CYP52*	*CYP53*	*CYP54*	*CYP55*	*CYP56*	*CYP57*	*CYP58*
S. cerevisiae	51					56		
S. pombe	51							
C. tropicalis	51	52A1, A2, 52A6–8, 52B1, C1						
C. albicans	51							
I. orientalis	51							
C. krusei	51							
T. glabrata	51							
U. maydis	51							
Penicillium	51							
C. maltosa		52A3–5 52A9–11 52C2 52D1						
C. apicola								
Aspergillus niger		52E1, E2	53A1					
Rhodotorula minuta			53B1					
Neurospora crassa				54				
Fusarium oxysporum					55A1			
Cylindrocarpon tonkinese					55A2, A3			
Nectria haematococca							57A1, A2	
Fusarium sporotrichioides								58

Species	*CYP59*	*CYP60*	*CYP61*	*CYP62*	*CYP63*	*CYP64*	*CYP65*	*CYP66*
Aspergillus nidulans	59	60A2, B1		62				
Aspergillus parasiticus		60A1			63			
Aspergillus parasiticus						64		
Fusarium parasiticus							65	
Agaricus parasiticus								66

Reference: Nelson et al., 1996.

Notes:
CYP51 is a lanosterol 14α-demethylase, and CYP51 genes are also present in human and rat. CYP55A1 from *Fusarium oxysporum*, formerly known as $P450_{nor}$, has had its crystal structure determined but is unusual since it functions as a nitric oxide reductase.

Table 2.14 Plant P450 families CYP71–CYP79

Species	*CYP71*	*CYP72*	*CYP73*	*CYP74*	*CYP75*	*CYP76*	*CYP77*	*CYP78*	*CYP79*
Persea americana	71A1								
Solanum melongena	71A2, 71A3, 71A4				75A2	76A1,76A2	77A1,77A2		
Nepeta racemosa	71A5, 71A6								
Catharanthus roseus	71A7, 71D1, 71D2	72A1	73A4						
Mentha piperita	71A8								
Thlaspi arvense	71B1								
Arabidopsis thaliana	71B2, 71B3, 71B4 71B5, 71B6, 71D3	72A2	73A5	74		76C1			
Zea mays	71C1, 71C2, 71C3 71C4, 71C5		73A6–8					78A1	
Helianthus thuberosus			73A1			76B1			
Phaseolus aureus			73A2						
Medicago sativa			73A3						
Pisum sativum			73A9						
Petunia hydrida					75A1, 75A3				
Phalaenopsis sp. (orchid)								78A2	
Sorghum biocolor									79

References: Nelson et al., 1996; Bozak et al., 1990.

Notes:
CYP73 is a cinnamate 4-hydroxylase.
CYP74 is an allene oxide synthase (hydroperoxide lyase).

All plant P450 genes are from angiosperm species (flowering plants).

Table 2.15 Plant P450 families CYP80–CYP99

Species	*CYP80*	*CYP81*	*CYP82*	*CYP83*	*CYP84*	*CYP85*	*CYP86*
Berberis stolonifera	80						
Zea mays		81A1–4					
Helianthus tuberosus		81B1					
Pisum sativum			82				
Arabidopsis thaliana				83A1, B1	84		86A1, A2
Lycopersicon esculeatum						85	
Brassica campestris							86A3
	CYP87	*CYP88*	*CYP89*	*CYP90*	*CYP91*	*CYP92*	
Zea mays		88				92	
Helianthus annus	87						
Arnbidopsis thaliana				90	91A1, A2		
Vicia sativa			89				

Reference: Nelson et al., 1996.

Note:
CYP85 is involved in gibberellin biosynthesis.

Table 2.16 Bacterial CYP families CYP101–CYP120

Species	*CYP*	*Old name*	*Function*
Pseudomonas putida	101	cam	camphor 5-exo hydroxylase
Bacillus megaterium	102	BM-3	fatty acid ω–2 hydroxylase
Agrobacterium tumefaciens	103	PinF1	
Agrobacterium tumefaciens	104	PinF2	
Streptomyces griseolus	105A1	SU1	sulfonylurea methyl hydroxylase
Amylcolata autotrophia	105A2	VD25	vitamin D-3 25-hydroxylase
Streptomyces griseolus	105B1	SU2	sulfonylurea methyl hydroxylase
Streptomyces sp.	105C1	ChoP	
Streptomyces griseus	105D1	SoyC	guaiacol demethylase
Rhodococcus fasciens	105E1	-	
Bacillus megaterium	106A1	BM-1	fatty acid hydroxylase
Bacillus megaterium	106B1	meg	steroid hydroxylase
Saccharopolyspora erythraea	107A1	eryF	6-deoxyerythronolide hydroxylase
Streptomyces thermotolerans	107B1	–	
Streptomyces antibioticus	107C1	–	
Streptomyces antibioticus	107D1	–	
Micromosospora griseorubida	107E1	mycG	
Streptomyces griseus	107F1	–	
Streptomyces hygroscopicus	107G1	–	
Pseudomonas spp.	108	terp	α-terpineol 4-methyl hydroxylase
Bacillus subtilis	109	–	
Anabaena sp.	110	–	
Pseudomonas incognita	111	lin	linolool 8-methyl hydroxylase

Table 2.16 (continued)

Species	*CYP*	*Old name*	*Function*
Bradyrhizobium japonicum	112	BJ-1	
Saccharopolyspora erythraea	113A1	eryK	erythromycin C-12 hydroxylase
Streptomyces fradiae	113B1	–	
Bradyrhizobium japonicum	114	BJ-3	
Bradyrhizobium japonicum	115	BJ-2	S-ethyl dipropylcarbamothioate (herbicide) degradation
Rhodococcus sp.	116	–	
Bradyrhizobium japonicum	117	BJ-4	
Mycobacterium leprae	118	–	
Archaebacteria (Sulfolobus)	119	–	
Cyanobacteria	120	–	

References: Munro and Lindsay, 1996; Nelson et al., 1996.

Note:
Additional bacterial P450s comprising families CYP121–132 are from actinobacteria.

Table 2.17 Nematode P450s from *Caenorhabditis elegans*

CYP
13A1–13A10, 13B
14A1–14A4
16
22
23
25
31
32
33A, 33B, 33C, 33D
34
35A, 35B, 35C, 35D
36
37A, 37B

Table 2.18 Insect P450s

CYP
4C, 4D, 4E, 4G, 4H, 4J, 4K, 4L, 4M, 4N, 4P, 4Q, 4R, 4S
6A, 6B, 6C, 6D
9A, 9B, 9C
12A, 13B
15A
18A
28A
29A

Table 2.19 Mollusc P450s

CYP
10
30

References: Nelson et al., 1996; Nelson, 1998.

Table 2.20 New and old names of selected mammalian P450s

Current CYP name	*Previously used names*	*Species*
1A1	c, βNF-B, P_1, MC, P-448, LM_6	Rat, human, mouse
1A2	d, LM_4, P_3, ISF, P_2, P-448	Rat, rabbit, human, mouse
2A1	a1, a, 7α, RLM2	Rat
2A2	a2, RLM2, M-2	Rat
2A3	a3	Rat
2A4	15α OH-1	Mouse
2A5	15α OH-2, coh	Mouse
2A6	IIA3	Human
2B1	b, PB-B, PB-4, PBRLM5	Rat
2B2	e, PB-D, PB-5, PBRLM6	Rat
2B3	IIB3	Rat
2B4	LM2	Rabbit
2B5	B2	Rabbit
2B6	LM2	Human
2C1	PBC1	Rabbit
2C2	PBC2	Rabbit
2C3	PBC3	Rabbit
2C4	PBC4	Rabbit
2C5	Form 1	Rabbit
2C6	PB1, k, PB-C, RLM5a	Rat
2C7	f, PBRLM5b	Rat
2C8	IIC2, MP-12, MP-20	Human
2C9	MP-1, MP-2	Human
2C10	MP-8	Cloning artefact
2C11	2c, h, 16α, UT-A, RLM5	Rat
2C12	i, 15β, 2d, IT-I, fRLM4	Rat
2D1	db1, UT-7, UT-H	Rat
2D2	db2	Rat
2D3	db3	Rat
2D4	db4	Rat
2D5	db5	Rat
2D6	db1	Human
2E1	j, 3a, RLM6	Human, rabbit, rat
3A1	pcn1, PCNa, 6b-4	Rat
3A2	pcn2, PCNb/c, PCN-E, 6β-1/3	Rat
3A3	HLp	Human
3A4	h PCN1, nf-25, nf-10	Human
4A1	LAω, P452	Rat
4A2	IVA2, K-5	Rat
4A3	IVA3	Rat
4A4	p-2, PGω	Rabbit

References: Nelson et al., 1993; Ryan and Levin, 1990.

Chapter 3

The P450 catalytic cycle

For a state to exist it must first be observable.

(Lord Kelvin)

3.1 Introduction

In essence, the P450-mediated reaction (Figure 3.1) involves the combination of oxygen with the organic substrate (RH) to produce a molecule of water and a mono-oxygenated metabolite (ROH) according to the following scheme:

$$RH+O_2 \xrightarrow[2e^-]{2H^+} ROH+H_2O$$

The two reducing equivalents are supplied by NADH or NADPH, depending on the type of P450 system, and these are transferred in two consecutive stages via the mediation of either one or two redox partners, which may include either an iron-sulphur redoxin, a flavoprotein or cytochrome b_5, depending on the P450 enzyme concerned (reviewed in Lewis and Hlavica, 2000; Degtyarenko, 1995; Degtyarenko and Archakov, 1993).

It is possible to regard the P450-mediated reaction formally as involving the reduction of molecular dioxygen to the level of peroxide as follows:

$$O_2 + 2H^+ + 2e^- \xrightarrow{P450} H_2O_2$$

where the P450 system acts as a catalyst, and it would be more accurate to describe this process as occurring in two stages, namely

1. $O_2 + H^+ + e^- \longrightarrow \underset{\text{hydrosuperoxide}}{HO_2^\bullet} \rightleftharpoons \underset{\text{superoxide}}{O_2^{-\bullet}} + H^+$

2. $HO_2^\bullet + H^+ + e^- \longrightarrow \underset{\text{hydrogen peroxide}}{H_2O_2} \rightleftharpoons \underset{\text{hydroperoxide}}{HO_2^-} + H^+$

If one regards hydrogen peroxide as the precursor of the ultimate oxygenating species, then it is reasonable to assume that the molecule could either (1) cleave

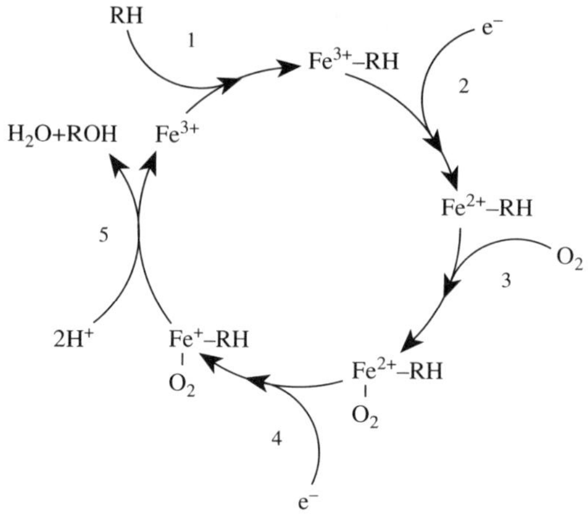

Figure 3.1 The reaction cycle and enzymatic intermediates known in P450 reactions (Lewis, 1996a).

Note: The various stages whereby a substrate (RH) becomes oxygenated via the activation of molecular oxygen, mediated by P450, are shown together with the heme iron redox state at each stage.

An interpretation of the spin-state equilbria in ferrihemoproteins has been provided by Otsuka (1970) based on theoretical considerations.

symmetrically to form two hydroxyl radicals or (2) break unsymmetrically forming a bare oxygen atom and water, as shown in the following equations:

1. $H_2O_2 \longrightarrow 2OH^{\bullet}$
2. $H_2O_2 \longrightarrow H_2O + [O]$

The former process is energetically more favourable, and it can thus be envisaged that one hydroxyl radical reacts with the substrate to form a hydrogen radical which would then, in turn, combine with the second hydroxyl species to give rise to a water molecule, as follows:

$$OH^{\bullet} + RH \longrightarrow ROH + H^{\bullet}$$

$$H^{\bullet} + OH^{\bullet} \longrightarrow H_2O$$

However, it is likely that a concerted mechanism occurs in the following manner:

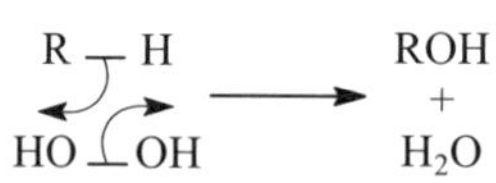

or

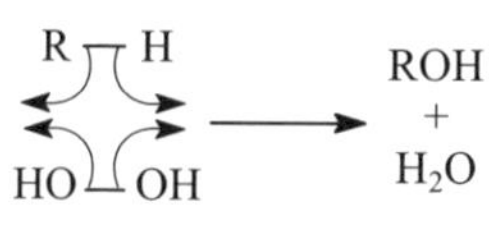

Alternatively, if reaction (2) were to be invoked, then one would have to consider the possibility of hydride ion transfer, as follows:

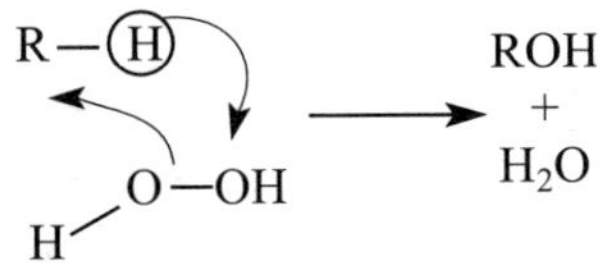

It is possible, however, that the exact mechanism is determined by the nature of the R (alkyl or aryl) group, i.e. depending on whether it is electron-donating or electron-withdrawing, although the type of P450 could ensure that the reaction proceeds preferentially via a specific route or pathway (reviewed in Lewis and Pratt, 1998).

A key to the likely mechanism of oxygenation may be provided by consideration of the electronic factors of substrates which correlate with rates of P450 reactions (Cho and Miwa, 1974; Jones and Korzekwa, 1996; Lewis et al., 1999b; Lewis, 2000a). For example, there are many cases where the ionization energy (IE) of the substrate correlates with the logarithm of the P450-mediated reaction rate (Lewis and Pratt, 1998) and, furthermore, the log rate constant for reaction with hydroxyl radicals also exhibits a parallelism with IE (reviewed in Lewis et al., 1999b). Obviously, the presence of the heme group will affect the electronic state of the bound oxygen ligand (Harcourt, 1977; Sawyer, 1987) in its varying changes throughout the cycle. However, the full effects of heme ligation on the electronic properties of the oxygen molecule are difficult to evaluate precisely, although recent molecular orbital calculations (de Groot et al., 1998; Segall et al., 1998) point the way forward to understanding such aspects of the catalytic cycle. Possibly, the effect of the heme-thiolate combination on the bound oxygen molecule is to bring about activation of this normally inert species (Hawkins and Dawson, 1992) but in a controlled manner involving the input of two reducing equivalents (two electrons and two protons) while the enzyme's active site residues maintain the position of the substrate relative to the active oxygenating species (Schlichting et al., 2000).

The thermodynamics of the P450 reaction can be studied, both experimentally and theoretically, where it can be appreciated that the enzyme dramatically lowers (Ruckpaul et al., 1989) the uncatalysed activation energy of a typical reaction, such as:

$$\text{Cyclohexane} + O_2 \longrightarrow \text{Cyclohexanol} + H_2O$$

This is achieved by binding the substrate at the active site and facilitating the mono-oxygenase reaction by activating molecular oxygen to the level of hydrogen peroxide by the consecutive input of two reducing equivalents. The general equation for this is as follows:

$$O_2 + 2H^+ \xrightarrow{2e^-} H_2O_2$$

where the four stages may be written as follows:

$$O_2 + e^- \longrightarrow \underset{\text{superoxide}}{O_2^{-\bullet}} \xrightarrow{H+} HO_2^{\bullet} \xrightarrow{e^-} \underset{\text{hydroperoxide}}{HO_2^-} \xrightarrow{H+} \underset{\text{peroxide}}{H_2O_2}$$

where oxygen is an electron-deficient π-acid, superoxide is a strong base and strongly nucleophilic, whereas hydroperoxide is weakly basic but also a strong nucleophile (reviewed in Lewis, 1996a; Lewis and Pratt, 1998).

The reduction of dioxygen by the P450 system is generally carefully regulated by the various components of the system, namely the redox partner which transfers reducing equivalents, the medium which may be cytosolic or membrane-bound (i.e. mitochondrial or microsomal), the nature of the active site and its degree of solvation, together with the influence of the heme-thiolate group itself (reviewed in Bernhardt, 1995). When this degreee of regulation is not tightly controlled uncoupling of the reaction occurs, leading to the formation of reactive oxygen species (Bast, 1986; Imai et al., 1989; Kappus, 1993; Martinis et al., 1989; Yeom et al., 1995; Yeom and Sligar, 1997). It is clear from accumulated experimental evidence that redox partner interactions play an important role in P450 catalysis (Daff et al., 1997; Hazzard et al., 1997; Hintz et al., 1982; Klein and Fulco, 1993; Koga et al., 1993; Lambeth et al., 1984; Munro et al., 1996a and b; Voznesensky and Schenkman, 1994; Strobel et al., 1989; Peterson and Mock, 1979; Peterson et al., 1997; Unno et al., 1996) and, apparently, both the type and efficiency of the P450 redox pathway have a strong bearing on reaction rate and catalytic properties of P450 systems in general (reviewed in Lewis and Hlavica, 2000; Lewis and Pratt, 1998).

3.2 Stages in the reaction

This can be appreciated by considering each sequential stage in the P450 catalytic cycle, starting with the enzyme in its resting state (Figure 3.2). Here, the heme iron is ferric low-spin and it is thought that the distal sixth ligand is either a water molecule or the hydroxyl ion, OH^- (Raag and Poulos, 1992) although consideration of the spectrochemical series for mixed ligand complexes of iron(III) tends to indicate the former (Shimura, 1988; Lewis and Pratt, 1998). The binding of a suitable substrate, such as camphor in the case of $P450_{cam}$ (CYP101), brings about a displacement of a number of bound water molecules present in the heme environment including the distal ligand, as is observed when one compares, for example, the substrate-bound and substrate-free CYP101 crystal structures (Raag and Poulos, 1992). Apparently, the displacement of this distal water ligand brings about a modulation in the heme iron spin-state equilibrium, which moves from predominantly low-spin to high-spin ferric (Sligar, 1976) and, moreover, alters the redox potential of the heme moiety such that it becomes less negative. It is this modification in the heme iron redox potential which facilitates the initial reduction of substrate-bound P450 by its redox partner. For example, in the $P450_{cam}$ (CYP101) system, binding of the camphor substrate gives rise to an alteration in the iron redox potential from –303 mV to –173 mV (Sligar, 1976). It is thus energetically

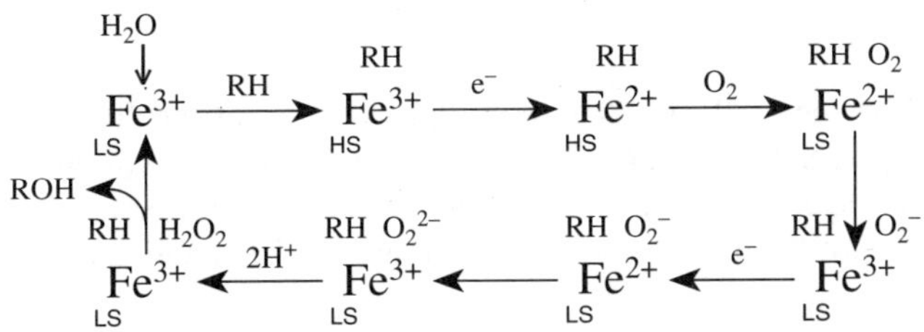

Figure 3.2 The P450 catalytic cycle in schematic form showing the heme iron spin-state changes and oxygen species (Lewis, 1992a; 1999a).

Note: The mechanistic sequence of substrate (RH) binding, P450 reduction, dioxgen binding and activation, together with generation of the oxygenated metabolite (ROH) has been adapted from Lewis and Pratt (1998).

feasible for the reductant, putidaredoxin, with a redox potential of –240 mV, to transfer electrons to $P450_{cam}$ once the camphor substrate is bound (reviewed in Lewis and Pratt, 1998), because electrons will then be able to flow down a potential gradient from the more redox-negative putidaredoxin $(E^{o'} = -240\ \text{mV})$ to the P450 hemoprotein $(E^{o'} = -173\ \text{mV})$.

Once the redox partner provides a reducing equivalent to the substrate-bound P450, the ferric iron of the hemoprotein moves to the ferrous state while still remaining high-spin. It is thus in a sitation whereby molecular oxygen is able to bind avidly to P450 via heme ligation, due to a high affinity ($K_D = 0.6\ \mu$M), when one considers the $P450_{cam}$ system (Archakov and Bachmanova, 1990), between the electron-deficient dioxygen molecule in its triplet ground-state and the electron-rich Fe(II) in its high-spin state with a single negative charge overall due to the heme-thiolate moiety. There is also evidence to suggest that oxygen binding affinity for hemoproteins is linearly related to their redox potential (Addison and Burman, 1985). In fact, reduced P450 is also able to bind carbon monoxide with a K_D of 0.1 μM in the case of $P450_{cam}$, and the resulting complex is relatively stable such that it is readily observable spectroscopically, and substrate-mediated changes in electronic states of the complex have been reported (Greschner et al., 1993). The characteristic absorption maximum in the UV at around 450nm for the Fe(II)P450.CO complex, where there is a prominent Soret peak in the UV spectrum, greatly facilitates identification of the enzyme and was largely responsible for its discovery in the 1950s (see Section 1.1). In fact, the position of the absorption maximum in the UV spectrum of CO-bound P450 has been used as a marker for identifying the particular form of the enzyme, as the precise location of the Soret peak is dependent on the heme environment. Consequently, the active site region of an individual P450 will have an effect on the electronic transition $(S_p \rightarrow \pi^*)$ which is associated with the band around 450 nm (reviewed in Lewis, 1996a).

Returning to the stage in the catalytic cycle when dioxygen binds, it is thought that the Fe(II)O_2 complex is in equilibrium with another state involving electron transfer from iron to oxygen, which leads to the formation of a ferric-superoxide species, Fe(III) $O_2^{-\bullet}$, according to the following scheme:

$$Fe^{2+}\ O_2 \longrightarrow Fe^{3+}\ O_2^{-\bullet}$$

The ferric-superoxide complex will be favoured by protonation of the superoxy anion and, therefore, the presence of exchangeable protons in the heme environment will facilitate its formation. Superoxide is also a fairly strong base ($pK_a = 4.5$) and the equilibrium constant for its reaction with protons ($K = 10^{12}$) indicates that the process:

$$O_2^{-\bullet} + H^+ \rightleftharpoons O_2H^\bullet$$

will be thermodynamically favourable (Chin et al., 1982). No state after this one has been observed experimentally, however, in either the bacterial or microsomal P450 systems (Martinis et al., 1991; Guengerich et al., 1976) and, consequently, the following stages should be regarded as hypothetical and, therefore, speculative despite the fact that they are consistent with known evidence and sound chemical principles (see Lewis and Pratt, 1998 for a fairly recent review of this area).

Resonance Raman spectroscopy of the $P450_{cam}$ system under catalytic conditions shows the presence of an oxygen species consistent with the formation of superoxide (Egawa et al., 1991) and it has been demonstrated that the addition of superoxide to reduced $P450_{cam}$ produces a catalytically competent enzyme (Kobayashi et al., 1994). It would appear, therefore, that the second reduction phase of the P450 catalytic cycle may give rise to the formation of a ferrous superoxy anion, complex Fe(II) O_2^- which, via the same argument outlined above, is likely to become protonated to form a relatively stable species. However, it may require an electron transfer from the heme iron to superoxide for the production of the active oxygen species, and this implies that ferric peroxide (or a protonated form) could represent a reactive intermediate which is capable of oxygenating the substrate (reviewed in Lewis and Pratt, 1998). It is known that peroxide in the presence of ferric iron is able to react with typical P450 substrates and form the expected oxygenated products but, in the enzyme's active site, there would be some degree of control exerted by the surrounding amino acid residues (see Mueller et al., 1995, for example). Moreover, the degree of solvation of the heme pocket by water molecules is likely to have an effect on the outcome of such oxidations (Oprea et al., 1997). This may explain the findings that different P450s are able to form various reaction products with the same substrate and, obviously, orientation of the molecule within the active site plays a key role in this situation. Nevertheless, one can write an equation describing the formal reduction of superoxide to peroxide, as follows:

$$O_2^{-\bullet} + H^+ + e^- \longrightarrow HO_2^-$$

where the electron is supplied either directly from the redox partner or via the transfer from P450 in its reduced ferrous form. The mechanism by which protons are acquired by the oxygen species is not entirely clear but may involve the mediating influence of protein-bound water molecules (Di Primo et al., 1992; Deprez et al., 1994; Griffin and Peterson, 1975; Yoshikawa et al., 1992; Helms et al., 1996; Hasinoff, 1985; Wade, 1990) that are present in the active site, and it appears that these are likely to be hydrogen-bonded such that proton transfer could occur along a conduit of water molecules occupying channels in the essentially hydrophobic environment encountered in the vicinity of the heme group (Oprea et al., 1997).

Consequently, the hydroperoxide species (presumably iron-bound) would be expected to be in equilibrium with solvated protons in the P450 active site, although the extent of hydration of this region will depend upon the occupancy by substrate, the molecular size of the substrate and, also, any conformational changes which may occur as a result of both substrate and redox partner binding (reviewed in Lewis and Pratt, 1998; Lewis and Hlavica, 2000). It is possible, therefore, to describe the reduction of dioxygen to the formal state of peroxide, brought about by the sequential addition of two electrons, as being in equilibrium with hydroperoxide and hydrogen peroxide, as follows:

$$O_2^{2-} + H^+ \rightleftharpoons HO_2^- + H^+ \rightleftharpoons H_2O_2$$

It is known that peroxide is a weak base (i.e. weaker than superoxide, for example) but it is a strong nucleophile when in its anionic state as HO_2^-, particularly with respect to reactions with 'soft' electrophiles such as the carbonyl group (Fleming, 1980). This is a consequence of the so-called α-effect whereby mixing of frontier molecular orbitals from oxygen alters the energy level of the HOMO when one compares peroxide with hydroxide (Fleming, 1980). However, the reaction with H_3O^+ is considerably slower than for OH^- because the solvated proton is a 'hard' electrophile and, consequently, will react more favourably with hydroxide than hydroperoxide (Fleming, 1980). Assuming that an iron-bound hydroperoxide is a possible active oxygenating species in P450 reactions, one might expect that it would act as a nucleophile by abstracting protons from the substrate, or undergo direct oxygen insertion for those substrates containing a carbonyl group (reviewed in Lewis and Pratt, 1998). In fact, it is likely that the P450-mediated oxygenation of aldehydes occurs via peroxy species, and this type of intermediate has been proposed for other P450 oxidations (Pratt et al., 1995). It would also follow that the likelihood of nucleophilic attack on a substrate could depend on the LUMO energy level, and several series of compounds exhibit a trend in ΔE (a quantity determined by the difference between the LUMO and HOMO energies) towards their relative metabolic clearance. However, even reactions regarded as electrophilic attack by a positive species, such as aromatic substitution, appear to show a correlation between rate and LUMO energy (Lewis, 1999a). Nevertheless, the majority of P450-mediated reactions seem to involve electrophilic attack by an oxygen species and the general trend for rate of substrate oxidation is associated with the substrate HOMO energy, or the equivalent expression involving ionization potential, IP (where IP = $-E_{HOMO}$ by definition).

It is possible that there are at least two mechanisms of oxygenation which may occur in P450 reactions depending on the nature of the substrate and, in some cases, the P450 enzyme involved (reviewed in Lewis and Pratt, 1998). There is, however, some circumstantial evidence for the involvement of hydroxyl radicals as active oxygen species in P450-mediated reactions and, of course, one can view $OH^{\bullet}$ as being the product of protons combining with a bare oxygen atom. Many workers in the P450 field favour the presence of an iron oxene in the P450 catalytic cycle and, formally, this would imply that the oxidation state of oxygen is zero in this species. If $OH^{\bullet}$ is the reactant species, then a simple radical-type reaction would occur whereby a proton is abstracted from the substrate to form the hydroxylated product, leaving a bare proton available for combination with a second hydroxyl radical which will then give rise to a water molecule. This could proceed from the symmetric cleavage of hydrogen peroxide via the following sequence of reactions:

$$H_2O_2 \longrightarrow 2OH^{\bullet}$$

$$RH + OH^{\bullet} \longrightarrow ROH + H^{\bullet}$$

$$H^{\bullet} + OH^{\bullet} \longrightarrow H_2O$$

Such a scheme can readily explain the formation of the typical products for P450-mediated oxygenation of a hydrocarbon, RH, and one could also envisage the reaction proceeding via a concerted mechanism, possibly mediated by the ligated iron and nearby threonine residue distal to the heme, as uncoupling of the normal reaction can occur when this conserved threonine is mutated to other amino acid residues (reviewed in Mueller et al., 1995). In fact, it is found that the rate constant for reaction with $OH^{\bullet}$ radicals is directly proportional to the ionization potential for about 50 chemicals of diverse molecular structure (reviewed in Lewis, 1999a) and, for P450-mediated oxidation of alkenes, the rate constant is also well correlated with substrate ionization potential (Traylor and Xu, 1988). Furthermore, several other series of P450 substrates exhibit similar correlations between either ionization energy or E_{HOMO} and the logarithm of their rate constant for the P450-catalysed reaction (reviewed in Lewis and Pratt, 1998).

Product release, following oxygen insertion, represents a process which is not fully understood. However it is clear that P450 reaction products still possess affinity for the enzyme (compare, for example, the K_D value for camphor = 1.6 μM and the K_D value for 5-hydroxycamphor = 10 μM), albeit somewhat lower, but one difference is that the products are normally difficult to oxygenate and, of course, the enzyme will not be in a catalytically active state until another substrate binds following the egress of metabolite. Also, the higher polarity and aqueous solubility ensures that the product will diffuse from the enzyme's active site, thus returning P450 to its resting state following the influx of water molecules, possibly via a 'gate' mechanism (Oprea et al., 1997). It has been reported (Bell and Guengerich, 1997) that product release is rate-limiting with respect to CYP2E-mediated oxidation of ethanol, although the second reduction step often tends to represent the slowest stage in many P450 systems, such that this can generally be regarded as rate-determining (Imai et al., 1977).

The way in which the P450 enzyme activates molecular dioxygen, a relatively stable species, to a highly reactive oxygenating intermediate in a controlled manner can be regarded as a stepwise lowering of the oxygen–oxygen bond energy such that cleavage of the dioxygen bond can occur (reviewed in Lewis, 1996a; Lewis and Pratt, 1998). Table 3.1 shows the relevant bond length and bond energy data for oxygen, superoxide and peroxide, together with their bond stretching frequencies. From an inspection of this evidence, it is clear that the sequential addition of electrons to oxygen lowers the bond energy and increases the bond length. In addition, statistical analysis of the data shows that the O–O stretching frequency is inversely proportional to bond length; this information can be employed to determine the electronic nature of a particular oxygen species, when bound to the enzyme, from either the infrared or Raman spectra of P450 complexes (Egawa et al., 1991).

If one considers the electronic structures of superoxide, peroxide and hydroperoxide, for example, it is possible to demonstrate that the frontier orbital energies are of the same order as those of a typical P450 substrate, such as camphor. Table 3.2 shows the energies of various species, frontier molecular orbitals of relevance to the $P450_{cam}$ catalytic cycle, together with their dipole moments and binding affinities (in the form of K_D values). These data indicate that, in terms of optimum frontier orbital energy differences, either hydrogen peroxide or the bare oxygen atom [O] could interact with the frontier molecular orbitals on camphor to form the metabolite, 5-hydroxycamphor, although it is not known at present precisely what effect the heme iron atom is likely to have on the orbital energies of the bound oxygen species (see, however, recent theoretical findings reported by Segall et al., 1998 and de Groot et al., 1998). Nevertheless, it is possible to view the peroxide and oxene states as two extremes of a continuum between various active oxygen species in equilibrium, via the mediation of proton transfer (reviewed in Lewis and Pratt, 1998). It would then depend upon the nature of the substrate and P450 active site environment to determine the actual nature of the oxygenating intermediate, and it is possible that the P450 system could actually control this through conformational and other changes within the heme environment, such as the binding of redox partners and participation of the membrane phospholipid, etc. (reviewed in Bernhardt, 1995). This equilibrium between various active oxygen species can be formulated as follows:

$$O_2^{2-} \underset{}{\overset{H^+}{\rightleftharpoons}} HO_2^- \underset{}{\overset{H^+}{\rightleftharpoons}} H_2O_2 \underset{\text{cleavage}}{\overset{\text{bond}}{\rightleftharpoons}} 2OH^\bullet \underset{\text{transfer}}{\overset{\text{proton}}{\rightleftharpoons}} [O] + H_2O \underset{\text{transfer}}{\overset{\text{proton}}{\rightleftharpoons}} [O] + OH^-$$

where [O] denotes a bare oxygen bound to the heme iron as the oxene intermediate, which will be electron deficient and, consequently, electrophilic in nature (Coon et al., 1998). It is apparent that iron-bound peroxy species will be nucleophilic, however, although the full description of P450 catalysis in terms of mechanisms of oxygenation awaits further elucidation and is a topic of considerable current

Table 3.1 Dioxygen species bond data

Species	*Bond length (Å)*	*Bond energy (kJ mole^{-1})*	*Stretching frequency (cm^{-1})*
Oxygen ion O_2^+	1.12	N/A	1905
Dioxygen O_2	1.21	494	1580
Superoxide $O_2^{-\bullet}$	1.33	276	1140
Hydrogen peroxide H_2O_2	1.48	213	850
Peroxide O_2^{2-}	1.49	178	802

Reference: Lewis and Pratt, 1998.

Notes:
N/A = data not available
The O–O stretching frequency correlates with the reciprocal of bond length according to the following equation:

$$\text{stretching frequency} = 4943.7 \text{ bondlength}^{-1} - 2519.6$$
$$n = 5; \quad s = 38.66; \quad R = 0.997; \quad F = 597.8$$

interest (Akhtar et al., 1993, 1994; Newcomb et al., 1995; Champion, 1989; Lipscomb et al., 1976; Okazaki and Guengerich, 1993; Ortiz de Montellano, 1987, 1989, 1998; Patzelt and Woggon, 1992; Poulos and Raag, 1992; Vaz et al., 1991, 1996; Vaz and Coon, 1994; White, 1991; Blake and Coon, 1989; Groves and Watanabe, 1988; Karki et al., 1995; Coon et al., 1998; Schlichting et al., 2000). Some aspects of this process will now be discussed in terms of physicochemical quantities.

3.3 Thermodynamics of the P450 reaction, rates of electron transfer and kinetics

It is possible to consider the thermodynamics of a typical P450-mediated reaction such as the oxidation of cyclohexane, shown below:

$$\text{cyclohexane} + O_2 \xrightarrow[2H^+,2e^-]{P450} \text{cyclohexanol} + H_2O$$

in terms of the individual stages involved and their likely energy changes (Ruckpaul et al., 1989). In this case, there is the energy required to break the dioxygen bond (460.5 kJ mole^{-1}) and the C–H bond of cyclohexane (418 kJ mole^{-1}). However, energy is gained in forming water from its elements (–470.1 kJ mole^{-1}) and in the generation of cyclohexanol from cyclohexane (–393.9 kJ mole^{-1}), thus leading to a net gain in energy overall of –384.2 kJ mole^{-1} (Lewis, 1996a; Ruckpaul et al., 1989). Consequently, the P450-mediated process described above is thermodynamically favourable for the various reactants and products in their standard states at normal temperatures, although it is found in practice that the rates are quite slow in the mammalian system, for example, due to the relatively high activation energy required to break, primarily, the substrate's C–H bond (about

Table 3.2 Molecular orbital data on P450$_{cam}$ ligands

Molecule	E(*NHOMO*)	E(*HOMO*)	E(*LUMO*)	μ_{calc} (*D*)	μ_{expt} (*D*)	K_D (μM)
Camphor	–11.005	–10.046	1.039	2.84	2.90	1.6
5-OH Camphor	–10.901	–10.154	0.920	1.76	N/A	10.0
Dioxygen (O_2)	–19.195	–19.195	–5.493	0.0	0.0	0.6
Superoxide (O_2^-)	–7.039	–0.478	4.420	N/A	N/A	N/A
Hydrogen peroxide (H_2O_2)	–12.975	–11.447	2.857	1.42	2.13	N/A
Oxygen (O)	–13.532	–13.532	–2.562	0.0	N/A	N/A
Water (H_2O)	–14.952	–12.463	4.417	1.86	1.84	N/A

Notes:

N/A	=	data not available
K_D	=	dissociation constant for compound binding to P450$_{cam}$
μ_{calc}	=	calculated dipole moment (Debyes)
μ_{expt}	=	experimental dipole moment (Debyes)
E(NHOMO)	=	energy of next highest occupied MO (eV)
E(HOMO)	=	energy of the highest occupied MO (eV)
E(LUMO)	=	energy of the lowest unoccupied MO (eV)

418 kJ mole^{-1}). Faster rates have been observed in the bacterial P450 system, however, and the kinetics of substrate binding in $P450_{cam}$ have been studied in detail (Griffin and Peterson, 1972).

It has been estimated that, in the absence of a catalyst, the activation energy of the above process is between 418 and 460 kJ mole^{-1} but P450 is, in fact, able to lower this to between 38 and 71 kJ mole^{-1}. All enzymes are able to lower the activation energies of the reactions they catalyse but, in this example, P450 facilitates the process by reducing the dioxygen bond energy and by constraining the substrate molecule, possibly also with the introduction of some degree of bond strain in the substrate itself, via binding to active site amino acid residues (reviewed in Lewis and Pratt, 1998). However, the lowering of dioxygen bond energy, with concomitant increase in bond length, is perhaps the most important means by which the P450 system enables mono-oxygenation of substrates (reviewed in Lewis and Pratt, 1998). Essentially, this involves the consecutive addition of two electrons to molecular dioxygen to form, initially, the superoxide anion and then peroxide, according to the following process:

$$\underset{\text{oxygen}}{O_2} \xrightarrow{e^-} \underset{\text{superoxide}}{O_2^-} \xrightarrow{e^-} \underset{\text{peroxide}}{O_2^{2-}}$$

where the dioxygen bond length increases (see Table 3.1) within this series from 1.20 Å to 1.26 Å (1.33 Å) and then to 1.49 Å with energies, respectively, of 497 kJ mole^{-1}, 276 kJ mole^{-1} and 146 kJmole^{-1} (Lewis, 1996a; Lewis and Pratt, 1998). The way in which this process occurs has been described in some detail previously, however, and we shall now consider, therefore, another aspect of the catalytic cycle, namely, spin-redox coupling.

3.4 Coupling of spin and redox equilibria

A fundamental aspect of P450 reactions is the substrate-induced modulation of hemoprotein spin-state equilibria and it has also been established (Sligar, 1976) that this effect is coupled with the iron redox potential equilibria, such that substrate binding effects an initial reduction of the enzyme via interaction with its redox partner (Raag and Poulos, 1989a; Sligar et al., 1979, 1984). Additionally, it would appear that the extent of spin-state conversion has a bearing on the subsequent rate at which substrates are metabolized by the P450 enzyme itself, as has been shown for benzphetamine derivatives (Blanck et al., 1983). As far as spin-state coupling with substrate binding is concerned, however, it is possible to represent this in terms of a four-state model, as follows:

$$\begin{array}{ccc} P450_{HS} & \overset{K_4}{\rightleftharpoons} & P450_{HS}S \\ \Big\updownarrow K_1 & & \Big\updownarrow K_2 \\ P450_{LS} & \overset{K_3}{\rightleftharpoons} & P450_{LS}S \end{array}$$

where $K_1 - K_4$ are the equilibrium constants for the four processes, HS and LS refer to high-spin and low-spin ferric P450, whereas S denotes the presence of bound substrate. Spectrophotometric determinations of the UV spectral changes which occur accompanying substrate binding enable a monitoring of the hemoprotein spin-state equilibria perturbation in P450 systems (Gibson and Tamburini, 1984; Gibson and Skett, 1994). However, it should be pointed out that the situation in the microsomal system (Guengerich and Johnson, 1997) is not as clear-cut as that exhibited by $P450_{cam}$, for example, as some microsomal P450s appear to reside primarily in the high-spin state and, consequently, may not necessarily require the binding of substrates to modulate their spin and redox equilibria for the initiation of catalytic activity. Nevertheless, the general viewpoint is that substrate binding usually has a significant effect on both the spin-state and redox potential of most P450s, whereby the change from predominantly low-spin ferric to high-spin brings about a lowering of the hemoprotein redox potential (i.e. it becomes less negative in value), which may be associated with desolvation of the heme environment, such that the substrate-bound P450 can be more readily reduced (reviewed in Lewis and Hlavica, 2000) to the ferrous form by interaction with the relevant redox partner (i.e. via reductase or redoxin). This is an essential prerequisite for dioxygen binding and, subsequently, also for the activation of oxygen itself which is obviously necessary for metabolism of the bound substrate molecule.

Table 3.3 shows the redox potentials of various components in the P450 system, from which it can be appreciated that there is a potential gradient for electron transfer and, consequently, reduction of the P450 enzyme under consideration provided that a bound substrate is present. For example, in the $P450_{cam}$ bacterial system, the binding of camphor (the endogenous substrate) brings about a change in the P450 redox potential from –303 mV to –173 mV, this enabling a flow of electrons down a potential gradient from NADH to FAD in putidaredoxin reductase to putidaredoxin itself, which then reduces the P450 hemoprotein to the ferrous state (Archakov and Bachmanova, 1990). Similar situations exist in other P450 systems as can be appreciated from Table 3.3. Furthermore, the reduced form of P450 is then able to transfer an electron to oxygen, as the redox potential for the dioxygen/superoxide couple is –160 mV. In order for reduction of oxygen to the level of peroxide, however, the redox potential becomes +380 mV although this value may differ somewhat when the oxygen species is ligating the heme iron.

The redox potentials of hemoproteins vary quite considerably, and it is possible that one factor is the extent of heme surface exposed to the aqueous environment. Stellwagen (1978) has shown that there appears to be a linear relationship between redox potential and percentage heme exposure for a number of hemoproteins, and Table 3.4 presents the relevant data, including that of $P450_{cam}$. From this it can be appreciated that an increase in hydrophobicity corresponds to a rise in the redox potential of hemoproteins and this may partially explain the effect of substrate binding on P450 redox potential, although the transition to predominantly high-spin ferriheme is also substrate-dependent (Gibson, 1985; Blanck et al., 1983).

Table 3.3 Redox potentials (mV) of components in various P450 systems

Bacterial system		*Mitochondrial system*	
NADH	–320	NADPH	–324
FAD	–290	FAD	–290
Pdx	–240	Adx	–270
$P450_{cam}$	–303 → –173 (+ camphor)	$P450_{scc}$	–400 → –280 (+ cholesterol)
Bacillus megaterium system		*Microsomal system*	
NADPH	–324	NADPH	–324
FAD	–292	FAD	–290
FMN	–270	FMN	–270
$P450_{BM3}$	–370 → –235 (+ arachidonate)	$P450_{LM2}$	–300 → –237 (+ hexobarbital)

References: Lewis, 1996a; Munro et al., 1999; Lewis and Hlavica, 2000; Lewis and Pratt, 1998; Backstrom, et al., 1983; Guengerich, 1983

Notes:

1 Cytochrome b_5 is also a redox partner in the microsomal system and has a redox potential of about +25 mV.
2 Redox potentials of various oxygen species are tabulated in Lewis 1996a although the redox potential for dioxygen forming superoxide is –160 mV, and +380 mV for peroxide.
3 Substrate binding generally lowers the P450 redox potential, i.e. making it less negative, such that reduction by the relevant redox partner is facilitated.
4 The redox potentials of the iron-sulphur redoxins change somewhat when bound to the relevant P450 concerned, usually becoming less negative so that they may themselves be reduced more easily by their reductase partners.

Table 3.4 Redox partners and heme exposure

Hemoprotein	*% Heme exposure*	*Redox potential (mV)*	*Axial ligands*
Cytochrome c_2	6	320	His, Met
Cytochrome c	4	260	His, Met
Cytochrome c_{550}	5	250	His, Met
Hemoglobin α	14	113	His, His
Hemoglobin β	20	53	His, His
Myoglobin	18	47	His, His
Cytochrome b_5	23	20	His, His
Cytochrome $P450_{cam}$	43, 35[a]	–303, –173[a]	Cys, H_2O

References: Stellwagen, 1978; Lewis, 1996a; Lewis and Pratt, 1998; Lewis and Hlavica, 2000.

Notes:

a = Data for the camphor-bound enzyme
Values for heme exposure in $P450_{cam}$ were calculated using the equation derived from statistical regression analysis of the above data:

$$E^{o\prime} = 14.95\%\ \text{heme exposed} + 343.9$$

$n = 7$; $s = 37.36$; $R = 0.96$; $F = 47.15$

Using an expression originally formulated by Kassner (1973) and based on the Born equation for solvation energy, it is possible to rationalize the effect of substrate binding on the redox potential of $P450_{cam}$ based on the change in local dielectric constant in the vicinity of the heme group (Lewis and Hlavica, 2000). From the various characteristics of the P450 system, one can simplify the Kassner relationship to the following equation for the change in redox potential, ΔE^{o}, as shown below:

$$\Delta E^{o} = 7.2\left(\frac{1}{r_1} - \frac{1}{r_2}\right)\left(\frac{1}{D_N} - \frac{1}{D_P}\right)$$

where r_1 and r_2 are the radii of a heme moiety and the hemoprotein, respectively, and D_N, D_P are the dielectric constants for non-polar and polar environments surrounding the heme itself.

A calculated value for the situation in $P450_{cam}$ can be obtained using 16 Å as the approximate radius of the enzyme, 6.78 Å for the heme group and dielectric constants of 4 and 28, respectively, for the substrate-bound and substrate-free heme environment based on typical values for globular proteins (Lewis and Hlavica, 2000). Putting this information into the above equation leads to 131 mV for the expected change in redox potential when camphor binds to the P450, and this agrees well with the experimental value of 130 mV, although more precise data on dielectric constants would be desirable. The less-pronounced changes in redox potential observed when substrates bind to microsomal P450s are likely to be a result of the increased lipophilic environment encountered by the hemoproteins which are, in this case, embedded in a phospholipid bilayer; and this would be consistent with application of the Kassner equation to the microsomal system (Lewis and Hlavica, 2000). Consequently, some characteristics of redox changes in the P450 cycle can be explained fairly satisfactorily by the application of physicochemical principles to the system, and a recent report (Sato and Guengerich, 2000) underlines the importance of substrate redox potential to the P450 reaction mechanism.

3.5 Factors influencing the rate of substrate turnover

It is likely that the local dielectric constant also has an effect on the rates of electron transfer in P450 systems and, consequently, this will have an influence on the overall rate of substrate turnover (see Tables 3.5 and 3.6). It is generally accepted that, in both bacterial and microsomal systems, the second reduction represents the rate-limiting step for P450-mediated metabolism (Archakov and Bachmanova, 1990). Consequently, if all other factors are essentially the same, determination of the rate of electron transfer between redox partner and P450 will enable an estimation of the overall catalytic turnover exhibited by the enzyme system (reviewed in Lewis and Hlavica, 2000). An expression for the rate of P450 reduction in the microsomal system can be formulated as follows:

Table 3.5 Physicochemical data for P450 catalysis

Redox partner	*CYP*	*Pseudo first-order rate constant* (k_{app})	*Dissociation constant* (K_D)
Putidaredoxin	101	30–33 s^{-1} (1st and 2nd e^-)	5.4 μM (17 μM also reported)
Cytochrome b_5	101	3–4 s^{-1} (2nd e^-)	0.6 μM (0.65 μM also reported)
Cytochrome b_5	2B4	2–7 s^{-1} (2nd e^-)	0.02 μM(0.4 μM also reported)
Reductase	2B4/1	11.5–18 s^{-1} (1st e^-) 1.1 s^{-1} (2nd e^-)	0.07 μM (0.1 μM reported for 2B1)

References: Archakov and Bachmanova, 1990; Davies et al., 1990; Gray, 1992; Ortiz de Montellano, 1995; Ruckpaul et al., 1989; Ivanov et al., 1997; Munro et al., 1999.

Notes:
It is generally accepted that the overall rate-limiting step (Imai et al., 1977) for P450-mediated monooxygenase activity is determined by the second reduction process stage, and this is consistent with the information presented above for both bacterial and microsomal systems.

Putidaredoxin is reduced via putidaredoxin reductase, whereas cytochrome b_5 is reduced via cytochrome b_5 reductase; both FAD-containing reductases are NADH-dependent. The data for cytochrome b_5 binding to CYP101 have been reported for reconstituted systems, as opposed to the use of the natural redox partner for this bacterial system, namely, putidaredoxin.

The interactions between cytochrome b_5 and its reductase (Dailey and Strittmatter, 1979) and between adrenodoxin and adrenodoxin reductase (Geren et al., 1984) have been investigated using residue modification experiments.

It is also apparent that different P450s may be involved in the reductase complexation process (Backes et al., 1998) whereby heterooligomers of P450 enzymes can form clusters around a central P450 oxidoreductase in the microsomal system (Stier, 1976).

The mechanisms of 1- and 2-electron transfers by flavin-containing enzymes have been investigated (Iyanagi, 1987) although the modes of electron transfer may vary depending on the type of P450 system concerned (Munro et al., 1996b).

$$\text{reduction rate} = K_a k_{ET}\,[\text{P450}]\,[\text{Fpt}]$$

where K_a is the association constant between P450 and flavoprotein reductase of concentrations represented by the terms in brackets, and k_{ET} is the electron transfer rate constant (Archakov and Bachmanova, 1990). Typical values for rates of reduction, protein concentrations and dissociation constants for the microsomal system yield electron transfer rate constants in the order of several reciprocal seconds (i.e. 5 to 10 s^{-1}).

Although substrate binding has an effect on the apparent rates of both the first and second reductions, it has been shown that there is tight coupling between spin-state equilibria, substrate binding and rate of the first reduction step which also results in an effect upon catalytic turnover (Blanck et al., 1983, 1991; Petzold et al., 1985). However, in the hepatic microsomal system the second reduction step is mediated by cytochrome b_5 (Bell and Guengerich, 1997) and this may have a bearing on the relatively low rate constants observed, possibly due to the fact that cytochrome b_5 has a substantially different redox potential (+25 mV) from FMN (–270 mV).

Table 3.6 Kinetics and thermodynamics of stages in the catalytic cycle

A. *Rate constants*

Stage	*CYP101 – camphor*	*CYP2B4 – benzphetamine*
1. Substrate binding	47 s^{-1}	50 s^{-1}
2. First reduction	30–33 s^{-1}	18 s^{-1}
3. Oxygen binding	1.7×10^6 M^{-1} s^{-1}	> 10^6 M^{-1} s^{-1}
4. Second reduction	3–4 s^{-1} (via cytochrome b_5)	2–7 s^{-1}
Turnover number	1900 min^{-1}	24 min^{-1}

B. *Electron transfer stages in redox partners*

NAD(P)H → reductase	> 600 s^{-1}
Reductase → putidaredoxin	26 s^{-1}
NADH → cytochrome b_5	30 s^{-1}

C. *Association constants*

	K_D(μM)
Oxygen binding to CYP101	0.6
Carbon monoxide binding to CYP101	0.1
Camphor binding to CYP101	1–5

References: Lewis, 1996a; Archakov and Bachmanova, 1990; Ruckpaul and Rein, 1984; Griffin and Peterson, 1972.

This stems from a consideration of the Marcus equation (Marcus and Sutin, 1985) which has been formulated to explain the variation in rates of electron transfer for various redox systems (Veitch and Williams, 1992). The Marcus equation takes the following form:

$$\ln k = \ln \gamma_{ET} - \left(\frac{\lambda - \Delta E^{\circ}}{4 k_B \lambda T} \right)^2$$

where k is the rate constant for electron transfer with a limiting rate of γ_{ET} at infinite temperature T, and λ is the relaxation (or reorganization) energy of the reaction between an electron donor and acceptor couple of redox potential difference, ΔE°. The quantity k_{B} refers to the Boltzmann constant and relates the vibrational energy of a harmonic oscillator with its thermal energy at temperature T (Moser and Dutton, 1992). This quadratic relationship between log rate and difference in redox potential shows regions of approximate linearity within a certain range of values and, for a small number of P450 systems, there is a good relationship between rate of reduction (or substrate turnover) and redox potential differences (Lewis and Hlavica, 2000). Furthermore, it has been demonstrated (Moser and Dutton, 1992) that there is an inverse linear correlation between log rate of electron transfer and distance between the relevant interacting centres for certain biological systems. This arises from a special case of the Marcus-type expression where λ is approximately equal to ΔE° (or ΔG, the free energy change for the process) and, in this respect, the equation derived becomes much simpler as follows:

$$\log k_{ET} = 15 - 0.6R$$

where k_{ET} is the rate constant for electron transfer and R is the distance between the two redox centres. The intercept of this correlation conforms well with the diffusion-controlled limit at high temperature, whereas employing the rate constant for electron transfer in the $P450_{BM3}$ system produces an edge-to-edge distance of 21 Å which is in excellent agreement with crystal structure data for the FMN-heme domain complex in $P450_{BM3}$ (Sevrioukova et al., 1999) where a 20 Å gap between FMN and heme moieties is observed. However, from the de Broglie equation for the wavelength of an electron, one can also derive an expression for the quantum-mechanical expectation value of the rate constant for electron transfer which incorporates a consideration of the intervening dielectric (Lewis and Pratt, 1998). This inverse square relationship with respect to distance is also inversely linear for the dielectric constant (D) of the medium although, for a given value of D, there is a roughly linear portion of the curve which is analogous to the empirical expression noted previously.

Nevertheless, it is apparent from experimental observations that the P450 reduction rate constant is not only directly proportional to redox potential (Fisher and Sligar, 1985) it is also linearly related (see Table 3.7) to the percentage high-spin content of the P450 hemoprotein itself (Gibson, 1985) and, as mentioned previously, this is probably due to the fact that the redox and spin equilibria tend to be directly coupled (reviewed in Lewis, 1996a; Lewis and Pratt, 1998; Lewis and Hlavica, 2000). There is also a correlation between substrate-induced spin perturbation and tyrosine exposure (see Table 3.8) in $P450_{cam}$ (CYP101) which points to the key role of an active site tyrosine (Tyr96) in the heme pocket (Sligar and Murray, 1986; Atkins and Sligar, 1988, 1990) and, therefore, it appears that important dynamic aspects exist in the P450 catalytic cycle with respect to protein conformational effects 'triggering' redox partner binding as the substrate enters the heme environment, which is possibly linked with desolvation processes in water channels close to the heme moiety (Oprea et al., 1997).

In conclusion, therefore, one can envisage that the rate of catalysis is fairly well regulated by the biophysical chemistry of the P450 system, but the redox partners and their cellular environment play a role in modulating the kinetics of P450 substrate turnover (Bernhardt, 1995). The tight regulatory control which the P450 system maintains upon oxygen activation and substrate metabolism may well have been the result of the evolutionary process itself, as it could be argued that the P450-mediated reactions capitalize on the latent potential of the normally stable dioxygen molecule for detoxifying lipophilic compounds and harnessing, for example, the hormonal possibilities inherent in steroidal substrates biosynthesized from cholesterol and other related plant products. Recently, each major stage in the catalytic cycle of $P450_{cam}$ (CYP101) has been characterized crystallographically for the isolatable complexes (Schlichting et al., 2000) such that our knowledge of this important enzymatic pathway is now becoming more

Table 3.7 Reduction rate constant and % high-spin content of phenobarbital-induced microsomal P450*

% high spin	$k(s^{-1})$
6.8	0.09
12.4	0.15
18.8	0.25
27.2	0.36
34.4	0.45

Reference: Gibson, 1985.

Notes:
* Benzphetamine employed as a substrate.
Correlation equation for the above data:

$$k = 0.013\%HS - 0.005$$
$$(\pm 0.003)$$
$$n = 5; \quad s = 0.073; \quad R = 0.999; \quad F = 1109.6$$

Table 3.8 Extent of tyrosine exposure and % high spin CYP101

Compound	*% high spin*	*Number of tyrosines exposed*
1. None	4	4.20
2. TMCH*	14	4.06
3. Norcamphor	45	3.92
4. Fenchone	46	3.95
5. Camphenilone	55	3.92
6. Thiocamphor	61	3.83
7. 5-Ketocamphor	64	3.83
8. 3-*endo*-Bromocamphor	67	3.83
9. Adamantane	79	3.76
10. 4-*exo*-Bromocamphor	81	3.78
11. Camphor	95	3.58
12. Adamantanone	99	3.66

Reference: Sligar and Murray, 1986.

Notes:
Correlation equation for the above data:

$$\%\ \text{high spin} = 712.9 - 169.4\ \text{tyrosine exposure}$$

$$n = 12; \quad s = 6.753; \quad R = 0.975; \quad F = 174.7$$

*TMCH = tetramethylcyclohexanone

complete, and the biophysical chemistry more readily understandable. However, it will be useful to have structural evidence for microsomal P450s which show details of catalytic cycle intermediates, such that further insights may be obtained for eukaryotic P450 systems.

3.6 Appendix – Factors influencing rates and clearance in P450 reactions: QSARs

The rate constant or turnover number, k_{cat}, for an enzyme reaction is related to the maximal velocity, V_{max}, via the following equality:

$$k_{cat} = \frac{V_{max}}{[E_t]} \qquad \text{where } [E_t] \text{ is the total enzyme concentration}$$

Moreover, the initial velocity, V_o, is related to the substrate binding dissociation constant, K_m, as follows:

$$V_o = \frac{k_{cat}}{K_m}[E_t][S] \qquad \text{where } [S] \text{ is the substrate concentration}$$

P450-mediated reactions exhibit a wide range of turnover numbers, from as high as 5000 min^{-1} (or more) for arachidonate metabolism by CYP102 (where k_{cat} = 300,000 sec^{-1}) to as low as 6 sec^{-1} for hepatic microsomal P450s.

The catalytic efficiency (k_{cat} / K_m) of CYP102 (1.38×10^{11} M^{-1} s^{-1}) is close to the diffusion-controlled limit ($k_B T/h = 6.46 \times 10^{12}$ at 37 °C where k_B is the Boltzmann constant and h is Planck's constant) for enzyme catalysis. The intrinsic clearance, Cl_{int}, can be expressed as the ratio:

$$Cl_{int} = \frac{V_{max}}{K_m}$$

Consequently, the logarithmic transformation of this relationship becomes:

$$\log Cl_{int} = \log V_{max} - \log K_m$$

where the second term on the right-hand side of the equation is linearly related to the free energy of binding, ΔG_{bind}, via the expression:

$$\Delta G_{bind} = 2.3\,RT \log K_m$$

where R is the gas constant and T is the absolute temperature.

A high level of clearance is, therefore, expected for substrates with large negative binding energies which will be composed of favourable interactions between the substrate and P450 active site.

Essentially, clearance comprises two main components relating to the activation energy of the reaction (V_{max} term) and possibility (K_m term) of the substrate reaching the enzyme's active centre. By the same token, the rate constant for the P450-mediated reaction can be described via a relatively simple linear equation based on transition state theory. The Eyring formula, which stems from a quantum-chemical treatment of the original Arrhenius expression, leads to a thermodynamic rate equation as follows:

$$k = \frac{RT}{Nh}\exp\left(\frac{\Delta S^{\neq}}{R}\right)\exp\left(-\frac{\Delta H^{\neq}}{RT}\right)$$

where k is the rate constant, R is the gas constant, T is the absolute temperature, N is the Avogadro number and h is Planck's constant.

Expressed in logarithmic form, this equation becomes:

$$\ln k = \ln\left(\frac{RT}{Nh}\right) + \frac{\Delta S^{\neq}}{R} - \frac{\Delta H^{\neq}}{RT}$$

where $\Delta S^{\neq}$ and $\Delta H^{\neq}$ are, respectively, the entropy and enthalpy changes associated with the formation of the transition state.

As the first term on the right-hand side will be constant at a given temperature, the factors determining the rate of these reactions for various substrates centre around the entropic and enthalpic components, $\Delta S^{\neq}$ and $\Delta H^{\neq}$, respectively. These terms can be estimated from an experimental study of P450 reaction kinetics under varying temperature conditions or evaluated via quantum-chemical calculations on the substrates themselves, although these would be based on certain assumptions about the nature of the transition state. In general, it is found that the energy change associated with hydrogen abstraction of the substrate gives a good correlation with log rate constant for various series of P450 substrates, although the ionization potential (or energy of the highest occupied molecular orbital, E_{HOMO}) usually improves these correlations significantly. In fact, it is possible that the IP (or E_{HOMO}) can become the dominant factor depending on the nature of the P450 substrates being investigated. Furthermore, additional terms may also be required to explain fully the variation in either rate or clearance data for a series of P450 substrates and it can be shown that transport factors, such as log $D_{7.4}$ or log P, often have a bearing on metabolic clearance in various cogeneric series of compounds catalysed via P450 enzymes. The relative magnitude of these terms and contribution made by each component can be evaluated from QSAR analyses conducted on the P450 substrates in question when these have been docked within the enzyme's active site or, possibly, from first principle structural calculations on the molecules themselves. Such considerations apply primarily to the hydrogen bond, ion-pair, π-π and hydrophobic terms which go to make up the overall binding free energy, ΔG_{bind}, which can be equated with the K_m value as follows:

$$\Delta G_{bind} - RT \ln K_m$$

where R is the gas constant and T is the absolute temperature.

However, the importance of transport through cell membranes is underscored by the frequent occurrence of log P and/or log $D_{7.4}$ as QSAR descriptor variables in the explanation of rate and clearance data for P450 substrates.

In this respect, the entropic term may be represented by the lipophilicity parameter (log P) or some aspect of the substrate's molecular size, such as the solvent-accessible volume or surface area. This would, therefore, relate to the desolvation aspect of substrate binding which is usually the major component of the overall binding free energy. For ionizable molecules, it is usual for the log $D_{7.4}$ value to be employed instead of log P because this is equivalent to an ionization-corrected log P at pH 7.4; D is the distribution coefficient whereas P is the partition coefficient, normally taken as between n-octanol and water (Lewis, 2000a and b). An example of clearance being related to log $D_{7.4}$ is afforded by a series of benzodiazepines and sodium channel blockers (Lewis, 2000b). In this case, a combination of log $D_{7.4}$ and ionization potential (IP) of 12 substrates gives a correlation of 0.91 with the log of clearance according to the equation:

$$\log \mathrm{Cl} = 0.64 \log D_{7.4} - 0.98\ \mathrm{IP} + 9.33$$

$$n = 12; \quad s = 0.498; \quad R = 0.91; \quad F = 21.4$$

However, an improved correlation ($R = 0.95$) is obtained by replacing the log $D_{7.4}$ term with the number of hydrogen bond (HB) donors and, also, by including a term describing the internal energy of the molecule. In this case, the QSAR expression is as follows:

$$\log \mathrm{Cl} = 0.05\ \mathrm{energy} - 0.53\mathrm{HB} - 0.95\ \mathrm{IP} + 10.63$$

$$n = 12, \quad s = 0.382; \quad R = 0.95; \quad F = 26.6$$

Essentially, it would appear that the polarity of the molecules is better described by their hydrogen bond donating capacity than log $D_{7.4}$, with the total internal energy of the molecule providing a measure of the entropic/desolvation component to the binding affinity of these compounds. For a series of p-substituted toluenes, it is found that log k_{cat} correlates well with $\Delta H^{\neq}$ particularly in combination with IP, according to the following equation:

$$\log k_{cat} = 19.97 - 0.024\ \Delta \mathrm{H}^{\neq} - 0.95\ \mathrm{IP}$$

$$n = 8; \quad s = 0.081; \quad R = 0.99; \quad F = 92.98$$

In this series, it is likely that the entropy change is roughly constant as the compounds are of similar structure and, therefore, the rate of hydroxylation is determined by the enthalpic component which appears to be composed of two factors, namely, the energy required for hydrogen abstraction and the electron-donating ability of the molecule as measured by IP.

As intrinsic clearance is comprised of the ratio between rate and binding affinity, it follows that a low K_m will necessarily give rise to a relatively high metabolic clearance. Consequently, a description of the factors which contribute to the substrate binding energy is important for predicting the likely extent of a particular compound's clearance. Of course, the specific interaction between the substrate molecule and the P450 active site will provide a means of estimating the binding affinity because this depends on the key contacts with amino residues within the heme environment. In general, these can be summarized as being a combination of electrostatic, hydrogen bond, π–π stacking and hydrophobic interactions, although the loss in substrate molecular degrees of freedom (i.e. translational and rotational entropy) upon binding can also be taken into consideration. For example, in a study of 10 CYP3A4 substrates it is found that the free energy of binding is well described ($R = 0.96$) by a combination of frontier orbital energies, number of hydrogen bonds formed and number of π–π stacking interactions present in the active site-docked substrate complex. Furthermore, for CYP2D6 inhibitors (some of which may also act as substrates) one can demonstrate that the variation in inhibitory potency (pK_i value) is related ($R = 0.96$) to relative molecular mass, log P , number of hydrogen bond acceptors and number of basic nitrogens. For CYP2C9 substrates, however, a very good correlation ($R = 0.99$) is produced with a combination of log P, log $D_{7.4}$, pK_a and number of hydrogen bond donors (Lewis, 1999b). In fact, log P appears to represent an important component of the binding affinity for substrates of CYP2B and CYP2E inhibitors. However, in most of these series of compounds, a quadratic expression in log P is the recurrent feature in QSARs. For example, in a series of primary aliphatic amines, the binding to CYP2B4 can be described by a parabolic expression in log P, as shown below:

$$\mathrm{p}K_1 = 0.88 + 0.93P - 0.19 \log P^2$$
$$n = 8; \quad s = 0.216; \quad R = 0.99; \quad F = 114.5$$

where K_1 is the high affinity binding constant and, in this example, log P is the experimental value at pH 7. There is also a similar relationship for barbiturates binding to microsomal P450 where the dissociation constant, K_D, is dependent on a quadratic in log P according to the following expression:

$$\log K_D = 1.39 \log P - 0.22 \log P^2 - 0.50$$
$$n = 10; \quad s = 0.120; \quad R = 0.95; \quad F = 65.2$$

where the log P values were experimentally determined in phosphate buffer, and K_D is the reciprocal of the K_S value obtained from UV difference spectra. Moreover, the K_S values of haloalkanes binding to phenobarbital-induced hepatic microsomal P450 follow a quadratic in log P when combined with ΔE values determined via electronic structure calculation. In this example, the QSAR expression is as follows:

$$\log K_S = 0.057\Delta E + 0.70 \log P - 0.34 \log P^2 - 0.44$$
$$n = 11; \quad s = 0.235; \quad R = 0.93; \quad F = 15.5$$

where the log P values were determined experimentally, although calculated log P gives a similar QSAR equation (Lewis, 1999a). For p-substituted toluenes (White and McCarthy, 1986) binding to phenobarbital-induced microsomal P450, the variation in K_m values can be explained (Lewis et al., 1995a) via a straightforward linear combination of log P and ΔE , as follows:

$$-\log K_m = 43.27 - 4.03\,\Delta E - 0.60 \log P$$
$$n = 8; \quad s = 0.394; \quad R = 0.95; \quad F = 23.7$$

where it would appear that the employment of a bilinear expression in log P is not required in order to give an adequate description of the binding data. Similarly, the binding free energy using experimental K_D values is closely related to an equivalent expression involving the molecular volumes and E_{HOMO} values for the same series of compounds, and this leads to the following QSAR equation:

$$\Delta G_{bind} = -32.28 - 2.66\, E_{HOMO} - 0.20\, MV$$
$$n = 10; \quad s = 0.912; \quad R = 0.98; \quad F = 112.4$$

where K_D values were available for an additional two compounds in the p-substituted toluenes series.

With inhibitors of CYP2E1, it is also found that quadratics in log P provide a good explanation of the data for compounds which may also act as substrates for this enzyme, such as alcohols and carboxylic acids. For 10 primary aliphatic alcohols, the K_i data can be explained by a parabolic expression in log P as follows:

$$-pK_i = 0.98 \log P - 0.20 \log P^2 - 1.79$$

$$n = 10; \quad s = 0.218; \quad R = 0.96; \quad F = 89.4$$

Whereas, in the case of long-chain carboxylic acids the relevant data conforms to a similar quadratic equation in log $D_{7.4}$, as the log P values have to be ionization-corrected using the pK_a of the carboxylic acid, yielding the following:

$$-pK_i = 1.05 \log D_{7.4} - 0.24 \log D_{7.4}^2 - 2.66$$

$$n = 11; \quad s = 0.169; \quad R = 0.97; \quad F = 119.0$$

These quadratic expressions all give rise to a maximum value depending on the series of compounds under consideration and the type of P450 involved. Consequently, one can use this as a guide to estimating an optimal log P value for substrates or inhibitors of the P450 in question. Interestingly, from a study of over 100 P450 substrates and inhibitors, it can be shown that specific substrates or inhibitors of individual P450 enzymes are very close to the average log P value for substrates (or inhibitors) of each one and, furthermore, compare favourably with the maximal values obtained for QSAR analyses such as those described above.

It is possible that this type of approach can assist in understanding the structural requirements for P450 enzyme selectivity and, therefore, aid in the design of specific substrates and inhibitors for human P450s, for example. It is also found that working with homologous series of substrates helps to define the dimensions of the P450 active site, as can be appreciated from consideration of a small series of *N*-alkyl 2,4-dichlorophenoxyl *N*-methylethylamines. In this example, it can be demonstrated that the rate constant for N-demethylation varies parabolically with both surface area and chain length for the reaction catalysed via CYP2B. The following QSARs were obtained by analysis of the rate data for five compounds:

$$\log k_{2B} = 3.50 \text{ length} - 0.13 \text{ length}^2 - 23.90$$

$$n = 5; \quad s = 0.153; \quad R = 0.98; \quad F = 20.6$$

Whereas, for the same N-demethylation mediated by CYP3A the rate constant exhibits a simple linear relationship with compound surface area, as follows:

$$k_{3A} = 0.0005\, SA - 0.52$$

$$n = 5; \quad s = 0.022; \quad R = 0.97; \quad F = 41.4$$

where k is the rate constant for N-demethylation attributed to the involvement of CYP3A. These findings indicate that there is a difference between the size of the active site in CYP3A and CYP2B isoforms with the former exhibiting a larger overall size, and this is also confirmed by the types of substrates known to be selective for the two P450s (Li et al., 1995; Guengerich, 1999; Ekins et al., 1999a and b; Lewis, 2000a and b). Drug clearance in *Homo sapiens* can also be predicted from *in vitro* and *in vivo* data derived from animal studies (Schneider et al., 1999) which utilize multivariate techniques of analysis that are similar to those adopted for QSAR. It is clear, therefore, that structure–activity relationship studies in P450 systems are able to provide satisfactory explanations of the variation in kinetic and clearance data, thus leading to the possibility of using such information in a predictive manner.

Chapter 4

Substrate selectivity and metabolism

> All science is either physics or stamp collecting.
>
> (Lord Rutherford)

4.1 Introduction

One of the characteristics of P450s is the large number and wide variety of structural classes of their substrates, diversity of reactions and, in general, their selectivity (Groves, 1997) and even specificity in some cases for certain substrate molecules (Coon et al., 1996; Porter and Coon, 1991; Juchau, 1990) as summarized in Table 4.1. This is especially true for the mammalian microsomal P450 enzymes associated with the oxidative metabolism of drugs and other exogenous compounds (reviewed in Lewis, 1996a). In contrast, the mitochondrial P450s and most bacterial enzymes are characterized by a general high degree of selectivity, even bordering on specificity, in their substrates (Munro and Lindsay, 1996; Sariaslani, 1991). Both of these aspects of P450 will be explored in this chapter under the broad headings of endogenous and exogenous metabolism, followed by aspects of binding and QSARs although defects in metabolism will also be touched on, together with some mention of inhibitors which can also act as substrates; more information is provided in Chapter 5.

4.2 Endogenous substrates

Many P450 enzymes catalyse pathways of endogenous metabolism, and this function is apparent across species and biological kingdoms in general (Nelson, 1999; Mansuy, 1998; Stegeman and Livingstone, 1998; Nebert et al., 1989a). However, as a generality it is found that the P450s present in non-animal species have a role in the metabolism of endogenous substrates primarily, although various exceptions are known (reviewed in Lewis, 1996a). In animal species, such as mammalia, for example, endogenous functions for P450 enzymes include steroid hormone biosynthesis, prostaglandin and fatty acid metabolism, lipid oxidation and vitamin D_3 hydrolysis (reviewed in Lewis, 1996a and b, 1997a). Other endogenous roles of mammalian P450s have been suggested, including those P450s

Table 4.1 Reactions catalysed by P450 enzymes

Aromatic hydroxylation
Aromatic epoxidation
Aliphatic hydroxylation
Alkene epoxidation
N-dealkylation
O-dealkylation
S-dealkylation
N-oxidation
N-hydroxylation
S-oxidation
Aldehyde oxidation
Androgen aromatization
Halothane oxidation
Halothane reduction
Arginine oxidation
Cholesterol side-chain cleavage
Dehydrogenation
Dehalogenation
Azoreduction
Deamination
Desulphuration
Amide hydrolysis
Ester hydrolysis
Peroxidation
Denitration

References: Smith, 1988; Omura et al., 1993; Gibson and Skett, 1994;Bernhardt, 1995; Lewis, 1996a.

Notes:
Over 40 types of reaction are known to be catalysed by P450s and over 1000 substrates have been documented. Of the 750+ P450s sequenced to date, around 50 have a known physiological role and it is anticipated that over 30 individual P450 genes are present in each animal species (Coon et al., 1996).

normally associated with exogenous metabolism as shown in Table 4.2. The biosynthesis of steroids (reviewed in Lewis and Lee Robichaud, 1998) and their metabolism appears to be a major endogenous function of P450s across most species, and it is likely that this role developed relatively early in the evolution of eukaryotes (Lewis et al., 1998b; Nelson, 1999). For example, the 14α-demethylase of lanosterol is a reaction catalysed by P450s of the CYP51 family which are present in both fungi and animals, including man (Omura et al., 1993; Rozman and Waterman, 1998; Lewis et al., 1999a). Moreover, the 14α-demethylation of obtusifoliol in higher plants exhibits a similarity with that of the structurally related steroid, eburicol, which occurs in filamentous fungi; these two P450 substrates display a close analogy with the endogenous substrate lanosterol that is able to undergo 14α-demethylation in both yeast and mammalian species (Taton et al., 1994).

Table 4.2 Endogenous roles of xenobiotic-metabolizing P450s

CYP	*Endogenous function*
1A1	Heme catabolism
1A2	Estradiol metabolism
2A	Testosterone metabolism
2B	Testosterone metabolism
2C	Testosterone metabolism
2D	Catecholamine metabolism
2E	Gluconeogenesis
3A	Testosterone 6β-hydroxylation
Endogenous substrates	
Melatonin (CYP1A) also CYP2C and CYP2D isoforms are suspected of involvement	
Estradiol (CYP1A) and also metabolized via CYP1B1	
Testosterone (CYP3A) and also CYP2A, CYP2B and CYP2C isoforms	
Catecholamines, e.g. adrenaline (CYP2D) and serotonin	
Acetol (CYP2E) in a gluconeogenesis pathway	
Progesterone (CYP2C and CYP3A)	
Lauric acid (CYP2E and CYP4A)	
Arachidonic acid (CYP2E)	
Specific isoforms mediating endogenous metabolism include:	
Retinol and retinoic acid metabolism (CYP2C8)	
Estradiol 4-hydroxylation (CYP1B1)	
EDHF* synthesis (CYP2C8)	

References: Rendic and DiCarlo, 1997; Fisslthaler et al., 1999; Hakkola et al., 1998; Waxman, 1988; Zimniak and Waxman, 1993; De Matteis et al., 1991; Dey and Nebert, 1998; Amet et al., 1998; Chen et al., 1997.

Note:
* EDHF = Endothelium-derived hyperpolarizing factor

One of the differences between the yeast and mammalian lanosterol 14α-demethylase activity lies in the nature of the product, however, because the metabolite produced in yeasts is ergosterol whereas cholesterol is formed in mammalia (Omura et al., 1993). The generation of cholesterol via CYP51 represents the start of an important series of pathways leading to the formation of several classes of steroid hormones, including progesterone, pregnenolone, cortisol, testosterone and estradiol, as shown schematically in Figure 4.1 which represents the major steroid biosynthetic pathways (Takemori and Kominami, 1984; Lambeth, 1990). This scheme represents the pathways occurring in mammalia and involves P450s present both in the adrenal corticoid mitochondria (CYP11) and hepatic endoplasmic reticulum (namely CYP17, CYP21 and CYP19) although the aromatase activity of CYP19 (Chen et al., 1989; Cole and Robinson, 1991) is also present in other tissues, such as the placenta and gonads (Schenkman and Griem, 1993).

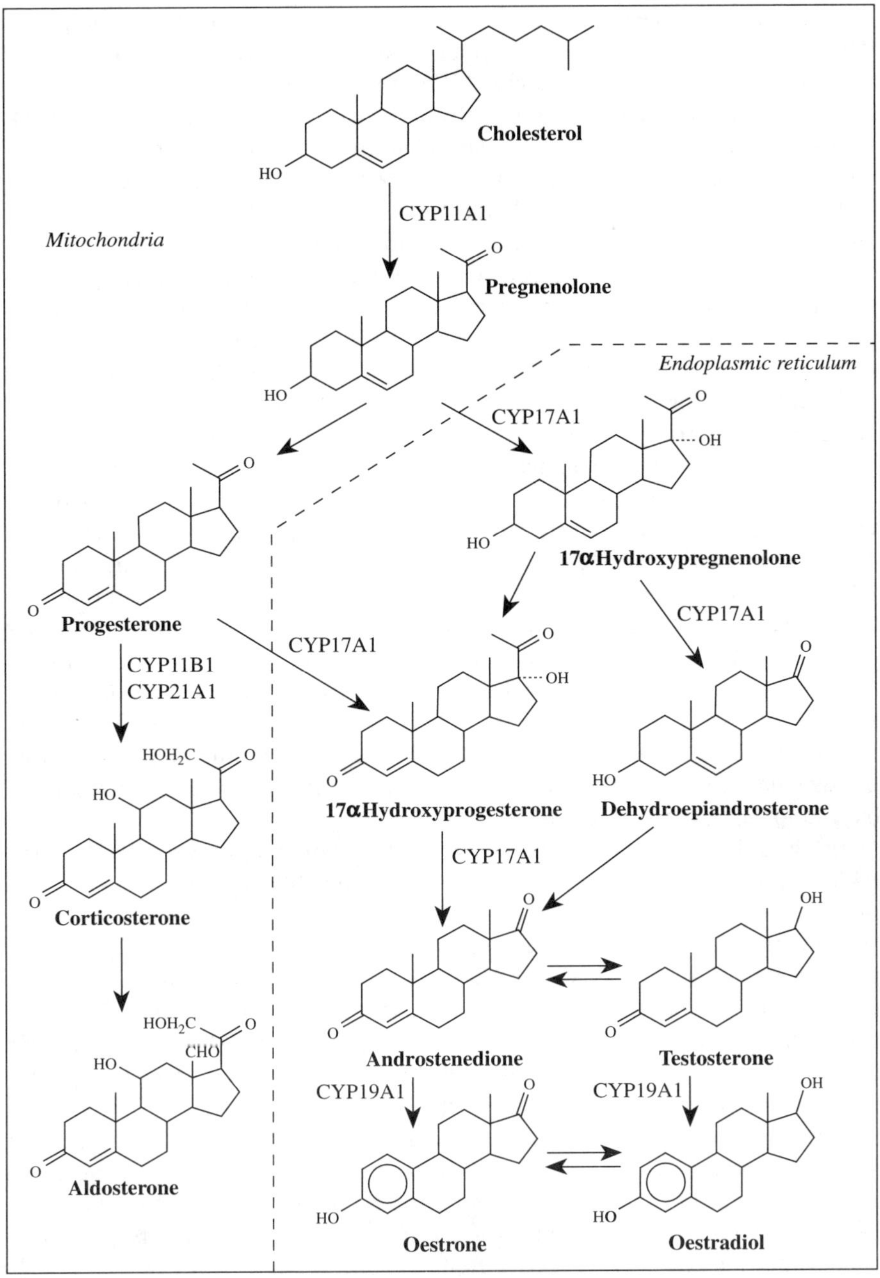

Figure 4.1 Steroid hormone biosynthetic pathways in the adrenal cortex showing reactions catalyzed by specific P450s (Lewis, 1997a). Aromatization of androgens to form estrogens is also catalyzed by CYP19 in the gonads.

A large number of endogenous P450 substrates have been found in plant species (Durst and Benveniste, 1993; Durst and O'Keefe, 1995; Durst, 1991) and these are summarized in Tables 4.3 and 4.4. From inspection of these tables it can be appreciated that the majority of the physiological activity for microsomal plant P450s comprise pathways of secondary plant metabolism (Harborne, 1993) and involve hydroxylation, epoxidation, isomerization, ring migration, cyclization, carbon–carbon coupling and oxidative desaturation (Durst et al., 1994; Chapple, 1998; Pinot et al., 1999). Coumaric acid, for example, is biosynthesized from cinnamic acid and is itself a precursor of the furocoumarin, 8-methoxypsoralen, which is the plant toxin (known as xanthotoxin) that is produced to protect certain plant species from predation (Guengerich, 1993). As xanthotoxin is metabolized by some animal species, such as the black swallowtail butterfly, which utilize P450 enzymes for detoxication (namely, CYP6B1 in the case of the black swallowtail butterfly) it would appear that P450s can play a crucial role in both biosynthesis and metabolism, depending on the type of species involved.

Fungi and bacteria also biosynthesize several different classes of compounds via P450 isoforms, and a selection of their various substrates is presented in Table 4.5. For example, ergosterol biosynthesis from lanosterol in yeasts is mediated by P450s of the CYP51 family (Yoshida and Aoyama, 1991) whereas the hydroxylation of alkanes is catalysed by CYP52 family enzymes in various species of yeast (Müller et al., 1991). In addition, CYP55 in the fungus *Fusarium oxysporum* is associated with nitric oxide reductase activity, and the crystal structure of $P450_{nor}$ (CYP55) which has been reported fairly recently (Park et al., 1997) appears to be related to the bacterial Class I P450s although, in this case, the substrate is nitric oxide and (unusually) the CYP55 enzyme does not require oxygen to catalyse the reaction.

Bacterial P450s are involved in the biosynthesis of erythromycin ($P450_{eryF}$ and $P450_{eryK}$), in the oxidative metabolism of terpenoids like camphor ($P450_{cam}$) and α-terpineol ($P450_{terp}$), and in the ω–2 hydroxylation of long-chain fatty acids such as palmitoleic, arachidonic and pentadecanoic acids ($P450_{bm3}$ or CYP102) as shown in Table 4.5. Fatty acids also constitute endogenous substrates for mammalian P450s of the CYP4 family, for example, and it is possible that there is a close evolutionary relationship between the unusual Class II P450, CYP102, and enzymes of the CYP4 family (Lake and Lewis, 1996; Lewis and Lake, 1999). Furthermore, the structural similarity between cholesterol and a C_{12} fatty acid, as shown in Figure 4.2, suggest that fatty acids of this type may have preceded the steroid hormones as substrates for P450s, with the possibility of a long-chain carboxylic acid being employed as a template for the steroid nucleus as far as the evolutionary development of the P450 superfamily is concerned in respect of its endogenous metabolic roles (Lewis et al., 1998b).

In mammalia, these endogenous pathways involve both steroid hormones, eicosanoids (Capdevila et al., 1992, 1995; Makita et al., 1996) and saturated fatty acids as substrates, together with prostaglandins (Kupfer, 1980) and vitamin D_3 as has been mentioned previously (reviewed in Lewis, 1996a). The various P450s

Table 4.3 Classes of plant P450 substrates

Classes	*Example*	*P450 enzyme*
Phenylpropanoids lignins pigments UV-protectants defence molecules	Cinnamic acid	CYP73
Terpenoids steroids hormones defence molecules aroma molecules/odorants	Geraniol	CYP71
Fatty acids cutin precursors suberin precursors defence molecules	Oleic acid	CYP94
Diverse substrates arginine opioid precursors xenobiotics	Tyrosine	CYP79

References: Durst and Benveniste, 1993; Durst and Nelson, 1995; Durst et al., 1994; Durst, 1991; Pinot et al., 1999.

Notes:
The majority of the above substrates are endogenous, although some plant P450s also play a role in the biotransformation of xenobiotics such as pesticides and environmental pollutants. Furthermore, there are species-specific metabolic pathways including arginine hydroxylation in cyanogenic plant species and carbon–carbon coupling in opioid-synthesizing species.

In addition, an unusual P450, CYP74, has been identified as an allene oxide synthase (Song et al., 1993) in flaxseed, whereas the CYP73 family appears to be associated with cinnamate hydroxylase activity. Moreover, the biosynthesis of dibenzylisoquinoline alkaloids (i.e. reticuline → salutaridine) in *Berberis stolonifera* is catalysed by a P450 (Ortiz de Montellano, 1995). A geraniol/nerol hydroxylase has been isolated and sequenced from *Catharanthus roseus*, and designated CYP71A5, whereas P450s associated with the biosynthesis of certain flower colours have also been characterized (Holton et al., 1993).

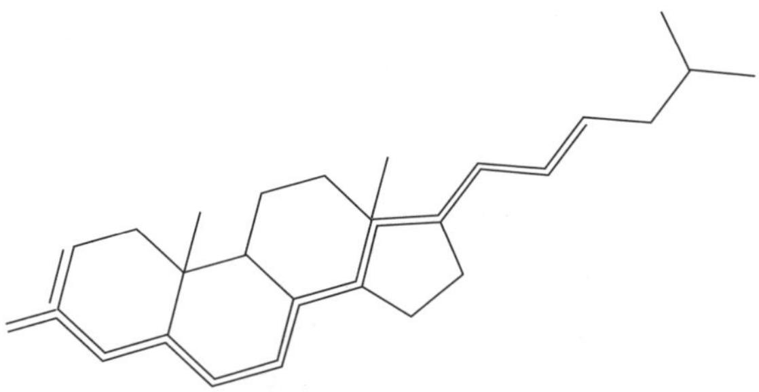

Figure 4.2 A structural overlay between lauric acid and cholesterol molecules showing the similarity between part of their carbon skeleton.

Table 4.4 Plant P450 functionalities

CYP	*Reactions catalysed*
71A	Nerol and geraniol hydroxylases
71C1	Indole hydroxylase
72	Geraniol 10-hydroxylase
73	Cinnamate hydroxylase
74A	Allene oxide synthase
74B	Fatty acid hydroperoxide lyase
75	Flavonoid 3′,5′-hydroxylase
79	Tyrosine *N*-hydroxylase
80	Berbamunine synthase
84	Ferulate 5-hydroxylase
88	Kaurene hydroxylase
90	Cathasterone 23-hydroxylase
94A1	Oleic acid 9,10-epoxidase

References: Chapple, 1998; Pinot et al., 1999.

Table 4.5 Fungal and bacterial P450 substrates

Substrate	*CYP*	*Reaction catalysed*
Lanosterol	51	14α-demethylation
Alkanes	52	ω-hydroxylation
Benzoic acid	53	4-hydroxylation
Nitrate	55	reduction
Nitrite	55	reduction
Pisatin	57	demethylation
Camphor	101	5-exo hydroxylation
Pentadecanoic acid	102	ω-2-hydroxylation
Palmitoleic acid	102	ω-2-hydroxylation
Arachidonic acid	102	ω-2-hydroxylation
Sulphonylureas	105	methyl hydroxylation
Fatty acids	106	hydroxylation
6-Deoxyerythronolide	107	hydroxylation
α-Terpineol	108	4-methyl hydroxylation
Linalool	111	8-methyl hydroxylation
Erythromycin	113	12-hydroxylation

References: Lewis, 1996a; Munro and Lindsay, 1996; Sariaslani, 1991.

which mediate these endobiotic pathways are summarized in Tables 4.6 and 4.7, from which it can be appreciated that a relatively small number of these enzymes are associated with the biosynthesis of endogenous compounds. However, it is found that those P450s normally involved in exogenous metabolism (i.e. the CYP1, CYP2 and CYP3 families) also catalyse oxidations of endogenous substrates, particularly steroids (Schenkman, 1992), although there is often regioselectivity of metabolism in many cases (reviewed in Lewis, 1996a). For example, CYP3A isozymes tend to mediate in the 6β-hydroxylation of testosterone and cortisol,

Table 4.6 Rat hepatic P450s (families CYP1–3) and endogenous steroid metabolism

CYP	*Testosterone metabolism*	*Gender selectivity*	*% P450*	*Inducibility by PB*
1A1	6β-hydroxylation	None	1	a
1A2	6β-hydroxylation	Female dominant	≤ 5	b
2A1	7α- and 6α-hydroxylation	Female dominant	12	2-fold
2A2	15α-hydroxylation	Male specific	3.5	–
2B1	16α-hydroxylation	Male dominant	≤ 1	40-fold
2B2	16β-hydroxylation	Male dominant	< 1	12-fold
2C6	2α-hydroxylation	None	20–30	2-fold
2C7	16α-hydroxylation	Female dominant	15	–
2C11	2α- and 16α-hydroxylation	Male specific	40	–
2C12	15β-hydroxylation	Female specific	27	–
2C13	15α-, 6β- and 7α-hydroxylation	Male specific	23	–
2D1	6β-hydroxylation	Male dominant	unknown	–
3A1	6β- and 7α-hydroxylation	Male specific	≤ 20	12-fold
3A2	6β-hydroxylation	Male dominant	unknown	3-fold

References: Schenkman and Griem, 1993; Lewis, 1996a.

Notes:
a, b – these enzymes are, respectively, 50-fold and 20-fold induced by 3MC or βNF.

Other P450s present in rat liver which are not associated with steroid metabolism include:

1 CYP2C22 which is male specific and non-inducible, but its involvement in testosterone metabolism has not been characterized to date.
2 CYP2E1 is female dominant (14% total P450 complement) and is 3-fold inducible by acetone.
3 CYP4A1 shows no sex differences; it is involved in ω-hydroxylation of lauric acid, and is inducible by clofibrate and other peroxisome proliferators.

whereas CYP1A tends to catalyse the hydroxylation of estradiol in the 2-position (Rendic and DiCarlo, 1997). In the CYP2 family, however, the situation is rather more complicated due to the fact that testosterone metabolism in the rat, for example, exhibits gender selectivity due to the presence of P450s which are either sex-specific or show markedly altered levels between male and female, especially with respect to CYP2A, CYP2B and CYP2C isoforms (Schenkman and Griem, 1993) as shown in Tables 4.6 and 4.7. In this, and for other areas of P450 catalysis, it is very likely that there is active site 'steering' of substrate regioselectivity which accounts for the observed findings, and this aspect will be discussed in the following section on the drug metabolizing enzymes (DMEs) within the P450 superfamily. The endogenous role of DMEs has been discussed by Waxman and colleagues as representing a means of regulating hormone levels (Zimniak and Waxman, 1993; Waxman and Chang, 1995; Waxman et al., 1985, 1991); and other isoforms have been associated with other activities such as gluconeogenesis (CYP2E), neurotransmitter metabolism (CYP2D) and fetal development (CYP1A). Moreover, androgenic imprinting in the noenate has been postulated as an endogenous function of DMEs involved in testosterone metabolism, such as those of the

Table 4.7 A summary of endogenous substrate metabolism by P450 enzymes

CYP	*Species*	*Functionality*
1A1	human	estrogen hydroxylation
1A2	human	estradiol 2-hydroxylation
1A3	human	estradiol 4-hydroxylation
2A1	rat	testosterone 5α-hydroxylation
2A2	rat	testosterone 7α-hydroxylation
2B1	rat	testosterone 16α-hydroxylation
2B2	rat	testosterone 16β-hydroxylation
2C7	rat	testosterone 16α-hydroxylation
2C11	rat	testosterone 2α-hydroxylation
2C12	rat	testosterone 15β-hydroxylation
2C13	rat	testosterone 6β-hydroxylation
2D9	mouse	testosterone 16α-hydroxylation
2E1	human	acetone and acetol oxidation
3A1	rat	testosterone 6β-hydroxylation
4A1	rat	lauric acid ω-hydroxylation
4A4	rabbit	prostaglandin E_2 ω-hydroxylation
4B1	human	lauric acid ω-hydroxylation
4F3	human	leukotriene B_4 ω-hydroxylation
5	human	thromboxane synthesis
6B1	butterfly	xanthotoxin metabolism
7	human	cholesterol 7α-hydroxylation
11A1	human	cholesterol side-chain cleavage
11B1	human	progesterone 11β-hydroxylation
11B2	human	aldocorticoid synthesis
17	human	pregnenolone 17α-hydroxylation
19	human	androgen aromatization
21	human	progesterone 21-hydroxylation
24	human	steroid 24-hydroxylation
27	human	cholesterol 27-hydroxylation
51	yeasts	lanosterol 14α-demethylation
52	yeasts	alkane ω-hydroxylation
55	fungi	nitric oxide reduction
71	plants	geraniol hydroxylation
72	plants	geraniol 10-hydroxylation
73	plants	cinnamic acid 4-hydroxylation
74	flaxseed	allene oxide synthesis
94	plants	oleic acid epoxidation
101	*P. putida*	camphor 5-exo hydroxylation
102	*B.megaterium*	arachidonic acid ω-2 hydroxylation
103	*A.tumefaciens*	N/A
104	*A.tumefaciens*	N/A
105	*S.griseolus*	sulfonylurea herbicide oxidation
106	*B.megaterium*	fatty acid and steroid hydroxylation
107	*S.erythraea*	erythromycin biosynthesis
108	*Pseudomonas spp.*	α-terpineol 4-methyl hydroxylation
109	*B.subtilis*	N/A
110	*Anabaena spp.*	N/A
111	*P. incognita*	linalool 8-methyl hydroxylation

continued...

Table 4.7 (continued)

CYP	*Species*	*Functionality*
112	*B.japonicum*	N/A
113	*S.erythraea*	erythromycin 12-hydroxylation
114	*B.japonicum*	N/A
115	*B.japonicum*	N/A
116	*Rhodococcus spp.*	herbicide degradation

References: Munro and Lindsay, 1996; Sariaslani, 1991; Nelson et al., 1996; Rendic and DiCarlo, 1997; Ruckpaul and Rein, 1991; Andersen et al., 1993; Boddupalli et al., 1992; Davis et al., 1996; Narhi and Fulco, 1986; Fulco, 1991; Fulco and Ruettinger, 1987; Miles et al., 1992; Shafiee and Hutchinson, 1987.

Notes:
* Some of the pathways listed above involve exogenous compounds, such as xanthotoxin, alkanes and herbicides which are degraded by the reactions catalysed respectively by CYP6B1, CYP52, CYP105 and CYP116. Also, $P450_{sca\text{-}2}$ from *Streptomyces carbophilus* is involved in pravastatin biosynthesis (Watanabe et al., 1995).

N/A = information not available

CYP2A, CYP2B and CYP2C subfamilies in particular (Zimniak and Waxman, 1993).

Most of the steroid hormone biosynthetic pathways in mammalia are mediated by P450 enzymes, and these are summarized in Figure 4.1 (Lewis and Lee-Robichaud, 1998). In particular, the CYP11 family enzymes are involved with side-chain cleavage of cholesterol (CYP11A1) and with the 11β-hydroxylation of progesterone (CYP11B1) which is converted to corticosterone via CYP21, whereas CYP17 is associated with 17α-hydroxylation of pregnenolone and progesterone (Takemori and Kominami, 1984). CYP19 represents the androgen aromatase which converts androstenedione and testosterone to estrone and estradiol, respectively (reviewed in Bernhardt, 1995). Of these pathways, the reactions mediated by CYP11 isoforms occur in the adrenal corticoid mitochondria while the remainder take place in the endoplasmic reticulum, although aromatase activity is also present in the placenta and gonads (Schenkman and Griem, 1993; Bernhardt, 1995). Other steroid metabolizing P450s in mammalia are those from the CYP7 family (7α-hydroxylation), CYP24 family (24-hydroxylation) and CYP27 family (25-, 26- and 27-hydroxylation); with the latter being associated with vitamin D_3 hydroxylase activity, in addition to oxidations of C_{27} steroids such as cholesterol (reviewed in Schenkman and Griem, 1993; Lewis, 1996a) which leads to the formation of cholic acid as a degradation product of cholesterol in mammalian liver (Okuda et al., 1993).

4.3 Exogenous substrates

In mammalia and other animal species, P450 enzymes from families CYP1, CYP2 and CYP3 catalyse the Phase 1 metabolism of the majority of foreign compounds to which the organism is exposed, and the liver is the main site of these enzymes in mammals although their activity has also been observed in other organs and tissues (reviewed in Lewis, 1996a; Guengerich, 1995a; Ioannides, 1996; Bernhardt, 1995; Omura et al., 1993; Schenkman and Griem, 1993; Gibson and Skett, 1994). For example, xenobiotic-metabolizing P450s are known to be present in the breast (Hellmold et al., 1998), feto-placental unit (Hakkola et al., 1996, 1998), lung (Arinc, 1993), prostate gland (Williams et al., 2000a), gastrointestinal tract (Strobel et al., 1993), olfactory tissue (Ding and Coon, 1993) and brain (Warner and Gustafsson, 1993), together with P450-dependent activity being observed in skin (Mukhtar and Khan, 1989), neural tissue (Strobel et al., 1997) and kidney (Miura et al., 1989; Honkakoski and Negishi, 1997). However, as the primary organ of exogenous P450 functionality is the liver, this section will focus on the metabolic transformations carried out by hepatic microsomal P450s from human and other mammalian species. Table 4.8 represents a summary of these P450s, whereas Table 4.9 describes those encountered in human liver, although it is important to note that the levels of individual P450s vary considerably between ethnogeographic groups (Shimada et al., 1994) and, within these populations, are also subject to variation depending on age (Faustman et al., 2000), diet and nutrition (Ioannides, 1999), sex (Horn et al., 1995), pathophysiological status (Morgan, 1997; Morgan et al., 1998) and the degree of medication (Gibson and Skett, 1994). Those interested in P450s present in non-mammalian animal species are referred to Schenkman and Griem (1993) and also to a special issue of *Comparative Biochemistry and Physiology* (Volume 121C, Nos.1–3, 1998) for details of substrate metabolism across various biological kingdoms.

4.3.1 CYP1 family substrates

The P450s which comprise this family include CYP1A1, CYP1A2 and CYP1B1 in mammalia and these enzymes are inducible by many planar polyaromatic hydrocarbons (PAHs) and their nitrogenous derivatives, together with environmental pollutants such as 2,3,7,8-tetrachlorodibenzo-*p*-dioxin (TCDD) which is a highly potent inducer of CYP1A1 in particular (Kawajiri and Hayashi, 1996). A substantial number of these inducers are also substrates for the CYP1 enzymes, although TCDD is an exception as it is refractory to oxidative metabolism due to the stabilizing effect of the chlorine substituents. Table 4.10 summarizes a representative number of the known substrates for CYP1 family enzymes present in mammalian species, including man (Lewis, 2000a) and a number of typical CYP1A2 substrates are shown in Figure 4.3. However, in contrast with the situation in most experimental animal species such as the rat and mouse, CYP1A1 is poorly expressed in human liver where the major isoform is CYP1A2, although the former is present in human lung tissue (Kawajiri and Hayashi, 1996).

Essentially, substrates of CYP1A1 are characterized as relatively planar PAH molecules comprised of two to four fused aromatic rings, which also tend to be rectangular in shape and are fairly hydrophobic with respect to those of CYP1A2 (Lewis, 1997b, 2000a). The selective substrate for this enzyme is 7-ethoxyresorufin

Table 4.8 Substrates, inhibitors and inducers of P450 families CYP1, CYP2, CYP3 and CYP4

CYP	*Substrates*	*Inhibitors*	*Inducers*	*Nuclear receptor involvement*
1A1	Aryl hydrocarbons	9-Hydroxyellipticine	TCDD, PAHs	AhR (+ARNT)
1A2	Arylamines	Furafylline	TCDD, PAHs	AhR (+ARNT)
2A	Coumarin	Pilocarpine	Phenobarbital	HNF-4
2B	Phenobarbital	Secobarbital	Phenobarbital	CAR
2C	Mephenytoin	Sulfaphenazole	Phenobarbital	RAR (and RXR?)
2D	Debrisoquine	Quinidine	None known	HNF-4
2E	4-Nitrophenol	4-Methylpyrazole	Ethanol	None known*
3A	Nifedipine	Ketoconazole	Dexamethasone	GR and PXR
4A	Lauric acid	10-Imidazolyl decanoate	Clofibrate	PPAR

References: Rendic and DiCarlo, 1997; Guengerich, 1999; Whitlock, 1999; Honkakoski and Negishi, 2000.

Notes:
* Regulation is probably not receptor-mediated.

TCDD	=	2,3,7,8-Tetrachlorodibenzo-*p*-dioxin
PAHs	=	Polyaromatic hydrocarbons
AhR	=	Aryl hydrocarbon receptor
GR	=	Glucocorticoid receptor
PXR	=	Pregnane-X receptor
PPAR	=	Peroxisome proliferator-activated receptor
CAR	=	Constitutive androstane receptor
ARNT	=	Ah receptor nuclear transporter
RAR/RXR	=	Retinoic acid receptor/Retinoid-X receptor
HNF-4	=	Hepatocyte nuclear factor 4

(Burke and Mayer, 1983) which contains four hydrogen bond acceptor atoms such that tight intermolecular binding with the active site is facilitated (Lewis et al., 1999c). For CYP1A2, the preferred substrates are usually heterocyclic amines and amides such as caffeine and the cooked food mutagens, IQ and PhIP, which tend to be more polar than CYP1A1 substrates, and it is likely that surface amino acid residues on the CYP1A enzymes are responsible for the recognition of, and selectivity towards, the two types of CYP1A substrates (Lewis et al., 1999c). The number of CYP1B1 substrates known is limited by the fact that this enzyme has only been characterized relatively recently, although it appears that 4-hydroxylation of estradiol is a characteristic activity and the metabolism of several heterocyclic amines has also been reported (Rendic and DiCarlo, 1997) together with an involvement in certain bioflavonoid oxidations (Doostdar et al., 2000). In contrast with CYP1B1, however, CYP1A2 preferentially catalyses the 2-hydroxylation of estradiol whereas the substrate showing particularly high binding affinity for human CYP1A2 is 7-methoxyresorufin, although caffeine N_3-demethylation is generally regarded as specific for this isoform (Rendic and DiCarlo, 1997). The strong

Table 4.9 Human drug-metabolizing P450s

CYP	*Level (%)*	*% Drug oxidations*	*Polymorphisms, etc.*	*Inducibility*	*Marker substrate*	*Selective inhibitor*
1A1	< 1	2.5	yes	inducible (AhR)	7-ethoxyresorufin	9-hydroxyellipticine
1A2	~ 13	8.2	variable levels	inducible (AhR)	caffeine	furafylline
1B1	< 1	?	not known	inducible (AhR)	estradiol	propofol
2A6	~ 4	2.5	2% Caucasians PMs	w.inducible (HNF-4)	coumarin	pilocarpine
2B6	< 1	3.4	not known	w.inducible (CAR)	4-trifluoromethyl-7-ethoxycoumarin	orphenadrine
2C8	?	?	not known		taxol	sulfinpyrazone
2C9	~ 18	15.8	1–3% Caucasians PMs	w.inducible via RAR/RXR and possibly by others, e.g. CAR	tolbutamide	sulfaphenazole
2C19	~ 1	8.3	3–5% Caucasians PMs 15–20% Asians PMs		(*S*)-mephenytoin	fluconazole
2D6	≤ 2.5	18.8	5–10% Caucasians PMs 1% Orientals PMs	non-inducible	debrisoquine	quinidine
2E1	≤ 7	4.1	yes	inducible	chlorzoxazone	4-methylpyrazole
3A4	≤ 28	34.1	variable levels	inducible (GR, PXR)	nifedipine	ketoconazole

References: Rendic and DiCarlo, 1997; Guengerich, 1995a, 1999.

Notes:
PMs = poor metabolizers
? = unknown

Table 4.10 CYP1A substrates

Compound	log *P value*
Acetanilide	1.16
2-Aminoanthracene	3.40
4-Aminobiphenyl	2.86
Benzo(*a*)pyrene	6.35
Caffeine	0.01
Dibenz(*a*,*h*)anthracene	6.75
Ellipticine (inhibitor)	4.80
Estradiol	2.69
3-Methylcholanthrene	6.75
β-Naphthoflavone	4.65
α-Naphthylamine	2.25
β-Naphthlyamine	2.28
Phenacetin	1.57
PhIP	2.23
IQ	1.84
Tacrine	2.71
7-Methoxyresorufin	3.15[c]
Furafylline (inhibitor)	0.12
Paracetamol	0.25
6-Aminochrysene	4.98

Reference: Lewis, 2000a.

Note:
c = calculated value

association between CYP1 and the activation of procarcinogens (Conney et al., 1994; Guengerich, 1988) is discussed in Section 4.5, although the role of CYP1A1 in carcinogenesis has been questioned (Beresford, 1993).

4.3.2 CYP2A subfamily substrates

Although primarily associated with the metabolism of endogenous steroids like testosterone in experimental mammalian species (Chang and Waxman, 1996; Honkakoski and Negishi, 1997), CYP2A enzymes also catalyse coumarin 7-hydroxylation (Yamano et al., 1990; Lake, 1999) in both mouse (CYP2A5) and *Homo sapiens* (CYP2A6), with the human isoform also being responsible for the metabolism of a number of drug substrates (Lewis et al., 1999d; Pelkonen et al., 2000) and some carcinogens, although these are probably activated more readily by other P450s such as CYP2E1 (Lewis et al., 2000a). Table 4.11 summarizes the main substrates oxidized by CYP2A, and it would appear that coumarin represents the selective marker substrate for the human enzyme CYP2A6 (reviewed in Chang and Waxman, 1996) whereas several known CYP2A6 substrates are shown in Figure 4.4. The general characteristics of CYP2A substrate selectivity exhibit some degree of overlap with those of CYP2B and CYP2E, although many of the former possess a ketonic function (or its nitrogenous equilavent, namely, the nitroso

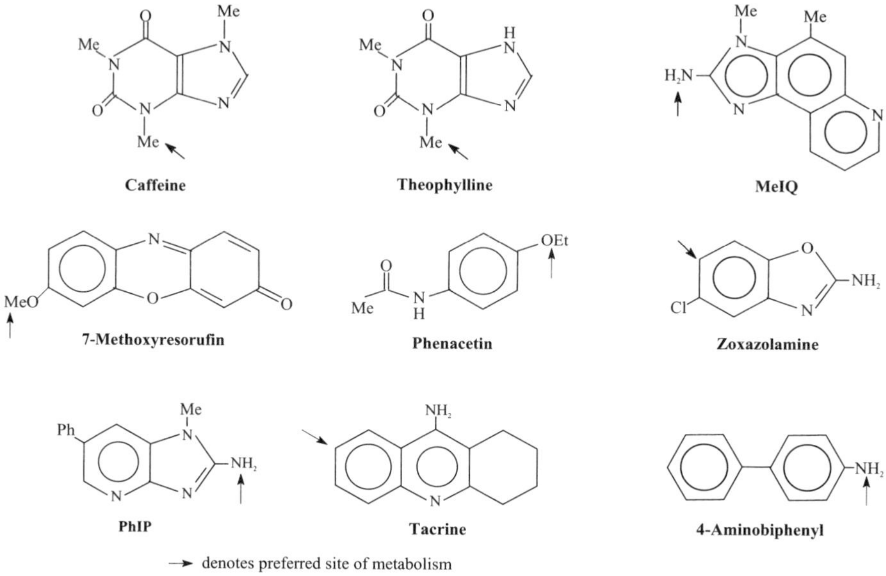

Figure 4.3 *A selection of human CYP1A2 substrates showing the preferred sited of metabolism.*

group) and are relatively polar medium to low molecular weight compounds, with hydrogen bond acceptor atoms fairly close to the preferred sites of metabolism (Lewis et al., 1999d). Coumarin 7-hydroxylase activity is generally linked with CYP2A6 levels in *Homo sapiens* which appear to exhibit a degree of variability, however, possibly being linked with those of CYP2B6, and there has been a suggestion that the two enzymes are coordinately regulated as their respective genes occupy the same chromosomal location in the human genome (Gonzalez, 1992).

4.3.3 CYP2B subfamily substrates

Being the main hepatic P450 isoform inducible by phenobarbital and other barbiturates in experimental animals, CYP2B-mediated biotransformations have been extensively studied in both small rodents and in the rabbit (Nims and Lubet, 1996). The preferred substrates of CYP2B enzymes (see Table 4.12) are phenobarbital and related barbital drugs, benzphetamine and other aliphatic amines, with 7-pentoxyresorufin representing a specific marker substrate in both rat and mouse. The human orthologue, CYP2B6, is somewhat poorly expressed in human liver and, consequently, its role in xenobiotic metabolism appears to be more limited than that of the other mammalian counterparts (Guengerich, 1995a). However, CYP2B6 is involved in the oxidation of cyclophosphamide (Rendic and DiCarlo, 1997; Ekins and Wrighton, 1999) and ifosfamide, together with

Table 4.11 CYP2A substrates

Compound	log *P value*
Coumarin	1.39
Losigamone	1.46
SM-12502	1.06
Fadrozole	2.79
Cotinine	0.07
534U87	2.20
Butadiene	1.99
Halothane	2.30
Methoxyflurane	2.21
Pilocarpine (inhibitor)	1.14

Reference: Lewis, 2000a.

Table 4.12 CYP2B substrates

Compound	log *P value*
4-Trifluoromethyl-7-ethoxycoumarin	3.31
Bupropion	3.64
Cyclophosphamide	0.51
Ifosfamide	0.86
7-Pentoxyresorufin	5.14
Antipyrine	0.23
Benzphetamine	2.27[e]
Cocaine	2.09
Nicotine	1.17
Phenobarbital	1.42
Allylisopropamide	1.14
Chloramphenicol	1.14
DDT	3.98
Heptachlor	5.05
Methoxychlor	4.30
Phenytoin	2.47
Phenylbutazone	3.25
Octylamine (inhibitor)	0.76
2-Phenylimidazole (inhibitor)	1.88
Orphenadrine (inhibitor)	3.71
SKF-525A (inhibitor)	4.65

Reference: Lewis, 2000a.

Note:
e = estimated value

being a mediator in the N-demethylation of mephenytoin, and its selective substrate is generally regarded as 4-trifluoromethyl-7-ethoxycoumarin (reviewed in Lewis et al., 1999e). Figure 4.5 shows the structures of selected CYP2B6 substrates as examples, and their sites of metabolism by this enzyme are also indicated. The

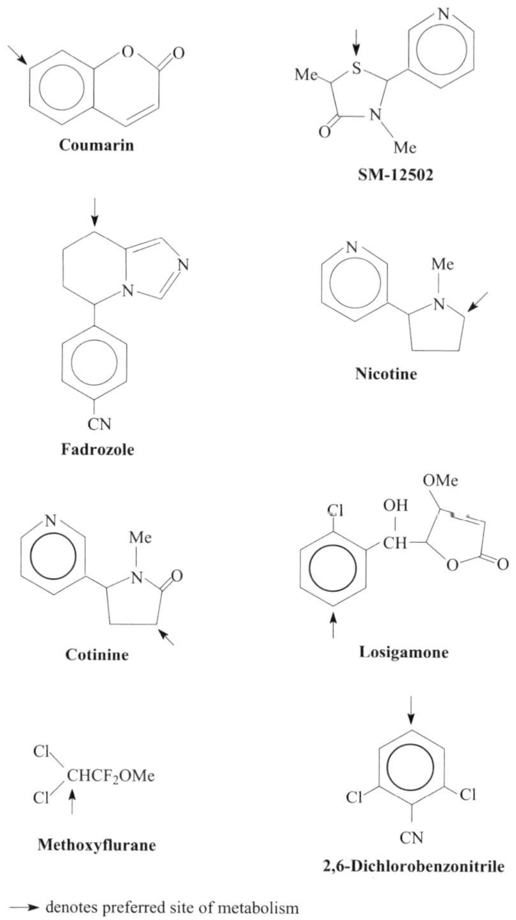

Figure 4.4 A selection of known CYP2A6 substrates showing the preferred sites of metabolism.

general characteristics of CYP2B substrates include non-planar molecular geometries coupled with a relatively high lipophilicity, together with the presence of one or two key hydrogen bond-forming groups and, usually, at least one aromatic ring adjacent to a tetrahedral carbon often imparting a V-shaped conformation to the structure (Lewis et al., 1999b and e). Enzymes of the CYP2B subfamily are induced by phenobarbital (PB) and other barbiturates (Nims and Lubet, 1996) and it is now established that an orphan nuclear receptor, namely, the constitutive androstane receptor (CAR) is involved in PB-type induction of CYP2B (Honkakoski and Negishi, 2000).

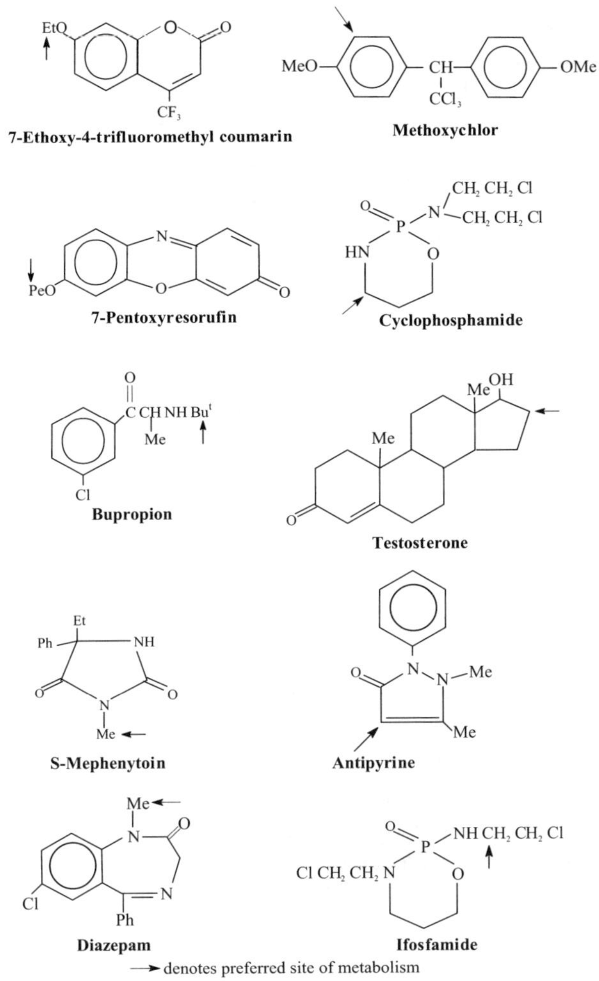

Figure 4.5 A selection of known CYP2B6 substrates showing the preferred sites of metabolism.

4.3.4 CYP2C subfamily substrates

There is a degree of overlapping substrate selectivity between CYP2C and CYP2B enzymes in experimental animal species, although the human CYP2C isoforms exhibit greater demarcation in their substrate selectivities (Lewis et al., 1998c) particularly with regard to CYP2C9 which is the major human orthologue (Richardson and Johnson, 1996). Table 4.13 summarizes information on typical substrates for enzymes of the CYP2C subfamily (Lewis, 1996a; Guengerich, 1995a) and many CYP2C9 substrates, such as diclofenac (Mancy et al., 1999) and tienilic acid (Lecoeur et al., 1994), contain carboxylic acid groups or amides which exhibit weakly acidic properties (Goldstein and de Morais, 1994; Lewis et al., 1998c). Figure 4.6 presents structures of representative CYP2C9 substrates as

Table 4.13 CYP2C substrates

Compound	log *P*	pK_a
Diazepam	2.86	4.75[b]
Hexobarbital	1.49	8.2[b]
Naproxen	3.18	4.15[a]
Ibuprofen	3.51	4.5[a]
Phenytoin	2.47	8.1[a]
Sulfaphenazole (inhibitor)	1.56	6.5[a]
Tolbutamide	2.34	5.43[a]
Warfarin	2.52	5.1[a]
Diclofenac	4.40	4.22[a]
Tienilic acid	3.15	4.8[a]
S-Mephenytoin	1.74	8.1[b]
R-Mephobarbital	1.86	7.8[b]
Proguanil	2.53	10.4[b]
Moclobemide	2.13	7.09[b]
Omeprazole	2.23	8.7[b]
58C80	5.18	5.0[a]

Reference: Lewis, 2000a.

Notes:
a = acidic   b = basic

examples, showing the common feature of an acidic grouping in the molecule. In contrast, the polymorphic human orthologue CYP2C19 (Wrighton et al., 1993) exhibits an altered substrate selectivity (see Figure 4.7 for typical examples) from CYP2C9, with (*S*)-mephenytoin representing a typical marker substrate for the former, and tolbutamide or diclofenac being preferred substrates for CYP2C9 (Richardson and Johnson, 1996). A third human CYP2C enzyme, CYP2C8, has a degree of overlapping substrate selectivity with CYP2C9, although taxol 6α-hydroxylation appears to be mediated primarily by the CYP2C8 isoform (Rendic and DiCarlo, 1997). Figure 4.8 shows some examples of known CYP2C8 substrates, together with the relevant sites of metabolism. Induction of CYPC enzymes may be mediated via the retinoic acid receptor (RAR) although there is also some evidence for other nuclear receptor involvement, including that of HNF-4 (Honkakoski and Negishi, 2000).

4.3.5 CYP2D subfamily substrates

Enzymes of the CYP2D subfamily tend to be associated with the metabolism of basic drugs and other xenobiotics containing a nitrogen atom which becomes readily protonated at physiological pH (G. Smith et al., 1998b). This property is reflected in the typical substrates for CYP2D isoforms which are listed in Table 4.14, the majority of which are metabolized by the polymorphic human form, CYP2D6 (Guengerich, 1995a; Gonzalez, 1996; de Groot et al., 1996, 1999). The marker substrates for CYP2D6 include debrisoquine (Ellis et al., 1992) and metoprolol (Ellis et al., 1996), although a substantial number of drugs in current

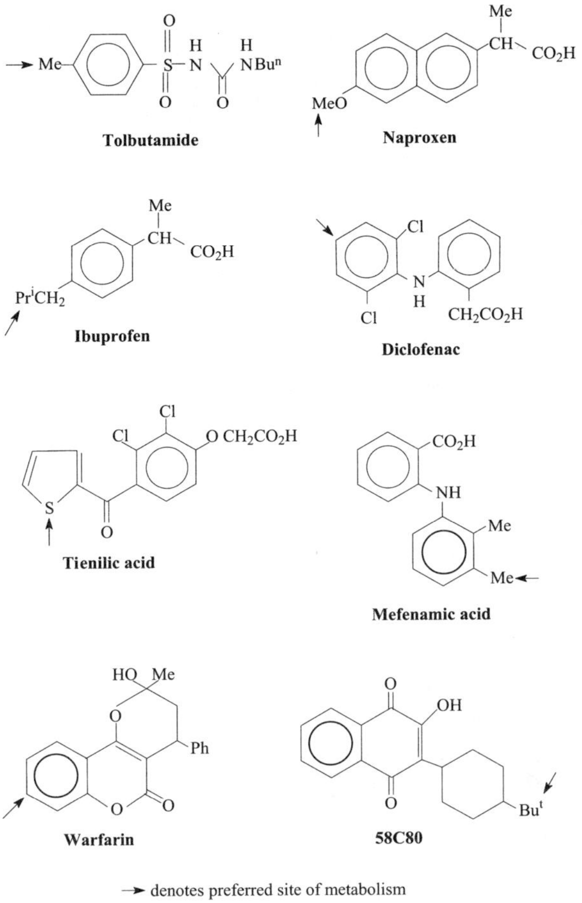

Figure 4.6 A selection of known CYP2C9 substrates showing the preferred sites of metabolism.

clinical use are also recognized substrates for this enzyme (Lewis et al., 1997a) Figure 4.9 provides some examples of typical CYP2D6 substrates, where sites of metabolism are also indicated. Although the rat orthologue CYP2D1 displays similar selectivity towards nitrogenous bases as CYP2D6, the mouse form, CYP2D9, does not appear to metabolize debrisoquine which is a preferred substrate for both the rat and human orthologues (Gonzalez, 1996). Species differences and regulatory aspects of the CYP2D subfamily have been reviewed (Gonzalez, 1996), together with a description of the pharmacogenetics associated with CYP2D6 polymorphism. The enzyme is constitutively expressed and it may be unlikely that nuclear receptors mediate its regulation, although there is evidence for the involvement of the estrogen receptor and the hepatocyte nuclear factor, HNF-4 (Honkakoski and Negishi, 2000).

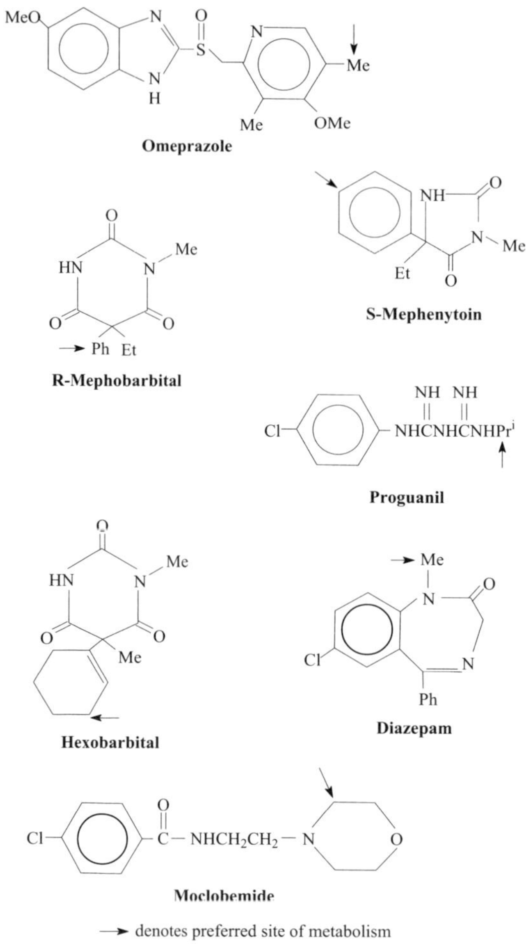

Figure 4.7 A selection of known CYP2C19 substrates showing the preferred sites of metabolism.

4.3.6 CYP2E subfamily substrates

Although structurally diverse in nature, substrates of CYP2E enzymes are characterized by their relatively low molecular weight (Lewis et al., 2000a) and many are fairly hydrophilic, with quite low log *P* values (Lewis et al., 1999b). Table 4.15 shows a representative number of known CYP2E substrates which include benzene, ethanol, acetone and aniline, with *p*-nitrophenol and chlorzoxazone being regarded as typical marker substrates (Ronis et al., 1996). The substrate selectivities of CYP2E enzymes appear to be largely conserved across mammalian species and the relevant proteins in rat, mouse, rabbit and human exhibit close sequence homology (Song et al., 1986; Lewis et al., 1997b).

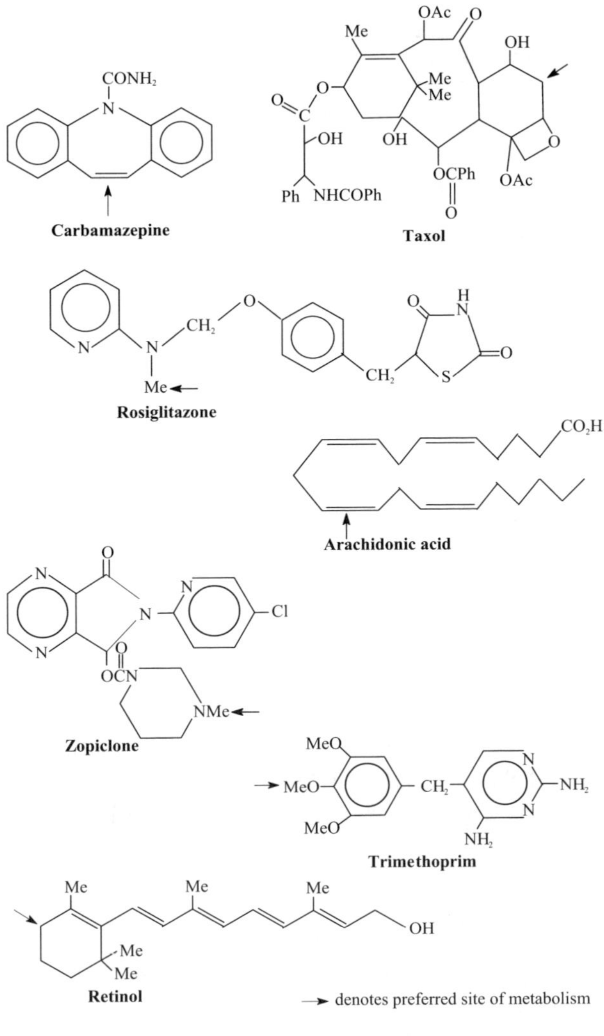

Figure 4.8 A selection of known CYP2C8 substrates showing the preferred sites of metabolism

CYP2E isoforms are associated with the activation of several carcinogens and other toxic chemicals, such as benzene, dialkylnitrosamines, halothanes (Elfarra, 1993; Lau and Monks, 1993) and paracetamol; the generation of oxygen radicals and other reactive oxygen species has also been linked with CYP2E induction (Ronis et al., 1996), probably due to a poorly coupled redox cycle. In *Homo sapiens*, CYP2E is involved in the metabolism of a relatively small number of pharmaceutical agents, including chlorzoxazone and paracetamol (Rendic and DiCarlo, 1997) although a substantial proportion of industrial solvents and petrochemicals are also metabolized via this enzyme. A selection of human

Table 4.14 CYP2D substrates

Compound	log *P*	pK_a
Debrisoquine	0.75	13.01[b]
Metoprolol	2.47	9.8[b]
Bufuralol	3.50	9.0[b]
Sparteine	2.13	11.8[b]
Codeine	1.07	8.2[b]
Dextromethorphan	3.36	8.3[b]
Oxprenolol	2.00	9.3[b]
Propranolol	3.37	9.5[b]
Citalopram	3.68	9.3[b]
Amitriptyline	5.04	9.4[b]
Chlorpromazine	5.40	9.5[b]
Desmethylimipramine	4.05	10.0[b]
Clomipramine	5.19	9.4[b]
Levopromazine	3.39	9.4[b]
Yohimbine	2.59	9.87[b]
Quinidine (inhibitor)	2.83	7.9[b]

Reference: Lewis, 2000a.

Note:
b = basic

CYP2E1 substrates, showing the relevant sites of metabolism, are presented in Figure 4.10. Regulation of CYP2E appears to be governed by post-translational modification, possibly mediated via protein kinases (Ronis et al., 1996), and where serine phosphorylation may play a role in enzyme degradation by loss of the heme moiety (reviewed in Lewis, 1996a).

4.3.7 *CYP3A subfamily substrates*

CYP3A substrates tend to exhibit structural diversity although, in contrast to those of CYP2E, such compounds are usually of relatively high molecular weight (Lewis et al., 1996; Li et al., 1995; Maurel, 1996; Smith et al., 1998; Guengerich, 1999). The major human isoform CYP3A4 is involved in the metabolism of a large number of drugs and other exogenous chemicals, and Table 4.16 indicates the broad substrate preferences for the human enzyme (Lewis, 2000a) and some selected examples of typical substrates are presented in Figure 4.11. It should be recognized also that CYP3A enzymes are primarily associated with detoxifying metabolism but may participate in the activation of a small number of pro-carcinogens and other toxic compounds, such as aflatoxin B_1 and pyrrolizidine alkaloids (Gonzalez and Gelboin, 1994). The levels of CYP3A isoforms are substantially lower in experimental rodent species than in *Homo sapiens*, where their association with drug oxidations is broadly commensurate with their proportion of the hepatic P450 complement, although there is significant variation between individuals and

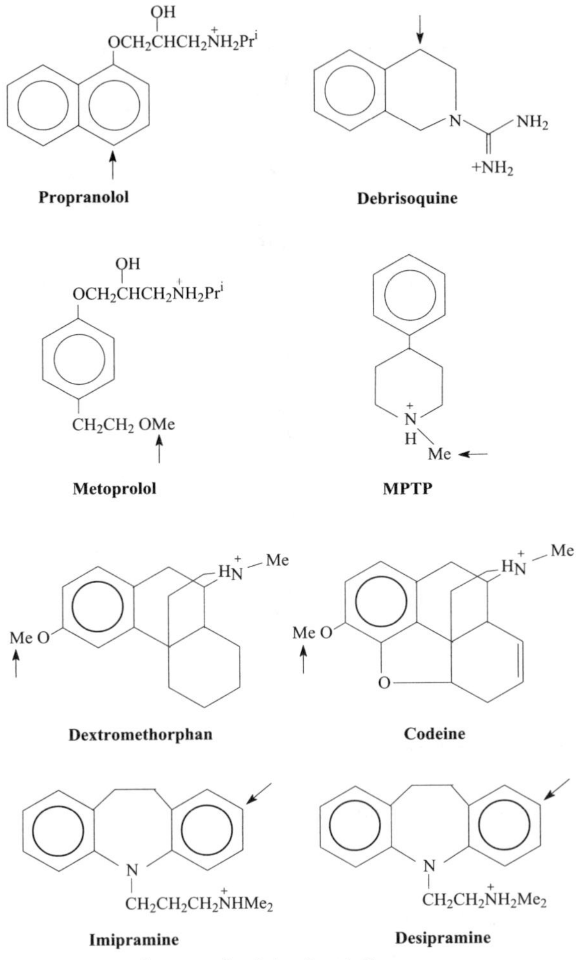

Figure 4.9 A selection of known CYP2D6 substrates showing the preferred sites of metabolism.

it is thought that CYP3A5 at least may be polymorphically expressed (Maurel, 1996). The specifically catalysed reaction of the major hepatic form, CYP3A4, is the N-oxidation of nifedipine (Rendic and DiCarlo, 1997) although other marker substrates, such as erythromycin, are also known. CYP3A genes are regulated by the glucocorticoid receptor (Guengerich, 1999), and via the pregnane-X receptor (PXR), although the possible involvement of HNF-4 has also been reported (Honkakoski and Negishi, 2000).

Table 4.15 CYP2E substrates

Compound	log *P*
Halothane	2.30
Azoxymethane	–0.94[c]
4-Nitrophenol	1.91
Carbon tetrachloride	2.83
Isoniazid	–1.14
Paracetamol (Acetaminophen)	0.25
Dimethylnitrosamine	–0.57
Nitrosopyrrolidine	–0.19
Dimethyl sulphoxide	–1.35
Aniline	0.90
p-Xylene	3.15
Pyridine	0.65
Pyrazole	0.13
Imidazole	–0.08
Acetone	–0.24
Benzene	2.13
Butan-2-ol	0.76
Pentane	3.11
Diethyl ether	0.83
Propan-2-ol	0.05
Acetonitrile	–0.34
Ethanol	–0.32
Chlorzoxazone	2.36
Diethylnitrosamine	0.48

Reference: Lewis, 2000a.

Note:
c = calculated value.

4.3.8 CYP4A subfamily substrates

This subfamily generally metabolizes endogenous fatty acid substrates such as lauric acid, palmitic acid, myristic acid and arachidonic acid (Lake and Lewis, 1996) but some exogenous substrates are also known, such as the phthalate ester, mono(2-ethylhexyl)phthalate (MEHP). The characteristics of CYP4A substrates include the presence of a carboxylic acid moiety situated about 14 Å from the preferred site of metabolism, which is usually at the terminus of an aliphatic chain (Lewis and Lake, 1999). These compounds also tend to be relatively hydrophobic due to the fact that they often contain a substantial number of hydrocarbon groupings, such as the methylene groups of long-chain carboxylic acids, for example. Regulation of the CYP4A subfamily appears to be controlled by an orphan nuclear receptor, termed the peroxisome proliferator-activated receptor (PPAR) and most inducers of these enzymes are also involved in peroxisomal proliferation (Lake and Lewis, 1996; Honkakoski and Negishi, 2000).

In conclusion to this section, it is also known that exogenous substrate metabolism occurs in plants, fungi and bacteria (reviewed in Schenkman and

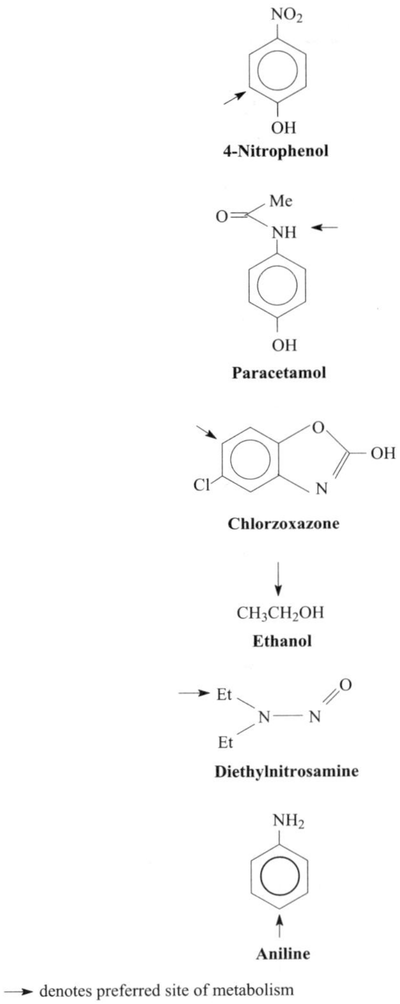

Figure 4.10 A selection of human CYP2EI substrates showing the preferred sites of metabolism.

Griem, 1993) together with xenobiotic metabolism being reported in birds, fish and arthropods (see, for example, the special issue of *Comparative Biochemistry and Physiology*, **121C**, Nos. 1–3, 1998). In some cases, these are compounds produced biosynthetically in other organisms, such as aflatoxin B_1, xanthotoxin and pisatin (Omura et al., 1993) although man-made chemicals like the sulphonylurea herbicides constitute substrates of certain bacterial P450s, and some examples of these are listed in Table 4.7. As far as mammalian P450 substrates are concerned, it appears that there are fairly well-defined ranges of log *P* values for different hepatic microsomal P450s (Lewis, 2000a) and selective inhibitors tend to have log *P* values close to the average shown for each type of P450. (The

Table 4.16 CYP3A substrates

Compound	log *P*
Dexamethasone	1.99
Pregenenolone 16α-carbonitrile	3.49[c]
Troleandomycin	4.78[c]
Erythromycin	2.48
Ethylmorphine	1.43[e]
Aflatoxin B_1	2.20[c]
Clotrimazole	5.71[c]
6-Aminochrysene	4.98
Rifampicin	1.29
Quinine	2.11
Nifedipine	2.82[c]
Testosterone	3.32
Ketoconazole (inhibitor)	3.90
Cyclosporin A	2.92
Diazepam	2.86
Lidocaine	2.26
Dapsone	0.97
Progesterone	3.87
Verapamil	3.79
Cortisol	1.61
Midazolam	1.53
Carbamazepine	2.28[c]

Reference: Lewis, 2000a.

Notes:
c = calculated value e = estimated value

lipophilic nature of a chemical can be evaluated from its log *P* value where *P* is the octanol/water partition coefficient (Pliska et al., 1996).)

4.4 Criteria governing substrate selectivity for human hepatic P450s

An understanding of the factors involved in human P450 substrate selectivity is of considerable current interest, particularly to the pharmaceutical industry (Smith et al., 1997a and b, 1998) where an early prediction of likely drug metabolism pathways in *Homo sapiens* can aid in the design and development of new chemical entities (NCEs). It is recognized, however, that some compounds are able to act as substrates for more than one P450 isoform because their structures fit the relevant enzyme active sites, although they may be rather more selective towards one P450 than another and the position of metabolism can also vary (see, for example, Wiseman and Lewis, 1996; Lewis and Lake, 1998a). For example, omeprazole undergoes 5-methyl hydroxylation mediated by CYP2C19, whereas its S-oxidation is catalysed primarily via CYP3A4 (reviewed in Lewis and Lake, 1998a). Other examples of such 'promiscuous' substrates include tamoxifen and trimethoprim

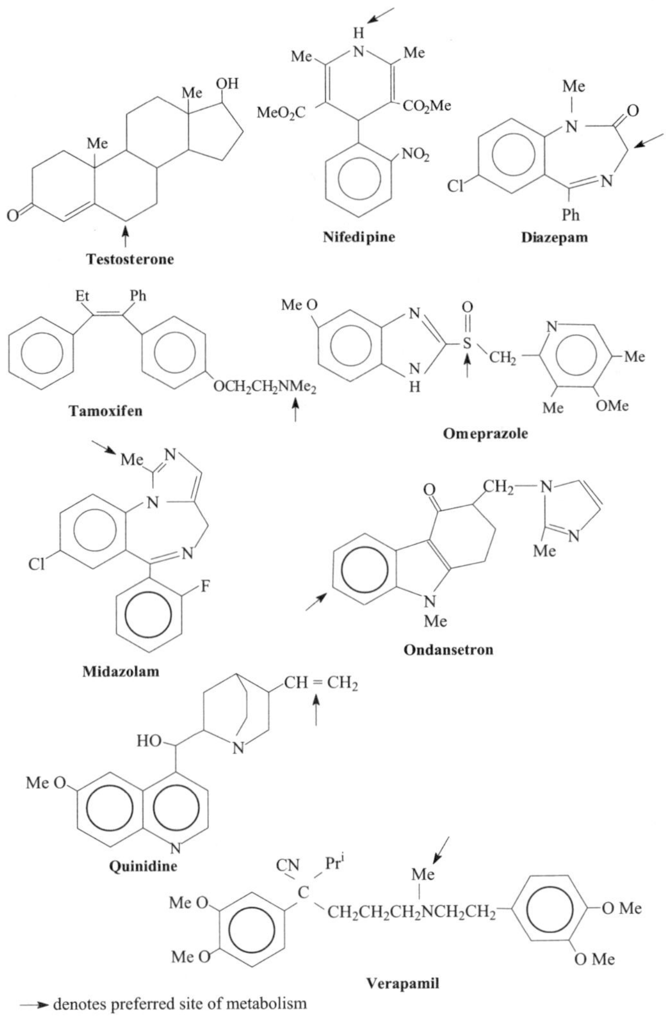

Figure 4.11 A selection of known CYP3A4 substrates showing the preferred sites of metabolism.

(Rendic and DiCarlo, 1997) and rationalizing their metabolism by human P450 enzymes represents a challenge for those involved in molecular modelling of P450–substrate interactions (see, for example, Lewis et al., 1999b).

One means of approaching the problem of satisfactorily explaining substrate selectivity of P450 enzymes involves an assessment of their molecular and physicochemical characteristics (Lewis et al., 1999b; Lewis, 1999b, 2000a and b). Table 4.17 presents the relevant data for substrates of CYP1A2, CYP2A6, CYP2B6, CYP2C9, CYP2C19, CYP2D6, CYP2E1 and CYP3A4 which have been

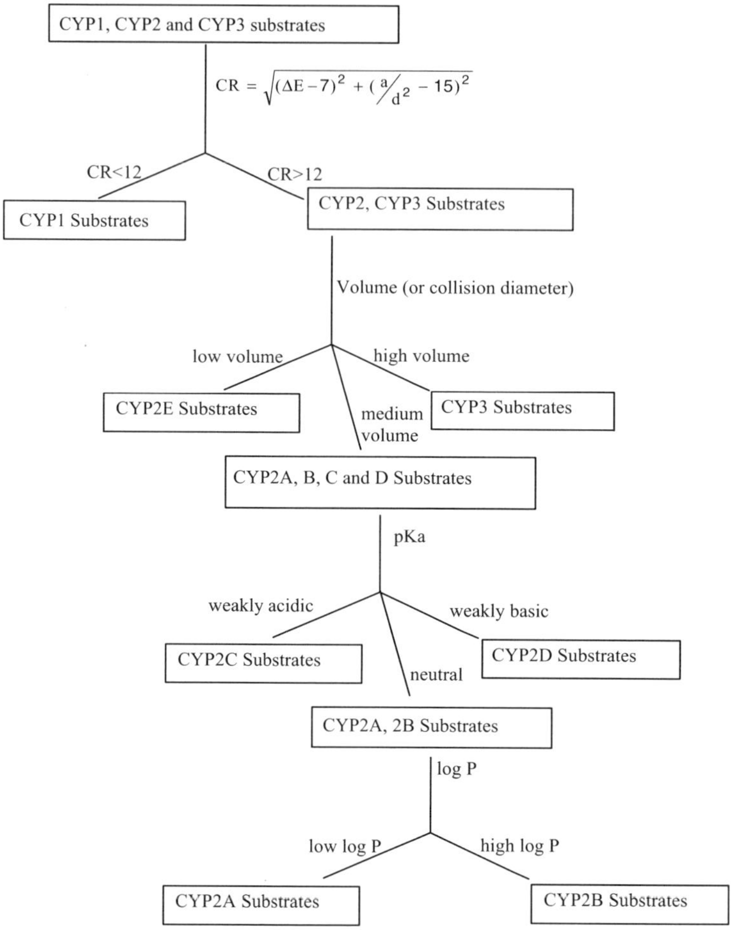

Figure 4.12 A decision tree scheme for differentiating between human P450 substrates from the CYP1, CYP2 and CYP3 families on the basis of their physicochemical and structural characteristics (Lewis, 2000b).

Notes:
1 By and large, the log P values of P450 substrates are greater than zero, although there are exceptions, such as some CYP2D and CYP2E substrates.
2 CYP2D6 substrates usually possess a protonable nitrogen atom 5–7Å from the site of metabolism.
3 CYP2C9 substrates generally possess a hydrogen bond donor/acceptor atom 5–8Å from the site of metabolism.
4 Some compounds can be substrates for more than one P450, especially where there is overlapping specificity.
5 The above scheme does not necessarily apply to steroids and other endogenous chemicals which can be substrates of several P450s.

Table 4.17 Structural descriptors for 48 substrates of 8 human P450s

	log P	pK_a	log $D_{7.4}$	a/d^2	l/w	SA	Vol	ΔE	E_{LUMO}	μ
1. CYP1A2										
Caffeine	0.08	neutral	0.08	5.106	1.139	190.22	174.33	8.6021	–0.3448	3.4490
PhIP	2.23	8.69[b]	0.92	5.148	1.952	228.55	213.59	8.4878	–0.2199	4.3344
7-Methoxyresorufin	3.15	9.94[b]	0.64	4.609	1.883	221.88	203.51	7.4266	–1.6707	4.3911
Phenacetin	1.57	neutral	1.57	5.118	2.017	208.79	183.91	8.7223	0.3635	3.0950
Tacrine	2.71	9.8[b]	0.36	3.183	1.459	212.84	203.69	8.1645	–0.2534	2.1640
IQ	1.84	9.18[b]	0.06	5.294	1.484	200.94	186.08	7.8889	–0.4215	3.4515
2. CYP2A6										
Coumarin	1.39	neutral	1.39	6.738	2.929	152.67	133.08	8.5287	–0.9313	4.2692
Losigamone	1.46	neutral	1.46	1.678	1.543	234.61	226.70	9.5150	–0.3150	4.0459
SM-12502	1.02	4.46[b]	1.02	1.595	1.489	209.34	204.27	8.5819	–0.4858	0.8560
Fadrozole	2.79	6.16[b]	2.77	1.835	1.480	232.79	227.83	8.2534	–0.7067	3.6289
Cotinine	0.07	4.44[b]	0.07	1.360	1.273	190.32	181.27	9.5498	–0.1467	1.6650
534U87	2.20	6.04[b]	2.18	1.863	1.707	237.70	226.17	8.0853	–0.4952	1.8609
3. CYP2B6										
4-Trifluoromethyl EC	3.31	neutral	3.31	3.083	1.379	229.86	212.00	8.1876	–1.3459	3.8658
Bupropion	2.54	8.35[b]	1.54	1.802	1.120	254.07	254.78	8.8666	–0.6235	2.1291
Cyclophosphamide	0.51	neutral	0.51	1.808	1.029	235.79	233.18	11.2819	0.6327	3.5430
Ifosfamide	0.86	neutral	0.86	2.620	1.003	238.76	232.48	10.9334	0.5199	3.1131
7-Pentoxyresorufin	5.14	10.06[b]	2.50	2.577	1.993	300.19	278.87	7.2014	–1.6870	4.5153
Antipyrine	0.23	neutral	0.23	2.011	1.562	203.62	188.71	8.6238	–0.1603	3.9330
4. CYP2C9										
Tolbutamide	2.34	5.43[a]	0.37	2.573	2.523	280.37	260.13	9.3253	–0.8189	7.0221
Diclofenac	4.40	4.22[a]	1.22	1.466	1.190	257.17	262.62	8.7552	–0.2647	1.6864
Tienilic acid	3.15	4.8[a]	0.55	2.635	1.922	276.19	272.17	8.8443	–0.8578	4.4238
Ibuprofen	3.51	4.5[a]	1.07	2.092	1.535	239.49	231.98	9.5994	0.2013	1.8386
S-Warfarin	2.52	5.1[a]	0.12	1.734	1.383	285.05	291.99	8.4476	–0.8577	3.6629
Naproxen	3.18	4.15[a]	0.33	1.845	1.620	241.82	228.33	8.2505	–0.3931	2.2335

Notes:
PhIP = 2-amino-1-methyl-6-phenylimidazo[4,5-*b*]pyridine IQ = 2-amino-3-methylimidazo[4,5-*f*]quinoline EC = 7-ethoxycoumarin

continued ...

Table 4.17 (continued)

	log *P*	pK_a	log $D_{7.4}$	a/d^2	*l/w*	*SA*	*Vol*	ΔE	E_{LUMO}	μ
5. CYP2C19										
S-Mephenytoin	1.74	8.1[b]	0.96	1.865	1.129	223.45	218.90	9.8861	–0.0608	2.5168
Hexobarbital	1.49	8.2[b]	0.63	1.287	1.217	230.09	234.13	9.7495	–0.1303	2.0137
R-Mephobarbital	1.86	7.8[b]	1.31	1.306	1.141	233.52	236.94	9.7388	–0.1986	1.9466
Proguanil	2.53	10.4[b]	–0.47	2.784	1.384	263.33	246.67	8.8773	–0.1612	2.1785
Moclobemide	2.13	7.09[b]	1.96	3.002	2.343	275.85	262.03	8.9332	–0.4901	2.4491
Omeprazole	2.23	8.7[b]	0.91	1.841	1.035	340.19	330.37	8.2444	–0.3993	3.2240
6. CYP2D6										
Debrisoquine	0.75	13.01[b]	0.75	2.005	1.488	194.46	181.19	9.5405	0.4072	1.5615
Metoprolol	2.35	9.68[b]	0.07	1.703	1.446	310.45	303.15	9.3137	0.1657	1.1873
Bufuralol	3.50	9.0[b]	1.89	2.196	1.573	292.55	290.87	8.7460	0.0606	1.9560
Sparteine	2.13	11.8[b]	–2.27	1.649	1.268	250.06	271.25	11.5799	2.8428	0.3968
Codeine	1.07	8.2[b]	0.23	1.343	1.327	273.96	297.89	8.9460	0.3400	2.0116
Dextromethorphan	3.36	8.3[b]	0.91	1.490	1.155	273.14	299.54	9.2159	0.6788	1.1944
7. CYP2E1										
4-Nitrophenol	1.91	neutral	1.91	6.009	1.363	139.82	119.01	9.0070	–1.0652	5.1168
Diethylnitrosamine	0.48	neutral	0.48	1.522	1.327	134.55	115.33	10.8885	0.9370	2.8324
Aniline	0.90	4.7[b]	0.90	5.349	1.219	121.11	99.74	9.1636	0.6359	1.2790
Ethanol	–0.32	neutral	–0.32	1.829	1.331	79.36	56.09	14.4412	3.5650	0.9200
Paracetamol	0.25	9.5[a]	0.25	3.957	1.603	169.03	146.50	8.7114	0.2508	2.3131
Chlorzoxazone	2.36	8.3[b]	1.41	5.711	1.635	155.70	134.95	8.8842	–0.5480	2.0617
8. CYP3A4										
Nifedipine	1.96	neutral	1.96	1.735	1.199	311.02	325.29	8.3144	–0.5322	2.6233
Erythromycin	2.48	8.8[b]	1.06	2.641	1.330	501.08	634.04	10.0472	0.8507	4.9621
Cyclosporin A	2.92	7.97[b]	2.25	2.896	1.181	864.35	1318.37	10.7803	0.8079	11.0130
Lidocaine	2.26	7.85[b]	1.68	2.244	1.343	260.66	272.09	9.2559	0.4132	2.8288
Dapsone	0.97	neutral	0.97	2.014	1.540	238.58	228.76	8.6327	–0.2536	5.1632
Midazolam	1.53	4.93[b]	1.53	1.988	1.076	290.11	293.91	8.1995	–0.7595	3.4458

References: Lewis, et al., 1999a, Lewis, 200b

Notes: a = acidic b = basic *Table key* (see next page)

Key to table 4.17

$\log P$	=	logarithm of the octanol/water partition coefficient
pK_a	=	negative logarithm of the acid/base dissociation constant
$\log D_{7.4}$	=	logarithm of the octanol/water distribution coefficient at pH 7.4
a/d^2	=	ratio of molecular area and depth squared
l/w	=	ratio of molecular length and width
SA	=	surface area of the solvent-accessible surface
Vol	=	volume of the solvent-accessible surface
ΔE	=	difference between the frontier orbital energies
E_{LUMO}	=	energy of lowest unoccupied molecular orbital
μ	=	molecular dipole moment

Table 4.18 Characteristics of human P450 substrates

CYP	*General structural and physicochemical characteristics*
1A2	Planar (poly)aromatic/heterocyclic amines and amides with 2 or 3 hydrogen bond acceptors
2A6	Compounds usually contain ketonic or nitroso groups, generally polar
2B6	Non-planar (often V-shaped) molecules, usually lipophilic with hydrogen bond donor/acceptors
2C9	Generally weakly acidic compounds with hydrogen bond donor/acceptors
2C19	Generally neutral or basic compounds with hydrogen bond donor/acceptors
2D6	Nitrogenous bases with sites of metabolism 4–7 Å from basic nitrogen
2E1	Structurally diverse generally neutral compounds of low molecular weight
3A4	Structurally diverse compounds of relatively high molecular weight

Reference: Lewis et al., 1999b.

collated from the literature (Lewis, 2000a). It is possible to show that a relatively small number of structural parameters is sufficient for discrimination between the 48 chemicals with a confidence of over 95% (Lewis, 2000b; Lewis et al., 1999b). Table 4.18 lists the general characteristics of these substrates which can aid in the identification of potential human P450 substrates and Figure 4.12 shows how a decision-tree approach may be employed for distinguishing between human hepatic P450 substrates (Lewis, 2000b). However, it is also important to demonstrate that molecular templates of superimposed substrates are able to fit the appropriate active sites of human P450 enzymes and some examples of these include substrates of CYP1A2, CYP2A6, CYP2B6, CYP2C9, CYP2C19, CYP2D6, CYP2E1 and CYP3A4, where 8 to 10 compounds can be superimposed on their positions of metabolism in each case, and where site interactions such as hydrogen bonding are reinforced (Lewis, 2000b). Although the criteria vary depending on the particular P450 concerned, some common factors emerge which point to a generalized view of substrate selectivity, and this is underlined by quantitative structure–activity relationship (QSAR) investigations on various series of P450 substrates as summarized in Table 4.19. The commonalities in these structural studies may be related to the way in which the various chemicals bind to the relevant P450 isoform, and one can formulate a general description of binding

Table 4.19 QSARs for human P450 substrates

CYP	*Descriptors*	*No.*	*Correlation*	*Biological data*
1A2	l/w, ΔE, N_{HB}, E_L	11	0.97	ΔG_{bind} (K_M) values
2B6	HA, HD, N_{HB}	10	0.96	ΔG_{bind} (K_M) values
2C9	log D, log P, pK_a, HD	8	0.99	ΔG_{bind} (K_M) values
2C19	M_r, log M_r, HA	10	0.96	pK_i values
2D6	M_r, log P, HA, NB	11	0.96	pK_i values
2E1	log P quadratic	10	0.96	pK_i values
3A4	E_L, E_H, N_{HB}, $N_{\pi-\pi}$	10	0.96	ΔG_{bind} (K_D) values

Reference: Lewis, 2000a.

Notes:
No. = number of compounds in the dataset
E_L = LUMO energy
E_H = HOMO energy
ΔE = $E_L - E_H$
l/w = length to width ratio
HD = hydrogen bond donors
HA = hydrogen bond acceptors
NB = number of basic nitrogens
N_{HB} = number of active site hydrogen bonds
$N_{\pi-\pi}$ = number of active site π-π stacks

energy contributions (Lewis et al., 1998d) based on the following expression for the free energy change (ΔG_{bind}) involved:

$$\Delta G_{bind} = \Delta G_{desol} + \Delta G_{ionic} + \Delta G_{hbond} + \Delta G_{\pi-\pi} + \Delta G_{T\&R} + \Delta G_{rot} + \Delta G_{confor} + \text{constant}$$

where the seven ΔG components refer, respectively, to the desolvation, ionic, hydrogen bond, π-π stacking, translational/rotational freedom, bond rotational and conformational energy contributions to the overall energy of binding between substrate and P450. Some of these terms are not always easy to estimate, however, but usually those are ones which constitute relatively minor factors for describing the total binding affinity such that their effect may be small (Lewis, 2000b). ΔG_{bind} can be measured experimentally as either the enzyme–substrate dissociation constant, K_D, or the apparent Michaelis constant, K_M, obtained from kinetics determinations on the purified P450 enzyme concerned (reviewed in Lewis et al., 1998d). The experimental ΔG_{bind} value is then derived from the relationship:

$$\Delta G_{bind} = RT \ln K_D$$

where R is the gas constant, T is the absolute temperature and K_D is the enzyme–substrate binding dissociation constant. Consideration of the above equation indicates that large negative binding free energies will be associated with relatively low K_D (or K_M) values, which are indicative of tight binding affinity between substrate and P450 enzyme (reviewed in Lewis et al., 1998d, 1999b). Analysis of several examples of P450–substrate interactions shows that the desolvation and electrostatic contributions are the major components of the binding energy (Lewis et al., 1999b). However, in some cases it is clear that the loss in substrate translational and rotational energy upon binding has to be taken into account, usually when the molecular weight of the compound is relatively high.

Hydrogen bonding and π-π stacking interactions between aromatic rings represent important contributors to binding free energies in most, if not all, examples of human P450–substrate interactions, although it should be recognized that the hydrogen bond component at approximately –2 kcal mole^{-1} is roughly twice the value of a typical π-π interaction (Lewis et al., 1999b). However, the important distinction between the P450 active sites, and that which has a major bearing on substrate selectivity and orientation for metabolism, is the number and disposition of hydrogen bond donor/acceptor groupings on amino acid residues lining the heme pocket (Lewis, 2000b). This is also reflected in the structures of selective substrates which exhibit high binding affinities towards a given P450 enzyme. For example, 7-methoxyresorufin is a high-affinity substrate for CYP1A2 (K_m value of 0.21 μM) and consideration of its putative active site interaction shows that a total of four hydrogen bonds are formed between the substrate and enzyme, which serve to position the molecule for O-demethylation (Lewis et al., 1999c). This orientation is augmented by the favourable π-π stacking interactions between the resorufin nucleus and two approximately parallel aromatic amino acid residues lining the CYP1A2 active site (Lewis et al., 1999c). Indeed, similar instances can represent explanations for the tight binding of selective substrates of CYP2A6, CYP2C9, CYP2C19, CYP2D6 and CYP3A4, thus leading to a general description of human P450 substrate interactions (Lewis et al., 1999b; Lewis, 2000b) which involves consideration of the disposition and number of hydrogen bonds formed between enzyme and substrate. In fact, such interactions will also direct the course of metabolism in addition to explaining P450 substrate selectivity (Lewis, 2000b).

4.5 Toxicity mediated by P450 enzymes

Foreign compounds constitute a potentially major environmental hazard to humankind. These xenobiotics may arise from both plant and animal sources, either naturally or via cooking processes, together with those produced via modern technology, e.g. pharmaceuticals, agrochemicals, petrochemicals, plastics and other industrial chemicals. Fortunately, during the course of animal evolution, biological defence systems have developed naturally to counteract the potentially harmful effects of plant and fungal toxins, in addition to the deleterious action of oxygen radicals and other reactive oxygen species (ROS) such as superoxide, peroxide, hydroxyl radicals, singlet oxygen, etc. (reviewed in Lewis, 1996a). Central to this process are the cytochromes P450, as representing the major 'frontline' of this defence, which also includes superoxide dismutase (SOD) and catalase for detoxifying the ROS entities O_2^- and O_2^{2-}, respectively.

Antioxidants such as glutathione and vitamins A, C and E, for example, are also associated with the amelioration of oxidative damage to cells and tissues which give rise to ageing, inflammatory diseases and cancer (Ioannides and Lewis, 1995). As has been stated previously, the P450s constitute mixed function oxidases associated with the Phase 1 metabolism of drugs and other foreign chemicals to which the body may be exposed or may ingest as food (Ioannides and Lewis, 1995). The levels of these enzymes in the liver and other organs are variable and are subject to chemical induction, genetic variation and other factors relating to lifestyle, dietary habits, age, gender and pathophysiological state, etc. (reviewed in Gibson and Skett, 1994). Although general detoxifying in nature, some P450s have been linked with the metabolic activation of carcinogens and other toxic chemicals (Gonzalez and Gelboin, 1994). These activating P450 enzymes include those from the CYP1 family and the CYP2E subfamily in particular, although others are also associated with some toxic activation whilst being primarily involved in detoxication (e.g. the CYP3 family enzymes). However, some dietary compounds are known to act either as inhibitors or inducers of these P450 isoforms (Ameer and Weintraub, 1997). For example, the selective CYP2E inhibitor diallyl sulphide is a natural constituent of garlic whereas indole-3-carbinol present in cruciferous vegetables is a CYP1 inducer (Ioannides, 1999). Consequently, the levels of P450 inducers and inhibitors in the diet can have significant effects on the potential cellular damage caused by oxygen radicals and food pyrolysis products (Ioannides and Lewis, 1995). Certain P450s are involved in the toxic activation of pro-carcinogens and other toxins, as shown in Table 4.20. In particular, it is clear that enzymes of the CYP1A and CYP2E subfamilies are specificially linked with the metabolic activation of carcinogens (Gonzalez and Gelboin, 1994). This has been observed in experimental animals, especially small rodent species, but human P450s are also known to catalyse these oxidations, and, consequently, there is a perceptible health risk from human exposure to such chemicals. Furthermore, the role of metabolic activation in drug toxicity has been emphasized (Nelson, 1982) and the link between P450-mediated metabolism and toxicity of several compounds has also been shown (Nyarko et al., 1997; Gonzalez and Gelboin, 1994) such that there is an emerging picture of the particular importance of especially CYP1 and CYP2E enzymes in toxic activation.

Human P450s, however, tend to exhibit different substrate selectivities, tissue levels and inducibilities relative to those of experimental animal species, including age and gender differences, mechanisms of regulation and tissue distribution (such as that reported in the rat, for example), thus rendering much animal data conducted in this area, especially in small rodents, to be of little relevance or value for human xenobiotic metabolism studies and risk assessment (Lewis et al., 1998a). However, those compounds which are activated to carcinogenic or reactive intermediates by P450s from the CYP1A or CYP2E subfamilies are likely to manifest their toxic effects in both experimental animals and in *Homo sapiens* because these enzymes are highly conserved across mammalian species. Where differences occur in respect of these chemicals and their P450 activators, one can anticipate that they will be due primarily to the significant variation in body weight between

Table 4.20 Human P450 enzymes involved in activation of carcinogens and toxins

CYP1A1	*CYP1A2*	*CYP2E1*	*CYP3A*	*CYP1B1*
PAHs, e.g.	PAH amines and	DMN	Aflatoxin B1	17β-estradiol
Benzo(*a*)pyrene (BP)	Heterocyclic amines, e.g.	DEN	Alfatoxin G1	
	PhIP	Benzene	Senecionine	
	Glu-P-1	Butadiene	*tris*(2,3-dibromopropyl	
	Glu-P-2	NNK	phosphate)	
	IQ	Styrene	17β-estradiol	
	MeIQ	Acrylonitrile	6-Aminochrysene	
	MeIQx	Ethyl carbamate	MOCA	
	Trp-P-1	Vinyl chloride	BP-7,8-diol	
	Trp-P-2	Trichloroethene	Sterigmatocystin	
	2-Naphthylamine	Trichloromethane		
	4-Aminobiphenyl	Tetrachloromethane		
	AAF	NNN		
	2-Aminofluorene	NNAL		

References: Guengerich, 1995a; Gonzalez and Gelboin, 1994; Rendic and DiCarlo, 1997.

Notes:
Some carcinogens activated by CYP2E1 are also metabolically activated via CYP2A6 to a lesser extent.
Although activated via CYP1B1, estradiol is detoxified by CYP1A1 and CYP1A2 through 2-hydroxylation (Spink et al., 1998; Hakkola et al., 1997).

Abbreviations

BP = benzo(*a*)pyrene
PhIP = 2-amino-1-methyl-6-phenylimidazo-[4,5-*b*]pyridine
PAHs = polyaromatic hydrocarbons
Glu-P-1 = 2-amino-6-methyl-dipyrido[1,2-*a*:3′,2′-*d*]imidazole
Glu-P-2 = 2-aminodipyrido[1,2-*a*:3′,2′-*d*]imidazole
IQ = 2-amino-3-methylimidazo[4,5-*f*]quinoline
MeIQ = 2-amino-3,5-dimethylimidazo[4,5-*f*]quinoline
MeIQx = 2-amino-3,5-dimethylimidazo[4,5-*f*]quinoxaline
Trp-P-2 = 3-amino-1-methyl-5*H*-pyrido[4,3-*b*]indole
2-AAF = 2-acetylaminofluorene
NNK = 4-(methylnitrosamino)-1-(3-pyridyl)-1-butanone
DMN = dimethylnitrosamine
DEN = diethylnitrosamine
NNN = nornitrosonicotine
NNAL = 4-(methylnitrosamino)-1-(3-pyridyl)-1-butanol
MOCA = 4,4′-methylene bis(2-chloroaniline)
DPP = dibromopropylphosphate

small laboratory-bred rodent strains and man, with diet representing an additional factor of relevance to such comparisons (Lewis et al., 1998a; Ioannides, 1999). In this respect, inter-species scaling methodologies are being developed for extrapolating metabolic data in experimental animals to the likely situation in human subjects (Campbell, 1998; Lin, 1998) and including the relating of microsomal studies to those conducted using liver slices or cells in culture (Houston and Carlile, 1997). It is also possible to predict the effects of drug–drug interactions on pharmacokinetics for P450-mediated metabolism in the liver (Ito et al., 1998). However, it is important to have information on both the chemistry and enzymology of reactions in order to predict the likely course of drug metabolism in man (Smith, 1994).

4.6 Quantitative structure–activity relationships (QSARs) in P450 substrates

One means of investigating the variation in binding affinity, catalytic rates and selectivity of substrates is via QSAR analysis (Hansch and Zhang, 1993; Gao and Hansch, 1996; Lewis, 2000a). In this respect, one can determine the likely factors which contribute to the variation in binding towards a given P450 for a series of compounds (which may be structurally related) by calculating a set of molecular and electronic parameters for the chemicals concerned, such that a relevant combination of descriptor variables are derived via statistical analysis; this constitutes a means of generating a QSAR model for those substrates (or inhibitors) and the P450 in question. For example, the inhibition of rat CYP2E1 by a series of primary aliphatic alcohols is readily described in terms of a simple quadratic expression in log P, the lipophilicity parameter (Lewis et al., 2000a) as follows:

$$-\mathrm{p}K_\mathrm{i} = 0.98 \log P - 0.20 \log P^2 - 1.79$$

where K_i is the inhibitory concentration expressed in μM for a series of 10 alcohols (methanol through to decanol) with a correlation coefficient of 0.96 for experimentally determined octanol-water partition coefficients (Lewis et al., 2000a). Although interesting, such expressions only provide a limited amount of information about the binding interactions between substrate or inhibitor and the P450 enzyme itself. However, studying homologous series of compounds can be helpful in establishing the optimal size requirements for occupancy of the heme environment in different P450s (see Lewis, 1997b, for example). In addition to the aforementioned work on CYP2E1-selective compounds, fatty acid substrates of CYP4A1 also exhibit a similar parabolic relationship between chain length and activity, with lauric acid occupying the maximum position with respect to both binding and catalytic turnover (Lewis and Lake, 1999).

In fact, it has been demonstrated that CYP4A11 is the major lauric acid ω-hydroxylase in human liver (Powell et al., 1996) whereas CYP2E1 (and, to some extent, CYP2C9) catalyses ω-1-hydroxylation of laurate in hepatic microsomes. These findings indicate the likely distance ranges between the heme moiety and active site basic residues in the different P450s concerned. Potentially more precise

Table 4.21 Structural data for selected CYP1A2 substrates

Compound	log P	pK_a	log $D_{7.4}$	M_r	log M_r	K_m	ΔG_{bind}
1. Caffeine	0.01	neutral	0.01	194.22	2.2883	180	–5.3118
2. PhIP	2.23	8.69[b]	0.92	224.29	2.3508	55	–6.0422
3. 7-Methoxyresorufin	3.15[c]	neutral	3.15[c]	227.23	2.3565	0.21	–9.4577
4. Phenacetin	1.57	neutral	1.57	179.24	2.2534	48	–6.1260
5. IQ	1.84	9.18[b]	0.06	198.25	2.2972	33	–6.3568
6. MeIQ	2.40[c]	9.77[b]	0.06[c]	212.28	2.3269	13	–6.9307
7. 4-Aminobiphenyl	2.86	4.61[b]	2.86	169.24	2.2285	30	–6.4156
8. 7-Ethoxyresorufin	3.61[c]	9.85[b]	1.19[c]	241.26	2.3825	1.7	–8.1839
9. Aflatoxin	2.20[c]	neutral	2.20[c]	312.29	2.4946	31	–6.2231
10. Theophylline	1.40[c]	8.8[b]	–0.02	180.2	2.2558	455	–4.7405
11. Tacrine	2.71	9.8[b]	0.46	198.27	2.2973	14	–6.8851
12. Estradiol	2.69	neutral	2.69	272.37	2.4351	N/A	N/A
13. Acetanilide	1.16	neutral	1.16	135.16	2.1308	N/A	N/A
14. Zoxazolamine	2.21	neutral	2.21	168.58	2.2268	N/A	N/A

Notes:
b = basic c = calculated value N/A=data not available

QSAR equation: $\Delta G_{bind} = 0.30\ \mu - 0.90\ l/w + 2.11\ \Delta E - 0.5\ N_{HB} - 22.41$

$n = 11$; $s = 0.42$; $R = 0.97$; $F = 22.3$

information can be obtained from conducting QSAR evaluations on somewhat more structurally diverse substrates, especially when a molecular template of the same compounds is docked with the relevant P450 binding site (Lewis, 2000b). An example of this is afforded by consideration of some human CYP1A2 substrates, and their relevant structural information is presented in Table 4.21 which includes binding affinities toward that enzyme. It is found that the free energies of binding for these substrates (ΔG_{bind}) is best described in terms of a linear combination of several parameters, yielding a QSAR of the following type:

$$\Delta G_{bind} = 0.3\mu - 0.9\ l/w + 2.1\ \Delta E - 0.5\ N_{HB} - 22.4$$

where μ is the AM1-calculated dipole moment and ΔE is the energy difference between the two frontier orbitals, $E_{LUMO} - E_{HOMO}$, l/w is the aspect ratio of the molecule expressed as the ratio between its length and width (this being as a measure of molecular rectangularity), and N_{HB} is the number of active site hydrogen bonds between the substrate and human CYP1A2 active site. The correlation between experimental binding energies and those calculated via the above expression was found to be as high as 0.97 and also reproduced the correct ranking order of affinities based on K_m values (Lewis, 2000c).

High affinities are found for both 7-methoxyresorufin and 7-ethoxyresorufin substrates; and it is interesting to note that these substrates formed the greatest number of hydrogen bonds with CYP1A2 active site residues. Furthermore, the difference between the binding energies of these two alkoxyresorufins can be explained in terms of the additional bond rotational energy for the methylene group being restricted on binding. Essentially, the high affinity of the two 7-alkoxyresorufins is due primarily to the combination of hydrogen bond and π-π stacking interactions with complementary amino acid residues located within the putative binding site of CYP1A2; several of which have been confirmed by site-directed mutagenesis experiments (reviewed in Lewis et al., 1999c). The binding of substrates to CYP3A4 also show a QSAR (Lewis, 2000a) which provides insight into the likely interactions encountered with the enzyme's heme pocket. In this case, one finds that the number of active site hydrogen bonds and aromatic π-π stacking interactions are also important, although there is only one main π-π interaction common to most CYP3A4 substrate interactions (Lewis et al., 1996). A QSAR equation incorporating these two descriptors gave a 0.96 correlation with experimental binding energies, although the two frontier orbital energies (E_{HOMO} and E_{LUMO}) were also included in the final four-variable expression, as shown below:

$$\Delta G_{\text{bind}} = 2.0\, E_{\text{LUMO}} - 3.0\, E_{\text{HOMO}} - 0.5\, N_{\text{HB}} + 3.5\, N_{\pi\text{-}\pi} - 35.1$$

where the free energies of binding were derived from experimental dissociation constants between CYP3A4 and the various substrates themselves (reviewed in Lewis, 1999b, 2000a). There are certain similarities between this relationship and that shown for substrates of CYP1A2, although the fewer hydrogen bonds and π-π stacking interactions present in CYP3A4 probably contribute to the somewhat lower binding affinities encountered in the latter.

QSARs have also been derived (see Tables 4.22 and 4.23) for substrates of CYP2C9 and CYP2D6, where the importance of both ion-pairing and hydrogen bonding are evidenced by the particular descriptor variables involved (Lewis, 1999b, 2000c). In addition, these are supported by the reinforced interactions (mainly electrostatic and hydrogen bond) shown by superimposed substrate templates in the CYP2D6 (Lewis et al., 1997a) and CYP2C9 (Lewis et al., 1998c) active sites. The findings indicate that, when a group of selective substrates is docked and overlaid within the enzyme active centre, there is a strong reinforcement of hydrogen bond, π-π stacking and ionic interactions (Lewis, 2000b). Moreover, the amino acid residues involved in these substrate interactions tend to be identical with those which have been probed via site-specific mutation (reviewed in Lewis, 1998a) or are close to those positions. QSAR evaluations have also been conducted on CYP2B6, CYP2C9, CYP2D6 and CYP3A4 substrates (Ekins et al., 1999a–c, 2000) using the CATALYST system, which appear to support the active site modelling studies described previously.

In conclusion to this section, it is clear that the differing substrate selectivities of human P450s involved in drug metabolism (see Table 4.24 for a summary) can be rationalized in terms of likely enzyme–substrate interactions, and that only a relatively small number of structural descriptors are often required to discriminate between substrates for the major human hepatic P450 enzymes (Lewis, 2000b). However, less specific factors describing rates of P450-mediated metabolism include carbon–hydrogen bond energies (Hollebone and Brownlee, 1995) and

Table 4.22 Structural data for selected CYP2C9 substrates

Compound	log *P*	pK_a	log $D_{7.4}$	M_r	log M_r	K_m(μM)	ΔG_{bind}
1. Phenytoin	2.47	8.1[a]	2.39	252.27	2.4019	45	−6.166
2. Tolbutamide	2.34	5.43[a]	0.37	270.35	2.4319	132	−5.503
3. Ibuprofen	3.51	5.2[a]	1.31	206.28	2.3145	52	−6.065
4. Diclofenac	4.40	4.22[a]	1.22	296.15	2.4715	6	−7.407
5. Warfarin	2.52	5.1[a]	0.12	308.33	2.4890	4	−7.657
6. Tienilic acid	3.15[c]	4.8[a]	0.55[c]	331.17	2.5201	6	−7.407
7. 58C80	5.18[c]	5.0[a]	2.78[c]	312.44	2.4948	141	−5.462
8. Naproxen	3.18	4.15[a]	0.33	230.263	2.3622	126	−5.535
9. Piroxicam	1.58	6.3[a]	0.45	331.345	2.5203	40	−6.238
10. Mefenamic acid	5.12	4.2[a]	2.0	241.289	2.3825	7	−7.6568

Notes:
a = acidic c = calculated value

QSAR equation: $\Delta G_{bind} = 8.62 \log D_{7.4} - 8.02 \log P - 6.26\ pK_a + 0.57\ HB_D + 42.74$

Table 4.23 Structural data for selected CYP2D6 substrates

Compound	log *P*	pK_a	log $D_{7.4}$	M_r	log M_r	K_m(μM)	ΔG_{bind}
1. Bufuralol	3.50	9.0[b]	1.89	261.36	2.4172	8.6	−6.9535
2. Codeine	1.07	8.2[b]	0.23	299.37	2.4762	15	−6.6218
3. Ondansetron	2.14	7.7[b]	1.30	293.40	2.4675	102	−5.4790
4. Imipramine	4.42	9.5[b]	2.52	280.41	2.4478	2.4	−7.7143
5. Desipramine	4.05	10.0[b]	1.45	266.39	2.4255	20	−6.4503
6. Nortriptyline	4.04	9.73[b]	1.71	263.38	2.4206	47	−5.9410
7. Amitriptyline	5.04	9.4[b]	2.50	263.38	2.4206	74	−5.6703
8. Debrisoquine	0.75	13.01[b]	0.75	175.33	2.2436	13	−6.7071
9. Propranolol	3.37	9.5[b]	1.18	259.35	2.4139	2.73	−7.6375
10. Dextromethorphan	3.36	8.3[b]	0.91	271.4	2.4336	2.76	−7.6310
11. Metoprolol	2.35	9.68[b]	0.07	267.4	2.4272	46 (K_S)	−6.4515
12. MDMA	2.28	10.04[b]	−0.27	193.27	2.2862	1.72 (K_m)	−7.913

Notes:
b = basic MDMA = methylenedioxy methylamphetamine

QSAR equation: $\Delta G_{bind} = 492.03 \log M_r - 5.08\ N_{HB} - 3.76\ N_{\pi\text{-}\pi} - 0.88\ M_r - 947.7$

hydrogen abstraction energies (Korzekwa et al., 1990) sometimes coupled with ionization potential (reviewed in Lewis et al., 1998d). In fact, the latter parameter often correlates with catalytic rate for many species of P450 substrates (Lewis and Pratt, 1998), thus indicating the fundamental importance of this quantity, possibly due to its relation with frontier orbital interactions (reviewed in Lewis, 1999a). Therefore, it is possible to obtain meaningful relationships between P450 substrate properties and their enzyme selectivity, together with being able to formulate expressions governing binding affinity and rate of metabolism based on structural factors.

Table 4.24 Substrates, inhibitors and inducers of the major human cytochrome P450s involved in xenobiotic metabolism

P450 enzyme	*Substrates*			
CYP1A2	Acetanilide	7-Ethoxyresorufin	Paracetamol	Theophylline
	Caffeine	Imipramine	Phenacetin	Verapamil
	Estradiol	Mianserin	Propafenone	Warfarin
CYP2A6	Coumarin	Nicotine	4-Nitroanisole	Testosterone
	Halothane	4-Nitrophenol		
CYP2B6	Cyclophosphamide	Ifosphamide	Mianserin	Testosterone
CYP2C8	Carbamazepine	Taxol	Tolbutamide	Warfarin
CYP2C9	Diazepam	Phenylbutazone	Proguanil	Torbutamide
	Diclofenac	Phenytoin	Tenoxicam	Trimethoprim
	Ibuprofen	Piroxicam	Tienilic acid	*S*-Warfarin
CYP2C19	Citalopram	Diazepam	Lansoprazole	Pentamidine
	Clomipramine	Hexobarbital	*S*-Mephenytoin	Proguanil
	Clozapine	Imipramine	Omeprazole	Propranolol
CYP2D6	Amitriptyline	Debrisoquine	Imipramine	Perhexiline
	Bufuralol	Desipramine	Metoprolol	Propafenone
	Cinnarizine	Dextromethorphan	Mexiletene	Propranolol
	Citalopram	Encainide	Mianserin	Sparteine
	Clomipramine	Flecainide	Nortriptyline	Thioridazine
	Clozapine	Fluoxetine	Ondansetron	Timolol
	Codeine	Fluphenazine	Paroxetine	Trifluperidol
CYP2E1	Aniline	Dapsone	Isoflurane	Pyridine
	Benzene	Enflurane	Methylformamide	Styrene
	Caffeine	Ethanol	4-Nitroanisole	Theophylline
	Chlorzoxazone	Halothane	4-Nitrophenol	Toluene
CYP3A4	Alfentanil	Diltiazem	Losartan	Teniposide
	Amiodarone	Diazepam	Lovastatin	Terfenadine
	Astemizole	Erythromycin	Midazolam	Tetrahydrocannabinol
	Benzphetamine	Ethinylestradiol	Nifedipine	Theophylline
	Budesonide	Etoposide	Omeprazole	Toremifene
	Carbamazepine	Ifosphamide	Paracetamol	Triazolam
	Cyclophosphamide	Imipramine	Quinidine	Troleandomycin
	Cyclosporin	Lansoprazole	Retinoic acid	Verapamil
	Dapsone	Lignocaine	Tacrolimus	Warfarin
	Digitoxin	Loratadine	Taxol	

continued ...

Table 4.24 (continued)

P450 enzyme	*Inhibitors*		*Inducers*
CYP1A2	Fluvoxamine		Charcoal-broiled beef
	Furafylline		Cigarette smoke
	α-Naphthoflavone		Cruciferous vegetables
			Omeprazole
			TCDD
CYP2A6	Diethyldithiocarbamate	Tranylcypromine	Barbiturates
	8-Methoxypsoralen		Dexamethasone
CYP2B6	Orphenadrine		Not known
CYP2C8	Sulphaphenazole	Quercetin	Not known
	Sulfinpyrazone		
CYP2C9	Sulphaphenazole	Tienilic acid	Rifampicin
	Sulfinpyrazone		
CYP2C19	Fluconazole	Tranylcypromine	Rifampicin
	Omeprazole	Warfarin	
CYP2D6	Ajmaline	Fluoxetine	Not known
	Amiodarone	Paroxetine	
	Clomipramine	Quinidine	
	Flecainide	Trifluperidol	
CYP2E1	3-Amino-1,2,3-triazole	Disulfiram	Acetone
	Diethyldithiocarbamate	Phenethyl isothiocyanate	Ethanol
	Dihydrocapsaicin	4-Methylpyrazole	Isoniazid
	Dimethyl sulphoxide		4-Methylpyrazole
			Pyridine
CYP3A4	Clotrimazole	Ketoconazole	Carbamazepine
	Ethinylestradiol	Miconazole	Dexamethasone
	Gestodene	Naringenin	Omeprazole
	Itraconazole	Troleandomycin	Phenobarbital
			Phenytoin
	Activator:		Rifampicin
	α-Naphtholfavone		Sulphadimidine
			Sulphinpyrazone
			Troleandomycin

Chapter 5

Regulation of P450 enzymes

Nature is waiting patiently for mankind to improve its wits.
(Maurice de Maeterlinck)

5.1 Introduction

The ways in which P450 genes are expressed in different species and in various tissues, together with how the corresponding enzyme levels become regulated are complex; but much work has been carried out in this area, such that our knowledge and understanding of the processes involved is becoming clearer (Okey, 1990; Weigel, 1996; Waxman, 1999; McKenna et al., 1999). The interested reader is referred to a recent review on this aspect, especially that concerning nuclear hormone receptor involvement (Honkakoski and Negishi, 2000). The levels of P450s in human liver (see Table 5.1) are subject to considerable interindividual variation (Table 5.2), and it is known that regulation of the corresponding P450 genes in this and other organs/tissues is governed by mechanisms which include nuclear receptors, together with other factors such as stabilization of the corresponding mRNA and post-translational modification. A summary of the nuclear receptors known to be associated with human P450 gene expression and induction of the relevant P450 enzymes is provided in Tables 5.3 and 5.4, where the latter refers to endogenous metabolic pathways.

In particular, it is well documented (Micka et al., 1997; Whitlock et al., 1996, 1997; Whitlock and Denison, 1995; Reyes et al., 1992; Gonzalez and Fernandez-Salguero, 1998; Larsen et al., 1998) that regulation of the CYP1 family is mediated by the Ah receptor (Ah = aryl hydrocarbon) and its nuclear translocator, ARNT (Ah receptor nuclear translocator), which are sequentially-related nuclear receptors containing both DNA-binding and ligand-binding domains (Hankinson, 1995; Hahn, 1998; Fujii-Kuriyama et al., 1992; Hines et al., 1994). Both of these regulatory proteins have had their amino acid sequences determined on the basis of their cDNAs (Burbach et al., 1992) and appear to belong to the Per-ARNT-Sim (PAS) gene family of nuclear transcription factors (Hahn, 1998). The most potent ligand for the Ah receptor is the environmental pollutant and carcinogen TCDD (2,3,7,8-tetrachlorodibenzo-*p*-dioxin) (Huff et al., 1994; Landers and Bunce, 1991) for

Table 5.1 Average levels of P450 enzymes (%) in human liver

CYP	Shimada et al., 1994	Rendic and DiCarlo, 1997	Edwards et al., 1998		Rodrigues, 1999
	Percentage of total P450 complement				
			A	B	
1A2	12.7	13	8.5	12.5	8.0
2A6	4.0	4	13	16.9	13.0
2B6	0.2	<1	ND	ND	7.0
2C8			15		12.0
2C9	18.2 (2C)	18 (2C8, 2C9)	14	15.7 (2C8, 2C9)	18.0
2C18		ND	ND	ND	<2.5
2C19		1	11	16.7	4.0
2D6	1.5	2.5	4	3.2	2.0
2E1	6.6	7	3	2.7	9.0
3A4	28.8 (3A)	28 (3A)	21 (3A)	18.4	20.0
3A5				ND	0.2
4A11	ND	ND	10	14	ND

References: Shimada et al., 1994; Rodrigues, 1999; Rendic and DiCarlo, 1997; Edwards et al., 1998.

Notes:
A = immunoquantification B = enzyme activity towards selective substrates ND = not determined

Table 5.2 Human P450s: genetics and variation in levels

CYP	*Polymorphisms and range of levels*	*Level of enzyme variability (fold)*	*Chromosome location on the Human Genome*	*Number of alleles known*
1A1	Variable levels, PMs known	100	15	?
1A2	Variable levels	40	15	?
1B1	PMs known	?	2	18
2A6	< 2% Caucasians, variable levels	30	19	6
2B6	Variable levels	50	19	?
2C8	?	?	10	6
2C9	1–3% Caucasians	25	10	6
2C19	3–5% Caucasians; 15–20% Orientals	?	10	11
2D6*	5–10% Caucasians; 1% Orientals	1000	22	65
2E1	Variable levels, PMs known	20	10	4
3A4	Variable levels, PMs known	20	7	2
3A5	Variable levels, PMs known	100	7	2

References: Guengerich, 1995a; Rendic and DiCarlo, 1997; Sata et al., 2000; Bourian et al., 2000; and P450 websites at: http://www.icgeb.trieste.it/P450 and http://drnelson.utmem.edu/CytochromeP450.html

Notes:
* Polymorphism in the rat orthologue CYP2D1 has also been reported (Vorhees et al., 1997).
? = data not available at present
PMs = poor metabolizers

Table 5.3 Receptors associated with P450 expression and induction

CYP	*Nuclear receptor involvement*	*Typical inducer*
1A1	AhR, ARNT	TCDD
1A2	AhR, ARNT	TCDD
1B1	AhR, ARNT	TCDD
2A	HNF-4	Phenobarbital
2B	CAR	Phenobarbital
2C	RAR	Retinoic acid
2D	HNF-4	None known†
2E1	None known*	Ethanol
3A	PXR, GR	Rifampicin
4A	PPAR	Clofibrate

References: Honkakoski and Negishi, 2000; Waxman, 1999; Prough et al., 1996; Tukey and Johnson, 1990; Koskela et al., 1999; Lewis, 2001b; Gonzalez and Lee, 1996; Gonzalez et al., 1989.

Notes:
* Regulated by post-translational modification
† Constitutively expressed

Table 5.4 Receptor and other factors associated with P450 induction of endogenous substrate metabolism

CYP	*Nuclear receptor/ factor involved*	*Trophic hormone as mediating factor*
7	FXR, LXR	–
11	SF-1	ACTH (adrenals), LH (testis)
17	SF-1	ACTH (adrenals), LH (ovary)
19	SF-1, ER	FSH (ovary)
21	SF-1	ACTH (adrenals)
26	RAR, RXR	–
27	VDR	PTH (kidney)

References: Honkakoski and Negishi, 2000; Waxman, 1999; Lewis, 1996a; Gibson and Skett, 1994; Simpson et al., 1997; Edwards and Ericsson, 1999; Kagawa and Waterman, 1995; Kato and Yamazoe, 1993.

Notes:
The steroidogenic P450 genes are positively regulated in response to trophic hormones via activation of cAMP, with SF-1 conferring a partial cAMP response.

FXR = farnesoid X receptor (down-regulation)
LXR = liver X receptor (up-regulation)
SF-1 = steroidogenic factor 1
ER = estrogen receptor
VDR = vitamin D receptor
ACTH = adrenocorticotrophic hormone
LH = lutenizing hormone
FSH = follicle stimulating hormone
PTH = parathyroid hormone

which it has a very high affinity (Okey and Vella, 1982) and is, furthermore, able to induce CYP1 levels substantially even at nanomolar concentrations (Whitlock, 1989; Poland and Knutson, 1982; Astroff et al., 1988) although other receptors such as the estrogen receptor (ER) and glucocorticoid receptor (GR) may also be affected by TCDD administration (Lin et al., 1991). Typical ligands for the cytosolic Ah receptor (AhR) include planar polyaromatic hydrocarbons and their halogenated derivatives (e.g. polychlorobiphenyls) where the degree of molecular planarity is directly proportional to their AhR binding affinity and CYP1 induction (Lewis, 1997b). It has also been reported that dietary flavonols are AhR ligands affecting CYP1A expression (Ciolino et al., 1999) although some flavonoids can also act as CYP1A inhibitors (Zhai et al., 1998).

In contrast, regulation of the CYP2 family involves other nuclear receptors (Honkakoski and Negishi, 2000) related to the steroid hormone receptor superfamily, such as the constitutive androstane receptor (CAR) and the retinoic acid receptor (RAR), for the CYP2B and CYP2C subfamilies respectively, as shown in Table 5.3. However, it appears that the hepatic nuclear factor HNF-4 is associated with regulation of CYP2A and CYP2D enzymes, whereas induction of CYP2E in not mediated by nuclear transcription factors as this is known to be regulated by post-translational modification (Porter and Coon, 1991).

The nuclear orphan receptor CAR activates the phenobarbital (PB) responsive enhancer module of the CYP2B gene (Honkakoski et al., 1998) via heterodimer formation with RXR. As PB induces CYP2B primarily (Waxman and Azaroff, 1992), although CYP2A and CYP2C are also PB-inducible, it is likely that PB acts as a ligand for the constitutive androstane receptor (Baes et al., 1994; Choi et al., 1997; Forman et al., 1998; Honkakoski et al., 1998) which probably regulates endogenous steroid hormone levels (e.g. testosterone and other androgens) in mammalia. A similar role for the RAR and CYP2C induction may also exist as regulation of CYP2C genes by RAR (and retinoid X receptor (RXR)) has also been reported (Honkakoski and Negishi, 2000). It may be of relevance to note that retinoic acid is a substrate of CYP2C enzymes (especially for CYP2C8) and that many CYP2C9 substrates are also acidic in nature (Rendic and DiCarlo, 1997; Lewis, 2000a). Consequently, the regulation of CYP2C subfamily genes by a receptor that is activated by an endogenous compound like retinoic acid would seem to be readily understandable. Furthermore, due to the sex differences in testosterone metabolism exhibited by CYP2C enzymes in the rat, where certain isozymes are either male or female specific/dominant and hydroxylate the androgen at differing positions, it is thought that this subfamily (possibly in conjunction with CYP2B and CYP2A) are associated with the homeostasis of steroid hormone levels in the adult, together with gender imprinting in the neonate (Zimniak and Waxman, 1993).

However, such maintenance and modulation of sex hormones may not be confined to enzymes of the CYP2 family, as there is evidence for a role in this respect for both CYP1 and CYP3 due to the fact that certain steroids, especially estrogen and androgens, respectively, are metabolized at specific positions by these enzymes (Martucci and Fishman, 1993; Schenkman et al., 1989). Furthermore, it is reported (Stegeman et al., 1996) that the Ah receptor and CYP1 module could play a role in fetal development, as both exhibit relatively high levels in chondrocytes (Iwata and Stegeman, 2000). Moreover, estrogen metabolism differs between members of the CYP1 family and these enzymes are induced to varying extents by typical AhR ligands such as TCDD (Astroff et al., 1988). In addition, there appears to be a modulation in CYP1A2 levels during the estrus cycle, and variation between parous and non-parous women (Horn et al., 1995) is indicative of some degree of cross-talk between the estrogen receptor (ER) and AhR, as has been documented elsewhere (Honkakoski and Negishi, 2000).

Although there is evidence for PB induction of CYP2A, the regulation of this family may be governed by specific tissue factors such as HNF-4 (Honkakoski and Negishi, 2000) and it is established that sex and tissue differences exist in the mouse CYP2A enzymes (Lindberg and Negishi, 1989) which are linked with testosterone metabolism in liver and kidney of male and female mice. The hepatocyte nuclear factor HNF-4 is also involved in CYP2D regulation, although CYP2E levels appear to be up-regulated via protein stabilization (Yang et al., 1990). In the latter, post-translation modification involving PKA-mediated serine phophorylation plays an important role in CYP2E induction (Jansson, 1993; Jansson et al., 1990) with additional regulation occurring by cytochrome b_5 (Epstein et al., 1989).

The levels of CYP3 family enzymes (Guengerich, 1999) are known to be regulated by the glucocorticoid receptor (GR) and also via the pregnane-X receptor (PXR), although the latter may be involved in CYP2C induction (Honkakoski and Negishi, 2000). The PXR is an orphan nuclear receptor which is activated by pregnanes (Kliewer et al., 1998a and b; Scheutz et al., 1998) and also via a wide variety of xenobiotics (Blumberg et al., 1998; Jones et al., 2000) which exhibits a particular signalling pathway for CYP3A induction (Bertilsson et al., 1998) whereas the GR is a steroid hormone receptor in the same superfamily (Laudet et al., 1999) with some degree of homology with ER (Krust et al., 1986).

Enzymes of the CYP4 family are regulated (Johnson et al., 1996; Gibson, 1992; Aldridge et al., 1995) via the mediation of another nuclear receptor, the peroxisome proliferator-activated receptor (PPAR), whereas CYP7 is regulated by two related receptors, the liver X receptor (LXR) and the farnesoid X receptor (FXR) which modulate bile acid biosynthesis (Schmidt et al., 1999; Wang et al., 1999; Lehmann et al., 1997; Forman et al., 1997; Janowski et al., 1996; Russell, 1999; Peet et al., 1998). The PPAR is a member of the steroid hormone receptor superfamily (Issemann and Green, 1990) although its endogenous ligand is likely to be an unsaturated fatty acid, and is activated by clofibric acid and other peroxisome proliferators (Lewis and Lake, 1998b; Kliewer and Willson, 1998; Willson et al., 2000; Escher and Wahli, 2000; Michalik and Wahli, 1999). The role of PPARs in adipose tissue development has been reported (Ailhaud, 1999; Uauy et al., 2000; Vamecq and Latruffe, 1999) and site-directed mutagenesis experiments (Causevic et al., 1999) have shown likely amino acid residues involved in ligand binding to the PPAR itself.

P450s associated with steroid biosynthesis (CYP11, CYP17, CYP19 and CYP21) have specific transcription factors, such as SF-1 (Morohashi and Omura, 1996) and the trophic hormones (FSH, LH, PTH and ACTH) are involved in their regulation (Honkakoski and Negishi, 2000). The likely role of certain nuclear factors and receptors in the up-regulation of P450 enzymes is implicated by the presence of specific response elements (Kemper, 1993) in the relevant P450 gene's regulatory region which lies prior to the coding region (reviewed in Lewis, 1996a). Table 5.4 shows a summary of the various receptors and other proteins involved in regulation of P450s from CYP7, CYP11, CYP17, CYP19, CYP21, CYP26 and CYP27 (Honkakoski and Negishi, 2000). In the case of CYP7, the FXR and LXR are associated with regulation, and CYP26 is regulated by RAR and RXR, whereas CYP27 utilizes the vitamin D receptor (VDR). FXR and LXR are also involved in the regulation of bile acid-forming P450s (Honkakoski and Negishi, 2000) with LXR being associated with up-regulation and FXR acting as a down-regulator of bile acid biosynthesis. Although the trophic hormones regulate levels of the steroidogenic P450s, the SF-1 nuclear factor is also involved and both are associated with a cAMP-activated response (Honkakoski and Negishi, 2000). Cyclic AMP is also involved in serine (or theonine) residue phosphorylation, which is a non-receptor mediated mechanism of P450 induction. In this case, protein kinase A (PKA) or protein kinase C (PKC) mediates a post-translational modification of the P450 enzyme by utilizing cAMP for the phosphorylation of amino acids containing hydroxyl groups. This mechanism is important in the case of CYP2E induction, and it appears that serine phosphorylaton at a particular site occurs which can be blocked by cytochrome b_5 (Jansson et al., 1990; Epstein et al., 1989).

5.2 Mechanisms of P450 regulation and gene expression

Table 5.5 indicates the various mechanisms involved in P450 induction, from which it can be appreciated that many of these require gene transcription and are mediated by various nuclear factors (as described above) including receptor proteins. It has been shown that phosphorylation of the steroid hormone superfamily of receptors is also important for activation (Weigel, 1996) although, in this case, tyrosine phosphorylation facilitates triggering of transcription via the intervention of a specific protein which links the bound receptor to the basal transcription element (BTE).

In the case of PB-type induction, it is thought that the ligand binding to a repressor protein facilitates P450 gene expression, and this could involve the release of a suppressing factor which normally binds close to the BTE. A simplified scheme for P450 gene expression is presented in Table 5.5, which shows the stages involved and indicates the various pathways for P450 induction and also degradation. As far as receptor-mediated P450 gene expression is concerned, the sequence of events is thought to be as follows:

1 The ligand binds to a cytosolic receptor protein which displaces the heat shock protein (HSP) complex.
2 Ligand binding triggers a conformational change in the receptor which ‘primes’ it for dimerization and activation.
3 The ligand-receptor complex translocates to the nucleus.
4 The complex dimerizes and binds as a dimer to palindromic hormone response element (HRE)/xenobiotic response element (XRE) in the regulatory region of the P450 gene.
5 Activation of the P450 gene occurs, possibly involving a curvature of the DNA and linkage to the BTE.

There are several other factors, including cis- and trans-acting elements, together with co-activators or repressor proteins, which modulate regulation of P450 gene expression depending on the species in question and also on the particular cell type. Many of these await further characterization although the structures of some

Table 5.5 Regulation of P450 genes

Mechanism of induction	*P450s known to be regulated at specific stages*
Gene transcription ↓	1A1, 1A2, 1B1, 2B1, 2B2, 2C7, 2C11, 2C12, 2D9, 2E1, 2H1, 2H2, 3A1, 3A2, 3A6, 4A1, 11A1, 11B1, 17A1, 21A1
mRNA processing ↓	1A2
mRNA stabilization ↓	1A1, 2B1, 2B2, 2C12, 2E1, 2H1, 2H2, 3A1, 3A2, 3A6, 11A1
Translation ↓	2E1
Enzyme stabilization	2E1, 3A1, 3A2, 3A6

Stages in P450 gene expression

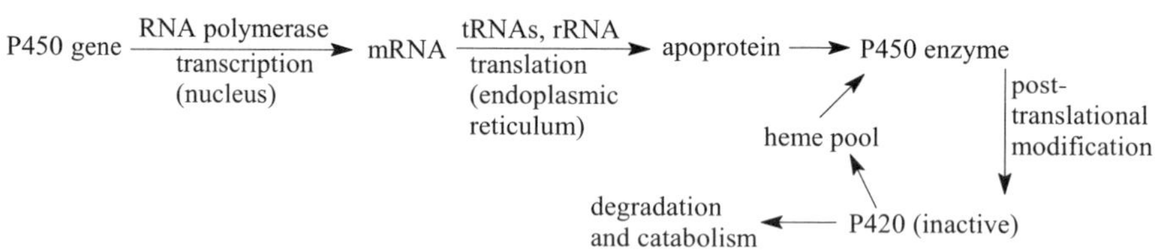

References: Lewis, 1996a; Gibson and Skett, 1994.

Notes:
1 The first stage in this process can be mediated by nuclear receptors.
2 Phosphorylation of serine or threonine residues via cAMP mediated by PKA or PKC is a common factor of post-translational modification of P450s (e.g. CYP2E1).

receptors are known, as listed in Table 5.6. The orphan nuclear receptors and steroid hormone receptors are apparently able to form both homodimers and heterodimers, usually with RXR. For example, CAR and RXR heterodimers appear to be involved in CYP2B induction activated by phenobarbital (Honkakoski and Negishi, 2000) and it is also reported that PPAR forms heterodimers with RXR, with specific response elements being present on the relevant P450 genes' regulatory regions (i.e. those of the CYP4 family). Co-activators, such as SRC-1 are involved in linking the nuclear receptor complex bound to its response elements with the transcription region (BTE/TATA box) thus triggering RNA polymerase to initiate gene transcription, and it is also apparent that phosphorylation of the ligand-bound receptor plays an important role in facilitating transactivation (Weigel, 1996). Part of the SRC-1 protein has been structurally characterized as a complex with the ligand-bound PPARγ which was recently resolved crystallographically (Nolte et al., 1998) and showing how a 'charge clamp' mechanism involving similar leucine motifs on the two proteins is associated with activation via the PPARγ C-terminus.

PB-type induction is thought to involve a PB-responsive enhancer module (PBREM) and a specific regulatory element, commonly referred to as the 'Barbie' box, appears to be present in the BTE regions of both PB-inducible mammalian P450 genes and also in that of CYP102 from *Bacillus megaterium* which is also known to be inducible by barbiturates (English et al., 1994, 1996) although this could represent a particular PB-deactivated repressor protein binding site. In some instances, there appears to be some degree of cross-talk between nuclear receptors, such as exists with CAR and PXR (Honkakoski and Negishi, 2000) and PB is a known ligand for both receptors, in addition to activating both CYP2B and CYP3A subfamilies. However, tissue- and sex-specific regulatory factors modulate the expression of mammalian P450 genes: the hepatocyte nuclear factor 4 (HNF-4) being involved in CYP2C subfamily regulation, for example; whereas sexual dimorphism in the CYP2A subfamily may be related to the presence of ERα levels and to the STAT5b transcription factor (Honkakoski and Negishi, 2000). Furthermore, the steroidogenic P450s that are cAMP-inducible appear to be controlled by SF-1 together with other factors such as the tissue-specific trophic hormones (see Table 5.4). Cyclic AMP-dependent pathways may also exist in other P450 genes due to the presence of cAMP response elements (CREs) in their regulatory regions (Kemper, 1993) although this remains to be fully elucidated. Cross-talk also seems to be of an indirect nature. However, it is known that CYP1A2 levels are modulated by estrus (Horn et al., 1995) and this enzyme is able to metabolize estradiol at the 2-position, which is also mediated by CYP1A1; whereas CYP1B1 appears to mediate the metabolic activation via 4-hydroxylation of estradiol (Rendic and DiCarlo, 1997) and is expressed at higher levels in breast cancer cell lines (Angus et al., 1999; McKay et al., 1996) although CYPA1 is also elevated in this tissue (Murray et al., 1991).

Although the AhR is involved in the regulation of all three of these CYP1 family enzymes, it has been reported (Astroff et al., 1988) that CYP1A1 is

Table 5.6 Nuclear receptors associated with P450 gene regulation

Receptor	*Level of structural characterization*	*Reference*
AhR	Model of LBD based on hERα LBD	Jacobs and Lewis, 2000b
CAR	Model of LBD based on hERα LBD	Jacobs and Lewis, unpublished
RAR	X-ray crystal structure of hRARγ LBD	Renaud et al., 1995
RXR	X-ray crystal structure hRXRα LBD	Bourget et al., 1995
PXR	Model of LBD based on hERα LBD	Jacobs and Lewis, 2000a
GR	Model of LBD based on hERα LBD	Lewis, 2000c
ER	X-ray crystal structure of hERα LBD	Brzozowski et al., 1997
PPAR	X-ray crystal structure hPPARγ LBD	Nolte et al., 1998

Notes:
1 The structures of the DNA-binding domains for GR and ER are also known.
2 It has been reported that there are similarities between the steroid-binding sites in both CYP11 and prostatic steroid-binding proteins (Gotoh et al., 1985) which could indicate a common evolutionary link.

LBD = ligand-binding domain   h = human

significantly more inducible by TCDD than CYP1A2. Moreover, the formation of heterodimers between AhR and ARNT appears to be prerequisite for XRE binding and CYP1 gene expression, with the activation of a whole gene battery (termed the Ah gene battery) being associated with the binding of typical ligands to the AhR (Nebert et al., 1990, 1991b). Consequently the initiation of specific P450 gene expression can lead to uncontrolled cell growth and tissue proliferation.

5.3 Toxic consequences of induction: predicting toxicity via the COMPACT method

Uncontrolled cell growth is obviously an undesirable side-effect from xenobiotic exposure and, consequently, its prediction from chemical structure represents an important area in the safety evaluation of NCEs for human risk assessment (Lewis, 1992b). It has been established that the toxic effects of carcinogens, for example, can fall into two main categories; namely, direct-acting (genotoxic) and epigenetic (non-genotoxic), although many hazardous chemicals can give rise to genotoxic intermediates as a result of metabolic activation, usually mediated by one or more of the P450 mono-oxygenases (Parke, 1987, 1994; Ioannides and Lewis, 1995). The direct-acting agents can, by and large, be readily identified on the basis of structurally alerting groups present within the molecule which, in general, tend to be electrophilic in nature (e.g. epoxide, nitrogen mustard, aziridine, etc.) and these are also associated with relatively low E_{LUMO} values, thus indicating a general ease of attack by nucleophilic agents such as may be encountered in DNA bases, for example (Lewis, 1995a, 1999a). However, an evaluation of potential toxicity arising from enzyme activation (Nelson, 1982) requires a more detailed study of the substrate selectivity and catalytic mechanism of P450s, together with an understanding of their general structure, particularly that of the active site region (Lewis et al., 1994a).

Due to the fact that CYP1 and CYP2E enzymes appear to represent the major activators of pro-carcinogens in both experimental animals and man (Gonzalez and Gelboin, 1994) a technique (COMPACT) has been developed which readily identifies potential selectivity for either of these P450s (Lewis et al., 1998e and f; Lewis, 2001a) such that the likelihood of carcinogenic reactive intermediates being formed as a result of enzyme activation can be predicted for untested chemicals. For example, substrates and inducers of CYP1 enzymes include planar polyaromatic hydrocarbons (PAHs) and their derivatives with their carcinogenicity and ability to bind to the Ah receptor linked with their overall molecular planarity (Lewis, 1997b); whereas the degree of planarity coupled with a relatively low activation energy (as determined by the ΔE value) provides both a ready means of discriminating CYP1-specific compounds from substrates and inducers of other P450 families, together with enabling an estimate of induction potency (Lewis, 1997b). PAHs, such as benzo(*a*)pyrene, are examples of 'complete' carcinogens (Ioannides and Parke, 1990, 1993; Ioannides and Lewis, 1995). Consequently, an evaluation of their relative potency from structural and electronic calculations represents a useful screening process in chemical risk assessment, and it can be shown that both molecular planarity (as measured by area/depth^2) and its combination with ΔE in the COMPACT radius (CR) correlate closely with CYP1 induction and Ah receptor binding (Lewis, 1997a; Lewis et al., 1998e).

Consideration of a third parameter (molecular diameter) for identifying CYP2E substrates and inducers facilitates a discrimination of both CYP1 and CYP2E selectivity from that of other P450s when combined with area/depth^2 and ΔE (Lewis et al., 1994a). However, for completely distinguishing between 8 human P450 selectivities in 48 compounds, a total of 5 structural/physicochemical descriptors are required (Lewis, 2000b) thus leading to a decision-tree approach for selecting likely substrate–enzyme combinations. When used in combination, the CR and diameter values enable a 70% accurate identification of rodent carcinogens from non-carcinogens (Lewis et al., 1998f). However the lipophilicity factor, log P, is also of importance (Lewis et al., 1998f) and, in the case of CYP2E-mediated toxicity which is largely due to the formation of oxygen radicals and other reactive oxygen species (ROS), it is found that both polarity (i.e. dipole moment) and rectangularity (length/width ratio) are also factors (Lewis et al., 2000a). Although many substrates of CYP2E are also inducers of the enzyme and tend to act as ROS generators, others are hepatotoxic (e.g. paracetamol) via the formation of genotoxic intermediates (Ioannides and Lewis, 1995), whereas some CYP2E substrates (e.g. acetone) are relatively non-toxic due to the fact that they are generally detoxified via this enzyme by forming polar metabolites which are readily excreted (reviewed in Lewis et al., 2000a). ROS generation has been associated with various degenerative disease states, cancer and ageing (Kehrer, 1993; Parke, 1987, 1994; Oberley and Oberley, 1995). Lifespan studies have indicated that oxygen radicals are involved in the ageing process (Orr and Sohal,

1994) although species body weight is an important factor in determining the magnitude of the potentially deleterious effect to the host organism (Martin and Palumbi, 1993; Lijinsky, 1993).

Induction of CYP2E potentiates ROS formation due to redox cycling (Zhukov and Ingelman-Sundberg, 1999) and, for poor substrates, uncoupling of the normal oxidative cycle leading to excess oxygen radical production occurs because the enzyme preferentially resides in the high-spin state (Guengerich and Johnson, 1997) and, consequently, single-electron transfer to bound oxygen will lead to the production of superoxide instead of substrate oxidation. The toxic effects of ROS generation are particularly apparent in small rodent species, which are especially susceptible to ROS injury due to rapid depletion of cellular glutathione levels (Lorenz et al., 1984; Parke, 1987, 1994) and it appears that caloric restriction (Seng et al., 1996) fasting and ether anesthesia can lead to the induction of CYP2E1 (Liu et al., 1993) with concomitant ROS production. It is possible that the heme moiety of CYP2E comprises predominantly high-spin iron(III) in its resting state (Guengerich and Johnson, 1997), because water molecules cannot readily ligate the heme iron due to the presence of several bulky amino acid residues in the active site (Lewis et al., 2000a) which sterically hinder access to the heme pocket for solvating waters. This would lead to facile reduction of the enzyme by reductase and subsequent superoxide formation, as mentioned previously.

However, induction of other P450s can also lead to redox cycling (e.g. CYP2B), whereas increased cell proliferation and tissue growth are features of CYP4 and CYP3 induction, respectively (Corton et al., 2000; Lake, 1995; Waxman, 1999). Many P450 substrates are also inducers of the same enzyme (Rendic and DiCarlo, 1997). For example, CYP1 enzymes are inducible by benzo(*a*)pyrene and 3-methyl cholanthrene which are substrates of this P450 family; phenobarbital induces CYP2B enzymes as well as acting as a substrate, and CYP3 family enzymes are readily induced by dexamethasone and rifampicin which are also metabolized by enzymes of this family (Rendic and DiCarlo, 1997). When one P450 is induced dramatically, the levels of others in the same tissue fall by the same extent due to competition for the heme pool (Ioannides and Parke, 1990, 1993). This can, therefore, lead to inadequate clearance of certain P450 substrates which may then build up cellularly, thus giving rise to potential toxicity to the host organism (Ioannides and Parke, 1990, 1993). Moreover, there is also a certain amount of competition between nuclear receptors for the same ligands, such as that reported for PXR and CAR (Moore et al., 2000a) although there is some degree of selectivity and, furthermore, PXR predominates over CAR for regulation of CYP3A4 expression (Honkakoski and Negishi, 2000; Waxman, 1999). Additionally, it is apparent that cross-talk exists between PXR and GR in competing for ligands which can induce CYP3A activity (Pascussi et al., 2000), thus leading to gene activation and cell growth. It can be demonstrated, furthermore, that CYP3A induction correlates closely ($R = 0.99$) with increased DNA synthesis and liver weight for steroids and their analogues (Lewis et al., 2000b). Table 5.7 provides some examples of P450 inducers which are either hazardous or potentially toxic, although it is important to quantify the relative degree of risk from P450 induction

Table 5.7 Examples of toxic or carcinogenic P450 inducers

Compound	*P450s involved*
TCDD, BP	CYP1A
Benoxaprofen	CYP1A and CYP4A
Phenytoin	CYP2B and CYP2C
Ethanol, benzene	CYP2E
MEHP, clofibrate	CYP4A
Hyperforin	CYP3A

References: Rendic and DiCarlo, 1997; Ioannides and Lewis, 1995; Lake and Lewis, 1996; Moore et al., 2000b.

Notes:
BP = Benzo(*a*)pyrene
TCDD = 2,3,7,8-Tetrachlorodibenzo-*p*-dioxin
MEHP = Mono(2-ethyl)hexylphthalate

when extrapolating data from experimental animal species to *Homo sapiens* (Paine, 1995). Clearly, the endocrine system is vulnerable to xenobiotic influence (Neubert, 1997) and it has also been established that P450 inducibility in some instances can be correlated with genetic differences in oxygen toxicity (Gonder et al., 1985). However, DNA strand cleavage represents a sensitive assay to hydroxyl radicals resulting from P450 induction (Kukielka and Cederbaum, 1994) whereas the use of P450 knockout mice strains points the way forward to formulating new toxicological models (McKinnon and Nebert, 1998; Nebert and Duffy, 1997).

5.4 QSARs in P450 inducers and nuclear receptor ligand binding

It is possible to derive quantitative structure–activity relationships for P450 inducers which provide some degree of rationalization for potency differences within groups of compounds acting at a common receptor (Lewis, 2000a). Table 5.8 shows a summary of QSARs generated for inducers of CYP1, CYP2, CYP3 and CYP4; some of these relationships refer to nuclear receptor binding in addition to P450 induction, however, although there is generally a correlation between the two events, such that it can be assumed that avidity of ligand binding tends to indicate induction potency (Safe et al., 1985).

For example, the Ah receptor binding affinity of diverse polyaromatic hydrocarbons and TCCD correlates closely ($R = 0.98$) with molecular planarity as measured by the ratio (a/d^2) between molecular area and the square of depth (Table 5.8, equation 1b). Figure 5.1 presents a plot of this relationship and the relevant data used to generate this QSAR is provided in Table 5.9. However, in a series of substituted dichlorodibenzo-*p*-dioxins (DCDDs) it is found that the AhR binding affinity correlates with the energy of the highest occupied molecular orbital (E_{HOMO}) as shown in Table 5.8, equation 1d (Lewis, 1997b). These two structural parameters (a/d^2 and E_{HOMO}) both figure in a QSAR for AhR binding within a series of 15 polychlorobiphenyls (PCBs) as shown in Table 5.8, equation 1a (Lewis,

Table 5.8 QSARs for P450 inducers and ligand-binding affinity

	n	*s*	*R*	*F*
1. CYP1A induction and AhR binding				
a) PCBs (Lewis, 1999a)				
$pEC_{50} = 0.33\ a/d^2 - 3.22\ E_H + 0.84\ \text{length} - 36.44$	14	0.308	0.95	31.5
$(\pm 0.04)\ (\pm 0.90)\ (\pm 0.26)$				
b) PAHs (Lewis and Dickins, 2001)				
$pEC_{50} = 0.44\ a/d^2 - 2.62$	7	0.230	0.98	112.6
(± 0.04)				
c) Diverse compounds (Lewis et al., 1998e)				
$\log \text{Induction} = 0.26 \log P - 0.15\ \text{CR} - 0.06\ \text{length} + 2.67$	12	0.203	0.99	105.1
$(\pm 0.05)\ (\pm 0.03)\ (\pm 0.02)$				
d) DCDDs (Lewis, 1997b)				
$pEC_{50} = 3.94\ E_H - 24.87$	8	0.550	0.87	19.3
(± 0.90)				
2. CYP3 induction				
a) Steroids (Lewis, 1999a; Lewis et al., 2000b)				
$\log \text{Induction} = a/d^2 - 0.37\ \Delta E + 6.75$	14	0.115	0.92	32.3
$(\pm 0.05)\ (\pm 0.05)$				
b) Diverse GR ligands (Lewis, 2000a)				
$\text{Fold induction} = l/w - 0.016\ M_r - 0.88\ \mu - 5.01 E_L - 4.58$	8	0.471	0.98	17.5
$(\pm 1.33)\ (\pm 0.002)\ (\pm 0.21)\ (\pm 0.74)$				
3. CYP4 induction				
a) Peroxisome proliferators (Lewis and Lake, 1998b)				
$\log \text{Relative Potency} = 0.008V - 0.22\mu + 0.38\ D^{acid}_{phenyl} - 2.95$	12	0.339	0.91	12.4
$(\pm 0.002)\ (\pm 0.06)\ (\pm 0.19)$				

Notes:

pEC_{50} = log EC_{50} for AhR binding

a/d^2 = ratio of area to depth-squared

$E_{H,L}$ = energy of HOMO, LUMO

ΔE = $E_L - E_H$

CR = $\sqrt{(a/d^2 - 15)^2 + (\Delta E - 7)^2}$

l/w = length to width ratio

M_r = relative molecular mass

μ = dipole moment

V = molecular volume

D^{acid}_{phenyl} = distance between acidic and phenyl groups in the molecule

1999a). In this case, however, the inclusion of molecular length provides a good agreement ($R = 0.95$) with compound potency, and this latter factor also appears in a QSAR describing CYP1 induction of diverse aromatic compounds (Table 5.8, equation 1c) which combines log P data for the 12 chemicals together with the COMPACT radius (CR). This factor is derived from a combination of a/d^2 and

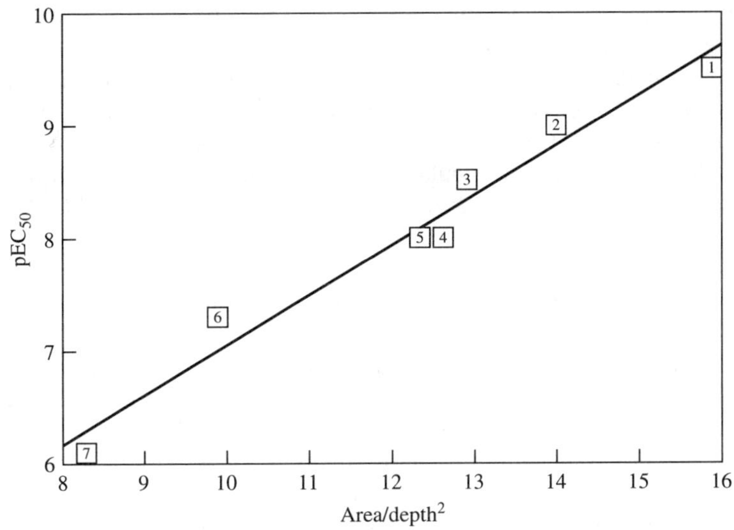

Figure 5.1 Graph showing the linear correlation (R = 0.98) between binding to the Ah receptor (pEC_{50}) and molecular planarity (area/depth^2) for 7 polyaromatic hydrocarbons.

Table 5.9 Ah receptor binding affinity and planarity of PAHs

	area/depth^2	pEC_{50}
1 TCDD	15.8450	9.5229
2 DB(*a*,*h*)anthracene	13.9903	9.0000
3 DB(*a*,*c*)anthracene	12.9125	8.5229
4 βNF	12.6053	8.000)
5 BP	12.3357	8.0000
6 BA	9.8858	7.3010
7 Anthracene	8.3161	6.0969

Reference to biological data: Okey and Vella, 1982.

Notes:
Figure 5.1 shows a linear plot between the data presented which exhibits a correlation of 0.983.

pEC_{50} = negative logarithm of effective concentration (*nM*) for displacement of 3MC
area/depth^2 = ratio between molecular area and square of depth
TCDD = 2,3,7,8-tetrachlorodibenzo-*p*-dioxin
DB = dibenzo
βNF = β-naphthoflavone
BP = benzo(*a*)pyrene
BA = benzanthracene
3MC = 3-methylcholanthrene

ΔE, which is the energy difference between the LUMO and HOMO (Lewis et al., 1998e). Although the compounds concerned are structurally diverse, and include both PAHs and certain drugs such as benoxaprofen and cimetidine, the correlation produced is excellent ($R = 0.99$) and could, therefore, be used predictively. These and other related studies on CYP1 inducers, such as aromatic amines (Lewis, 1999a) point to the importance of molecular planarity and frontier orbital energies for describing the compound potency differences in both AhR binding and CYP1 induction.

Very little QSAR work has been conducted on CYP2 family inducers, although a homologous series of 7-alkoxyresorufins exhibits a parallelism ($R = 0.85$) between length/depth and PB-induction (Lewis et al., 1995b). However, for CYP3 inducers, QSARs have been formulated within a series of endogenous and synthetic steroids (Lewis, 1999a; Lewis et al., 2000b) where induction appears to be related ($R = 0.92$) to a combination of area/depth^2 and ΔE, as shown in Table 5.8, equation 2a. Other additional factors may improve this correlation, and also it is not known whether the steroids concerned are ligands for the GR or PXR, although the latter may be more likely. Induction of CYP3A4 via GR mediation (Ogg, 1998) correlates well ($R = 0.98$) with a combination of several structural descriptors (Lewis, 2000a) as presented by Table 5.8, equation 2b. These parameters suggest that molecular size, rectangularity and polarity are important, together with the possibility of hydrogen bond formation, although only 8 compounds were subjected to QSAR analysis in this case. In contrast, several QSAR studies have been conducted on series of peroxisome proliferators, which are likely to be CYP4 inducers and bind to the PPAR, and equation 3a in Table 5.8 represents the most recent example (Lewis and Lake, 1998b). For these structurally diverse compounds, there is a good correlation ($R = 0.91$) between potency and a combination of molecular size, polarity and distance between two key groupings in the molecule, namely, the aromatic ring and carboxylic acid moiety which probably interact with the PPAR ligand-binding site (Lewis and Lake, 1998b). Consequently, although more work needs to be carried out in this area, a picture is steadily emerging as regards the structural factors responsible for nuclear receptor binding and P450 induction. In this respect, models for the ligand-binding sites of these receptors provide important information which can assist in the generation of useful QSAR data (Jacobs and Lewis, 2000a and b).

5.5 Inhibition of P450 enzymes

There are several ways in which chemicals can act as P450 inhibitors (Testa and Jenner, 1981; Testa, 1990; Murray 1992; Murray and Reidy, 1990; Ortiz de Montellano and Correia, 1995; Halpert, 1995). These include heme ligation, heme adduct formation, competition with a known substrate (competitive inhibition), acting as a substrate analogue, formation of a reactive intermediate (mechanism-based inhibition) or some combination of these (reviewed in Lewis, 1996a). Moreover, inhibitors may be selective (Halpert, 1995) for a particular P450 enzyme or non-selective to the degree in which they can inhibit any P450, such as the heme ligation shown by carbon monoxide, for example (Lewis, 1996a). In fact CO adduct formation is a major characteristic of P450s which can be used for identifying the presence of the enzyme due to the appearance of a strong absorption (Soret band) at around 450 nm in the UV spectrum (Omura et al., 1993). The crystal structure of $P450_{cam}$ (CYP101) with bound CO in the presence of substrate (Raag and Poulos, 1989b) shows that the carbon monoxide molecule ligates the heme iron and thus prevents access for an oxygen molecule. Apparently, the bond angle between CO and the heme plane in $P450_{cam}CO$ complexes displays a high correlation (R = 0.99) with the C–O stretching vibrational frequency in the IR region (Jung et al., 1992) and it is found that the crystallographic value of 14° for the CO-heme plane angle agrees well with the value of 16° estimated from the IR spectrum of $P450_{cam}CO$.

Small molecules like carbon monoxide (CO), nitric oxide (NO) and the cyanide ion (CN^-) are all able to act as non-selective inhibitors due to the fact that they possess π-acceptor properties and thus readily ligate the heme iron of P450 (Hill et al., 1970). The dissociation constant (K_D) for CO binding to $P450_{cam}$ (CYP101) is 0.1 μM whereas the K_D for oxygen (O_2) binding to the same enzyme is 0.6 μM. Consequently, CO competitively inhibits the stage in the P450 catalytic cycle where O_2 binds to the reduced enzyme (reviewed in Lewis, 1996a).

Other organic ligands containing sterically unhindered nitrogen or oxygen atoms with free valence electrons (i.e. lone pairs) tend to act as P450 inhibitors via reversible binding to the heme iron (Testa and Jenner, 1981). Examples of these include quinoline, furan, benzofuran, benzimidazole, pyridine, imidazole, pyrazole and triazole (together with their derivatives), some of which may be quite selective (Hargreaves et al., 1994). The crystal structures of several nitrogenous inhibitors bound to $P450_{cam}$ (CYP101) have been determined, and these include metyrapone and phenylimidazoles (Poulos and Howard, 1987; Poulos, 1988) where the average Fe–N distance is about 2 Å between heme iron and ligating nitrogen on the inhibitor. For a series of nitrogenous bases binding to iron porphyrin as a heme analogy, it is apparent that ligand basicity is a factor in binding affinity (Byfield et al., 1993) although the bond angle between ligand nitrogen and iron porphyrin ring nitrogen (N–Fe–N) is also important (Lewis, 1996a). In this respect, the decrease in log K (where K is the ligand-binding affinity) is attributable to the variation in N–Fe–N bond angle with high correlation (R = 0.97).

Some P450 inhibitors interact with the heme moiety as a result of some form of enzyme-mediated activation (Testa and Jenner, 1981). Such mechanism-based inhibitors can give rise the formation of a metabolic intermediate (MI) which then complexes with the heme (Murray and Reidy, 1990). For example, orphenadrine and its structural analogue SKF-525A (Rossi et al., 1987) are able to act in this way, together with the clinically used pharmaceuticals erythromycin and troleandomycin (Murray, 1992). The stages for metabolic activation of orphenadrine are as follows:

$$RNMe_2 \rightarrow RNHMe \rightarrow RNOHMe \rightarrow RNHOH \rightarrow RN{=}O$$

where *R* is an alkyl or aryl group and the nitroso product RN=O represents the heme ligand.

This scheme is common for the P450-mediated activation of other tertiary amines to form MIs containing the nitroso function which is then able to act as a heme ligand via the nitrogen lone pair (Murray and Reidy, 1990; Murray 1992). Orphenadrine and SKF-525A (proadifen) preferentially inhibit phenobarbital-inducible forms of P450, although neither of these are especially selective for the CYP2B subfamily (Murray and Reidy, 1990). The former is, however, a fairly selective inhibitor of the human orthologue, CYP2B6 (Rendic and DiCarlo, 1997). As is the case with several other inhibitors, SKF-525A is also able to act as a P450 inducer (Bornheim et al., 1983) and it is common that P450 inducers can become inhibitors at sufficiently high dose concentrations; for example, triacetyloleandomycin (TAO) is both an inducer (Lake et al, 1998) and inhibitor of CYP3A isoforms and also forms an MI complex with the enzyme. Other examples of groupings which are present in mechanism-based inhibitors include olefins, acetylenes, phenyldiazenes, benzothiadiazoles, benzotriazoles, hydrazines and methylenedioxybenzenes (Ortiz de Montellano and Correia, 1995). Of these, the phenyldiazenes are particularly useful as chemical probes for the active site topography of P450 enzymes (Ortiz de Montellano and Correia, 1995) due to the fact that, following *in situ* rearrangement of the phenyl-iron complex, the phenyl group of the phenyldiazene migrates onto one or more of the porphyrin ring nitrogens (Raag et al., 1990) depending on the degree of steric hindrance caused by neighbouring amino acid residues within the heme environment (Swanson et al., 1991, 1992; Tuck et al., 1992, 1993). For further details, the reader is referred to Lewis (1996a) which provides an overview of this area.

Some examples of selective inhibitors of various P450s are shown in Table 5.10, which includes K_i (inhibition constant) values against known substrates for each enzyme concerned. Many of these are competitive, often Type II ligands, which bind reversibly to the heme iron of the particular P450 enzyme. For example, ketoconazole is generally regarded as a selective inhibitor for CYP3A4 (K_i = 0.1 μM) and exhibits Type II binding spectra due to heme iron ligation via the imidazole ring nitrogen atom (reviewed in Lewis, 1996a). However, the inhibitor in this example is not specific for CYP3A4 as it is also able to inhibit CYP19 and CYP51 enzymes; and this non-specificity is often the case for other selective inhibitors, such as metyrapone (inhibits CYP2B and CYP11 as well as CYP101). Nevertheless, selective inhibitors are usually employed as diagnostic probes for individual P450s, and many of these can be substrate analogues, e.g. furafylline is analogous to the CYP1A2 substrate caffeine (Rendic and DiCarlo, 1997). In this instance, however, the mode of binding differs in that caffeine is preferentially N-3 demethylated via CYP1A2, whereas furafylline binds to the heme following 8-methyl oxidation (reviewed in Lewis et al., 1999c).

Occasionally, it is found that different compounds are able to act as selective inhibitors towards P450s within the same subfamily (see Table 5.10). For example, sulfaphenazole inhibits CYP2C9 activity preferentially (K_i = 0.2 μM), whereas

Table 5.10 Selective P450 inhibitors for different isoforms

CYP	*Inhibitor*	*Mode of inhibition*	K_i/K_m (μM)
1A2	furafylline	mechanism-based	0.7
2A6	pilocarpine	heme ligand	4.0
2B6	orphenadrine	mechanism-based	30.0
2C8	sulfinpyrazone	heme ligand?	17.0
2C9	sulfaphenazole	heme ligand	0.2
2C19	fluconazole	heme ligand	2.0
2D6	quinidine	heme ligand	0.06
2E1	3-amino-1,2,4-triazole, pyridine	heme ligand	0.4*
3A4	ketoconazole	heme ligand	0.1
4A1	10-imidazolyl decanoic acid	heme ligand	0.35
11A1	22-amino-23,24-bisnor-5-cholen-3β-ol	heme ligand?	0.025
17A1	4-cyclohexyl-2-methyl-2-(4-pyridyl)propanoate	heme ligand	2.2
19A1	4-hydroxyandrostenedione	heme ligand?	0.029
21A1	21,21-dichloroprogesterone	mechanism-based	12.0
51A1	itraconazole	heme ligand	0.022 (IC_{50} value)
101	metyrapone	heme ligand	0.0023 (K_s value)

References: Lewis, 1996a; Miners and Burkett, 1998; Murray and Reidy, 1990; Koop, 1990; Hargreaves et al., 1994; Ortiz de Montellano and Correia, 1995; Rendic and DiCarlo, 1997; Pelkonen et al., 1998; Halpert et al., 1989; Alterman et al., 1995; Peterson et al., 1971; Vanden Bossche et al., 1984; He et al., 1995.

Note:
* K_i value for pyridine

fluconazole is an inhibitor of CYP2C19 (Miners and Burkett, 1998; Rendic and DiCarlo, 1997). Sometimes, however, this difference in selectivity can relate to enantiomeric forms of the same chemical, such as that shown in the CYP2D subfamily where quinine is selective for the rat form CYP2D1 but its antipode, quinidine, is a selective inhibitor of the human orthologue, CYP2D6 (Rendic and DiCarlo, 1997; Boobis et al., 1990). It is possible that this enantioselectivity for inhibitors within the CYP2D subfamily is due to differences in active site amino acid residues for the two orthologous enzymes, and possible candidates in CYP2D6 include Thr107, Glu216 and Ser304 (Ellis, 2000), although the latter appears to be unlikely because the rat orthologue has this residue conservatively changed to threonine. However, Thr107 changes to phenylalanine in CYP2D1 and Glu216 becomes a valine residue in the same enzyme (see Lewis, 1996a for an alignment of CYP2D family enzymes), thus indicating that there could be a simple structural explanation for species differences in enantioselectivity towards CYP2D inhibitors in rat and human, which involves active site interactions between inhibitor and local amino acid residues.

In addition to their employment in the identification of particular P450s participating in drug oxidations, selective P450 inhibitors are also used as anticancer agents, in antifungal preparations and for the treatment of hyperaldosteronism (Correia and Ortiz de Montellano, 1993) together with a possible use in the alleviation of conditions related to defects in arachidonic and retinoic acid metabolism. Agrochemical uses of P450 inhibitors include plant growth regulation, fungicidal activity (Coulson et al., 1984) and for the development of novel antimicrobial agents and herbicides (Durst and Benveniste, 1993). Possible human health implications can arise from P450 inhibition, however, if there is dramatic down-regulation of an important P450 associated with drug metabolism, for example. Mibefradil, a calcium channel blocker, was withdrawn from clinical use due to significant inhibition of CYP3A4 (Prueksaritanont et al., 1999) and there is also some concern about flavonoid CYP3A4 inhibitors present in fruit drinks, such as grapefruit juice (Ameer and Weintraub, 1997). The antiestrogenic oral contraceptive drug, gestodene, is also a CYP3A4 inhibitor (K_i = 46 μM) and, consequently, its use could result in adverse drug interactions if taken in combination with other CYP3A4 selective compounds (Guengerich, 1990). Furthermore, P450 inhibition can occur in related areas, including the use of inhibitory antibodies for immunoquantification and detection (Edwards et al., 1995, 1998) and via the employment of reactive agents (reviewed in Lewis, 1996a) which selectively attack and bind covalently to certain types of amino acid residues, such as acidic amino acids (e.g. EDC; Strobel et al., 1989) and those with basic sidechains (e.g. MNT and FITC; Bernhardt et al., 1984, 1988).

In addition to organic compounds, metal ions can also act as P450 inhibitors (Testa and Jenner, 1981). Generally, this activity appears to be due to competition with iron for heme porphyrin binding, especially when the metal ion has the same ionic charge and a similar ionic radius to iron, such as that shown for Co^{2+}, Cd^{2+}, Mn^{2+} and Ni^{2+} (Testa and Jenner, 1981) although redox potential may also be a factor. The toxic metal cadmium inhibits CYP2E1 in the rat, but not CYP3A1, however (Alexidis et al., 1994), and it is thought that free radical generation could play a role in this selectivity as CYP2E1 is known to be associated with oxygen radical production (Ronis et al., 1996; Parke, 1987, 1994). Lead nitrate, however, preferentially inhibits CYP1A2 expression in the rat (Degawa et al., 1993) and, although the reason for this is not entirely clear, CYP1A2 was the most highly induced P450 in the study and, therefore, more likely to undergo down-regulation under the prevailing experimental conditions. Metal toxicity may be associated with P450 inhibition and a recent QSAR study (Lewis et al., 1999f) points to the involvement of redox potential, ionic radius (r) and charge polarization (Q/r^2) in the acute toxicity of 24 metals, showing a good correlation ($R = 0.92$) with toxicity data reported for intraperitoneal dosage of metal salts in the mouse. It may be possible to rationalize P450 inhibition by metal ions via a similar QSAR technique to that employed above for explaining potency differences in metal toxicity, as similar parameters may be involved in metal ion competition for iron in complexation with the porphyrin macrocycle (see Table 5.11 for inhibition data). In fact, it is possible to demonstrate that ionic radius plays a significant role in metal ion inhibition, as shown in Figure 5.2 which uses data presented in Table 5.11, although only 5 metals were included. However, if one considers the inhibition data for 9 metals, the relationship tends to indicate that a quadratic function of ionic radius is important.

Table 5.11 Inhibition of P450 by divalent metal ions

Metal/ion	*E°*(V)	*Ionic radius* (Å)	*Q/r*	*Q/r²*	*P450 loss* (%)
1 Co/Co^{2+}	–0.28	0.65	3.077	4.734	45
2 Cd/Cd^{2+}	–0.40	0.92	2.174	2.363	48
3 Mn/Mn^{2+}	–1.18	0.67	2.985	4.455	45
4 Pb/Pb^{2+}	–0.13	1.18	1.695	1.436	17
5 Fe/Fe^{2+}	–0.44	0.75	2.667	3.556	9
6 Cu/Cu^{2+}	+0.34	0.92	2.174	2.363	28
7 Ni/Ni^{2+}	–0.23	0.68	2.941	4.325	40
8 Zn/Zn^{2+}	–0.76	0.75	2.667	3.556	8
9 Cr/Cr^{2+}*	–0.89	0.73	2.740	3.753	7

Reference to inhibition data: Testa and Jenner, 1981.

Notes:
With the exception of Pb^{2+}, the inhibition data correlate ($R = 0.92$) with a quadratic expression in ionic radius, thus indicating that there is an optimal ionic site for heme binding which is close to that of Fe^{2+}. However, for five metal ions (including Pb^{2+}) there is a good linear relationship ($R = 0.987$) between inhibition and ionic radius (see Figure 5.2).

E° = metal/ion redox potential (volts)
Q = charge on ion (2 in each case)
r = ionic radius (Å)

* Although the inhibition data is for CrO_4^{2-}, it is assumed that reduction to Cr^{2+} occurs prior to incorporation into the heme porphyrin macrocycle as only divalent (or possibly trivalent in some cases) metal ions would readily form complexes with this tetradentate ligand.

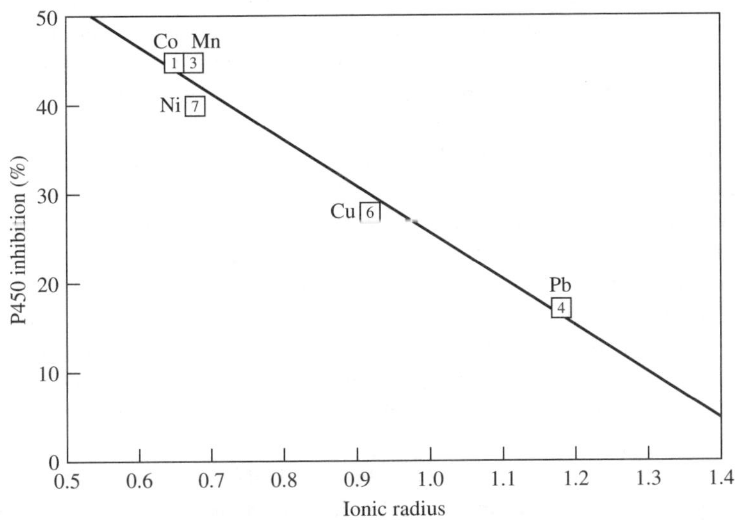

Figure 5.2 Graph showing a linear relationship ($R = 0.987$) between P450 inhibition (percentage loss of P450) and ionic radius (Å) for 5 divalent metal ions.

5.6 QSARs for P450 inhibitors

Quantitative structure–activity relationship (QSAR) analysis of P450 inhibition in a number of different gene families and subfamilies enables some degree of understanding about the possible molecular mechanisms involved in the process (Lewis, 2000a). For example, the results of a variety of separate QSAR studies are summarized in Table 5.12 (Lewis, 2000a). With inhibitors of CYP1 enzymes, it appears that inhibitory activity is primarily associated with the degree of molecular planarity as measured by area/depth^2 or area/depth in the respective cases of methylenedioxybenzenes (Table 5.12, equation 1a) and benzimidazoles (Table 5.12, equation 1b). In the latter example, however, compound polarity is an additional factor (Lewis, 1997b) and this indicates that there may be dipole–dipole interactions within the CYP1 active site region. This finding is supported by molecular modelling studies (Lewis et al., 1999c) which suggest that hydrogen bonding plays a role in both substrate and inhibitor binding to CYP1 enzymes.

QSAR evaluations conducted on a homologous series of primary amines acting as CYP2B inhibitors (Lewis, 1996a) shows the importance of lipophilicity as described by the log P value (where P is the octanol/water partition coefficient). In the example shown (Table 5.12, equation 2a) there is a clear quadratic expression in log P which suggests that the most potent inhibitor in the series (*n*-octylamine) possesses optimum hydrophobility for binding to the CYP2B active site, with aliphatic chain length being the most likely rationale for this effect. Similar expressions arise for aliphatic alcohols and carboxylic acids acting as inhibitors for CYP2E (Lewis, 2000a) as shown in Table 5.12, equations 5a and 5b; although, in these instances, the optimal carbon chain length appears to be a little greater than for amines binding to CYP2B. Desolvation of the P450 active site is also indicated in QSARs for CYP2C19 and CYP2D6 inhibitors (Table 5.12, equations 3a and 4a, respectively). In these cases, however, the molecular weight parameter (M_r) probably describes the relative degree of desolvation accompanying compound binding, with the number of hydrogen bond interactions also being important in both examples.

Finally, for series of antifungal imidazole and triazole inhibitors binding to CYP51 (Lewis et al., 1999a) the variation in compound potency is best described by a two-parameter expression (Table 5.12, equation 6a) involving molecular planarity (area/depth^2) and energy of the lowest unoccupied molecular orbital (E_L). This shows that both molecular shape and electron acceptance characteristics are relevant in combination for explaining the inhibition data, with the E_L descriptor possibly suggesting that hydrogen bond acceptor properties within the relevant molecules could play a role in binding to the heme environment in CYP51 (Lewis et al., 1999a). In summary it is apparent that factors describing P450 inhibition in QSAR studies relate to the key molecular recognition interactions between inhibitor and P450 active site residues in each case. This method of analysis (namely, QSAR) appears to be applicable to both inhibitors and inducers (Lewis, 2000a) of P450s, although one needs to investigate enzyme active site interactions in the former and receptor ligand-binding site interactions for the latter cases in order to obtain a wider perspective of the possible structural factors involved.

Table 5.12 QSARs for P450 inhibitors

	n	s	R	F
1. CYP1A (Lewis, 1997b)				
a) Methylenedioxybenzenes $pIC_{50} = 1.22\,a/d^2 - 5.68$ (± 0.14)	8	0.208	0.96	71.7
b) Benzimidazoles $pIC_{50} = 0.07\,a/d^2 + 0.01\mu^2 + 2.33$ $(\pm 0.01)\quad(\pm 0.003)$	8	0.127	0.98	53.0
2. CYP2B (Lewis, 1996a)				
a) Aliphatic amines $pK_i = 1.67 \log P - 0.17 \log P^2 - 2.15$ $(\pm 0.26)\quad(\pm 0.05)$	8	0.267	0.98	73.8
3. CYP2C19 (Lewis et al., 2001)				
a) Diverse substrates $pK_i = 0.03\,M_r - 19.13 \log M_r + 0.38\,HB_A - 0.28\,HB_D + 42.09$ $(\pm 0.01)\quad(\pm 2.93)\quad(\pm 0.05)\quad(\pm 0.11)$	10	0.172	0.98	38.4
4. CYP2D6 (Lewis, 2000a)				
a) Diverse compounds $pK_i = 0.01\,M_r - 0.48 \log P - 0.57\,HB_A + 1.79\,N_B - 3.56$ $(\pm 0.002)\quad(\pm 0.13)\quad(\pm 0.15)\quad(\pm 0.21)$	11	0.187	0.98	35.2
5. CYP2E1 (Lewis, 2000a)				
a) Primary alcohols $pK_i = 0.98 \log P - 0.20 \log P^2 - 1.79$ $(\pm 0.11)\quad(\pm 0.03)$	10	0.218	0.96	89.4
b) Primary carboxylic acids $pK_i = 1.05 \log D - 0.24 \log D^2 - 2.66$ $(\pm 0.10)\quad(\pm 0.02)$	11	0.169	0.97	119.0
6. CYP51 (Lewis et al., 1999a)				
a) Imidazoles $pIC_{30} = 0.58\,a/d^2 - 1.14\,E_L + 2.45$ $(\pm 0.10)\quad(\pm 0.41)$	7	0.084	0.95	18.6

Notes:

pIC_{50} = negative logarithm of the inhibitory concentration
pK_i = negative logarithm of the inhibition constant
a/d^2 = ratio of molecular area to depth-squared
μ^2 = square of the dipole moment
$\log P$ = logarithm of the octanol/water partition coefficient
M_r = relative molecular mass
$HB_{A,D}$ = number of hydrogen bond acceptors, donors
N_B = number of basic nitrogens
$\log D$ = logarithm of the distribution coefficient at pH 7.4
E_L = energy of the lowest unoccupied molecular orbital

Chapter 6

The structures of cytochromes P450

> Almost all aspects of life are engineered at the molecular level, and without understanding molecules we can only have a very sketchy understanding of life itself.
>
> (Francis Crick)

6.1 Introduction

It is probable that the cytochrome P450 system has been more extensively studied by various physical and chemical techniques than any other enzyme or protein (reviewed in Lewis, 1996a). The likely reason for this is due to the intense interest in its structure and function, which has been apparent over the last forty years or so since its discovery in the 1950s (Ruckpaul and Rein, 1984). Almost every type of spectroscopic method has been employed for achieving a detailed understanding of the P450 reaction mechanism, whereas X-ray crystallographic determinations of several bacterial P450 enzymes have had a major impact on our knowledge of how the P450 system catalyses the oxidation of carbon-based substrates (Peterson and Graham-Lorence, 1995; Graham and Peterson, 1999; Graham-Lorence and Peterson, 1996a; Poulos, 1996a and b). Moreover, quantum-chemical calculations have been carried out on many compounds metabolized by P450s, together with several theoretical investigations of the P450 heme-thiolate environment using molecular orbital (MO) procedures (de Groot et al., 1998; Segall et al., 1998; Lightfoot et al., 2000). Furthermore, homology modelling of microsomal P450s based on bacterial crystal structure templates has facilitated a deeper insight of substrate selectivity, and enabled evaluation of P450-mediated metabolism such that prediction of the metabolic fate of new chemicals can be achieved (Lewis et al., 1998d, 1999b). This chapter reviews the current status of these (and related) techniques for structural investigation of the P450 system, including aspects of redox partner interactions (see Lewis and Hlavica, 2000 for a review), together with summarizing the evidence obtained from site-directed mutagenesis, and other experimental methods, which help to build up an increasingly clearer picture of P450 structure and function.

6.2 X-ray crystallography

It is generally regarded that an X-ray crystal structure represents the most reliable model for describing the three-dimensional conformation of a protein, or enzyme, in the solid state (reviewed in Creighton, 1993) and the technique of X-ray diffraction has made an enormous contribution to our understanding of proteins and how they operate at the atomic level. An X-ray crystal structure is a 'frozen' moment in time, a molecular 'snapshot' reflecting a single frame of what is in reality a truly dynamic situation whereby, as far as biological macromolecular structures are concerned, a small effector molecule (e.g. substrate or ligand) binds to its biological receptor, or enzyme, in order to trigger a physiological process (see, for example, Branden and Tooze, 1991). Up until the early 1990s, the only P450 crystal structure available was that of cytochrome $P450_{cam}$ (CYP101) from *Pseudomonas putida* (Poulos et al., 1987). Since then, the structures of four more P450s have been determined, namely, $P450_{BM3}$ (CYP102), $P450_{terp}$ (CYP108), $P450_{eryF}$ (CYP107) and $P450_{nor}$ (CYP55). The protein databank currently indicates that there are about 30 separate crystal structures available for these five P450s, some of which contain either bound substrates or inhibitors (Brookhaven Protein Databank, http://www.rcsb.org/pdb). Table 6.1 provides the relevant information for some of these, together with data on a variety of redox partners such as cytochrome b_5, oxidoreductase and redoxins. These structurally characterized P450s primarily include those of Class I (namely CYP101, CYP107, CYP108 and CYP55) although one (actually, CYP102) is representative of a Class II P450. As $P450_{BM3}$ (CYP102) is the only one in Class II to have had its structure determined crystallographically, this constitutes the current template for modelling microsomal (and other similar eukaryotic) P450s by homology (Lewis, 1995b, 1998a–c); there is also a generally satisfactory sequence identity of around 20% between CYP102 and many other Class II P450s (Lewis, 1996a). The other bacterial P450s (namely CYP101, CYP107 and CYP108) of known structure are all of Class I and their homology with the microsomal enzymes is quite low (< 20%) such that their potential utility for modelling Class II P450s is somewhat questionable (Poulos, 1991). The fungal form (CYP55) also exhibits a structure and amino acid sequence which is more related to the Class I P450s, such as CYP101 (Park et al., 1997) and, therefore, it is unlikely that this bears much relevance to Class II P450 structures. However, various groups are currently engaged in the crystallization of mammalian P450s and, consequently, it is highly likely that several of these Class II enzymes will be resolved crystallographically in the near future. For the four bacterial P450s (CYP101, CYP102, CYP107 and CYP108) of known structure, it is clear that there is a similarity of protein conformation which includes a Rossmann domain, a Greek key helical bundle (E, F, G and I helices) and a globin fold, with four α-helices (D, E ,I and L) representing part of a common structural core (Poulos, 1986; Poulos et al., 1986, 1987). The substrate-binding site close to the distal heme face comprises helices B′, F and I, together with two β-sheet regions, which tend to remain structurally conserved despite primary

Table 6.1 Structurally characterized P450s and their redox partners

a) CYP	*Description*	*Iron state*	*PDB code*	*Resolution (Å)*	*References*
101	Camphor-bound $P450_{cam}$	Fe^{3+}_{HS}	2cpp	1.63	Poulos et al., 1987
101	$P450_{cam}$:CO complex[a]	Fe^{2+}_{LS}	3cpp	1.90	Raag and Poulos, 1989b
101	Substrate-free $P450_{cam}$[b]	Fe^{3+}_{HS}	1phc	2.20	Poulos et al., 1986
101	Adamantane-bound $P450_{cam}$[c]	Fe^{3+}_{HS}	4cpp	2.10	Raag and Poulos, 1989a
101	5-Hydroxycamphor-bound $P450_{cam}$[d]	Fe^{3+}_{LS}	1noo	2.00	Protein Databank, (http://www.rcsb.org/pdb)
102	Substrate-free $P450_{BM3}$	Fe^{3+}_{LS}	2hpd	2.00	Ravichandran et al., 1993
102	Substrate-free $P450_{BM3}$	Fe^{3+}_{LS}	2bmh	2.00	Li and Poulos, 1995
102	Palmitoleate-bound $P450_{BM3}$	Fe^{3+}_{HS}	1fag	2.70	Li and Poulos, 1997
102	Substrate-free $P450_{BM3}$:T268A	Fe^{3+}_{LS}	1fah	2.30	Raag et al., 1991
102	FMN and heme domains	Fe^{3+}_{LS}	1bvy	2.03	Sevrioukova et al., 1999
107	Erythronolide-bound $P450_{eryF}$	Fe^{3+}_{HS}	1oxa	2.10	Cupp-Vickery and Poulos, 1995
108	Substrate-free $P450_{terp}$	Fe^{3+}_{LS}	1cpt	2.30	Hasemann et al., 1994
55	Substrate-free $P450_{nor}$[e]	Fe^{3+}_{LS}	1rom	2.00	Nakahara et al., 1994

b) Redox partners	*Description*	*PDB code*	*Resolution (Å)*	*References*
Abbreviation				
Pdx	Putidaredoxin	1put	NMR	Pochapsky et al., 1994
Adx	Adrenodoxin	1ayf	1.85	Mueller et al., 1998
b_5	Cytochrome b_5	3b5c	2.0	Mathews et al., 1972
b_5 reductase	Cytochrome b_5 reductase	1ndh	2.4	Nishida et al., 1995
OR[f]	NADPH-dependent P450 oxidoreductase	1amo	2.6	Wang et al., 1997

Notes:
LS = low-spin
HS = high-spin

a Camphor moves about 1Å away from the heme relative to that in the enzyme–substrate complex.
b A water molecule occupies the 6th ligand distal to the heme, and there are six water molecules in the active site previously occupied by camphor in the 2cpp structure.
c Other substrates cited in this reference include adamantanone, camphane and norcamphor.
d Deposited by T.L. Poulos.
e A total of 32 individual P450 structures are accessible via the Protein Databank of which 22 are of $P450_{cam}$, 6 are of $P450_{BM3}$, 2 are of $P450_{nor}$, 1 is of $P450_{terp}$ and 1 of $P450_{eryF}$.
f Other heme-thiolate enzymes of known structure include chloroperoxidase and nitric oxide synthase. Recently, there has been an important crystallographic study of the $P450_{cam}$ catalytic cycle (Schlichting et al., 2000) which includes resolution of the oxygenated complex. The crystal structure of flavocytochrome b_2 shows that a bridging tyrosine residue is likely to be involved in electron transfer between flavin and heme domains (Xia and Mathews, 1990). This may be analogous to the conserved tryptophan residue, present in many eukaryotic P450s, which could represent a linkage between oxidoreductase and P450 facilitating electron transfer. Furthermore, the electron transfer complex between cytochrome c and its peroxidase has also been reported (Pelletier and Kraut, 1992).

sequence variations that reflect substrate selectivity differences (Peterson and Graham-Lorence, 1995).

The heme moiety is 'sandwiched' between two helical segments (I and L helices) where certain hydrophobic and π-π stacking interactions tend to bind the iron protoporphyrin IX complex (Poulos et al., 1995; Modi et al., 1995). The two propionate sidechains on this heme group are held by electrostatic interactions with basic amino acid residues that are present in all of the P450 crystal structures determined thus far (Peterson and Graham-Lorence, 1995; Ortiz de Montellano and Graham-Lorence, 1993). Usually, but not exclusively, these basic residues are conserved and, furthermore, there are additional basic amino acids which are almost certainly contact points for redox partner interaction that tend to be conserved, especially where the same type of redox partner (i.e. redoxin or reductase) is utilized for electron transfer (reviewed in Lewis and Hlavica, 2000).

One of the basic residues which can both bind to heme and act as a redox partner contact may be involved in facilitating the access of water molecules to the heme environment during the catalytic cycle (Oprea et al., 1997). It is suggested that a proximal arginine residue acts as a 'sluice gate' for allowing the egress of distal water molecules which are displaced following substrate binding; this may trigger the redox partner binding and reduction processes with the arginine once again allowing solvating water molecules to enter the functional water channel (or 'aqueduct') when the oxygenated metabolite leaves the heme pocket (Oprea et al., 1997). Another key residue which is conserved across the majority of microsomal P450s corresponds with tryptophan-96 (Trp96) in CYP102 which is also a Class II P450 and, like the microsomal enzymes, utilizes an NADPH-dependent oxidoreductase as a redox supply (Degtyarenko and Archakov, 1993). In the CYP102 hemoprotein domain, it appears that Trp96 may be involved in heme binding, possibly via a hydrogen bond to one of the heme propionates, and a role in electron transfer has also been postulated (Baldwin et al., 1991; Inouye and Coon, 1985; Munro et al., 1992, 1994a). Although this tryptophan is not conserved in the other bacterial P450s, it is interesting to note that, in the $P450_{cam}$ system, there is a C-terminal tryptophan on the putidaredoxin protein which does appear to participate in the electron transfer process (Sligar et al., 1974) and, consequently, it could be proposed that its incorporation into the Class II P450 sequences represents an evolutionary development which accompanied the change of redox partners from iron-sulphur redoxins to NADPH-dependent flavoprotein reductases (Baldwin et al., 1991).

Other features of the structurally characterized P450s include regions of the polypeptide that are predominantly helix-rich and those which can be regarded as essentially composed of β-sheets (Peterson and Graham-Lorence, 1995), leading to the designation of an α-domain and a β-domain in each bacterial crystal structure examined thus far (see Figure 6.1 as an example). The important review of these P450s by Peterson and Graham-Lorence (1995) shows the topologies of each bacterial form (i.e. CYP101, CYP102 and CYP108) where a clear similarity between these can be appreciated, including that of super-secondary structural motifs such as the Rossmann domain ($\alpha\beta\alpha\beta\alpha$) composed of repeating α-helical and β-sheet segments comprising the A, B and B′ helices and β_1 strands (Peterson and Graham-Lorence, 1995). Elements which contribute to maintaining the structural integrity of the single polypeptide chain in P450s, such that the tertiary fold tends to exhibit a similar shape, include a highly conserved tetrapeptide motif with glutamate and arginine residues at its termini (reviewed in Lewis, 1996a). This E*xx*R pattern (where *x* is any amino acid but usually a hydrophobic aliphatic residue) can be observed in virtually every P450 sequence across the entire superfamily, and it appears to lie close to the heme moiety in the known P450 crystal structures as part of the K helix (Hasemann et al., 1995). The possible role of this highly conserved ion-pair may be partially structural in nature to maintain the tertiary fold, and also to participate in both the redox partner interaction and heme binding (Peterson and Graham-Lorence, 1995; Hasemann et al., 1995).

Other salt bridges are known in P450 crystal structures, and these seem to be involved in retention of protein folding although a distal charge relay mechanism has been reported (Gerber and Sligar, 1992, 1994) whereby ion-pairing plays a role in hydrogen-bonded water conduits within the heme pocket which are crucial to the mono-oxygenase pathway (Mueller et al., 1995; Poulos et al., 1995). One distal helix residue appears to be particularly important in this respect, and this is a conserved acidic amino acid (aspartate or glutamate) which is located immediately prior to the highly conserved threonine found in many P450s sequenced to date (Nelson, 1995). Alteration to either of these functionally relevant residues, which are associated with an unusual deformation in the distal I helix brought about by an upstream glycine, leads to changes in the normal mono-oxygenase pathway of P450s that probably result from uncoupling of the typical dioxygen activation (Mueller et al., 1995). In fact, it is generally accepted that the conserved tetrapeptide G*x*D/ET (where *x* is any amino acid) represents the likely oxygen-binding site and point of access for the incoming dioxygen molecule (Poulos et al., 1995). Other commonalities exist between the various P450s of known structure but a full and detailed description is beyond the scope of this section, and the reader is referred to the original papers listed in Table 6.1, together with comparisons reported elsewhere (Hasemann et al., 1995; Peterson and Graham-Lorence, 1995; Poulos et al., 1995). For an account of the techniques involved in P450 crystal growing, Poulos (1996a) is very instructive, whereas the general principles of protein crystallography are usefully outlined in an excellent book by Thomas Creighton (1993). In this context, it should be recognized however that the results of an X-ray determination for a protein provides a structural description of a time-averaged minimum energy conformation in the solid state, and at a temperature where there is usually little thermal motion of the protein molecules in the crystal itself, although certain exceptions are known, such as in the CYP108 structure (Hasemann et al., 1995).

P450 redox partners have also been resolved crystallographically (see Table 6.1) and the redox interaction between the FMN and hemoprotein domains of CYP102 has been published recently (Sevrioukova et al., 1999). It is clear that eukaryotic P450 crystal structures are next on the list for X-ray crystallographic determination, and at least one may have been reported in the literature by the time this work is published. (The crystal structure of modified CYP2C5, a rabbit orthologue of the CYP2C subfamily, has recently been published (Williams et al., 2000b) and shows both similarities and differences with that of CYP102.) Due to the profound structural insights that this technique provides, one awaits with interest what the forthcoming years will bring to enhance our knowledge and understanding of P450 structure and function.

6.3 Spectroscopy

Almost every available spectroscopic procedure has been employed in order to probe the structure and function of P450 enzymes (reviewed in Lewis, 1996a). Indeed, it was one such technique (namely, UV spectrophotometry) which led to the discovery of cytochrome P450 and was also responsible for its name, as described in the Introduction. The major spectroscopic methods in the optical region (UV, CD, MCD and ORD), vibrational (IR and RR), magnetic resonance (NMR and ESR) and high energy areas (MB and EXAFS) have been detailed previously (Lewis, 1986, 1996a) and, therefore, these will only be summarized and briefly reviewed in this section. The interested reader is also referred to some excellent reviews by Hawkins and Dawson (1992), Hildebrandt (1992), Schenkman et al. (1981), Coon and White (1980), White and Coon (1980) and Lippard and Berg (1994) for further reading in this particular area.

Optical spectroscopy has been extensively used for the study of P450s and their substrate interactions, and it is recognized that absorptions in the UV/visible region are associated with electronic transitions within the heme unit (Hanson et al., 1976, 1977) which are exquisitely sensitive to the hemoprotein spin-state and also to the local heme environment itself (Schenkman et al., 1981). Thus the technique of UV spectrophotometry can be readily employed to probe the effect of substrate binding to various P450 enzymes, and also to monitor the extent (and form) of substrate binding in particular cases (Gibson and Tamburini, 1984). Different types of substrate-binding spectra have been established which can assist in identifying the specific mode of substrate interaction with the P450 heme pocket (Schenkman et al, 1981). In particular, these are classified as type I, type II and reverse type I according to the spectral change which occurs in the UV difference spectrum, and this relates to heme iron spin-state modulation together with ligation at the distal heme face (Gibson and Skett, 1994). For example, type I binding in the microsomal P450 system is regarded as being characterized by a shift in the spin equilibrium towards high-spin and this is evidenced by a decrease in the Soret band absorption maximum at 420 nm coupled with a concomitant increase of the 390 nm absorption in the ferric P450 UV spectrum. In contrast, type II binding is exemplified by a decrease in the UV absorption at 390–405 nm with an accompanying increase at around 425–435 nm such that there is a change to longer wavelengths associated with the spin-state population shift towards the low-spin ferric case (Gibson and Skett, 1994). The type II spectral change is indicative of direct heme iron ligation by the substrate and this represents a well-established mechanism of P450 inhibition, as discussed in Chapter 5. Reverse type I (sometimes referred to as modified type II) is typified by a UV spectrum which is essentially a 'mirror image' of the type I spectral change (Schenkman et al., 1981). Although this variety of substrate-binding spectrum resembles the type II spectrum, there is no accompanying shift in the absorption maxima which is a characteristic of the type II case and, consequently, it is thought that heme ligation does not occur in this instance (Schenkman et al., 1981).

The P450 system has also been studied using polarized optical spectroscopy including circular dichroism (CD), magnetic circular dichroism (MCD) and optical rotatory dispersion (ORD), with MCD representing the most extensively used technique of this type for investigating the P450 environment (reviewed in Lewis, 1996a). MCD spectroscopy has been applied by Dawson and colleagues who, in conjunction with other techniques such as ESR and EXAFS, have employed this physical method for elucidating the properties of P450 heme ligands such as the invariant cysteine proximal to the heme (Hawkins and Dawson, 1992; Dawson and Sono, 1987; Dawson, 1988; Dawson et al., 1982, 1986). In addition, Hanson and co-workers have utilized UV absorption spectroscopy with plane polarized light for an important study on the classification of electronic transitions in the P450 spectrum (Hanson et al., 1977).

Vibrational spectroscopy centres around the two largely complementary techniques of infrared (IR) and resonance Raman (RR) spectroscopy (reviewed in Lewis, 1996a). Of these, the latter has been more extensively used to study the P450 system and work in this area has been reviewed previously (Hildebrandt, 1992; Lewis, 1996a). Essentially, various types of vibrational modes of the heme moiety can be readily investigated via IR and RR spectroscopy, some of which give rise to characteristic marker absorption bands relating to the heme iron spin-state and redox state, with RR being particularly useful in this respect (Lewis, 1986, 1996a). Moreover, axial ligand stretching vibrations (including those involving the heme iron) may be monitored using both IR and RR, once again with the latter constituting the more commonly reported procedure (Lewis, 1996a). For example, the Fe–S stretching band at around 350 cm^{-1} in the RR spectrum is characteristic of the thiolate-ligated heme group in P450 enzymes (Champion et al., 1982; Munro et al., 1992, 1994b), whereas the presence of superoxide has been identified under catalytic conditions by an O–O stretching band at about 1140 cm^{-1} (Egawa et al., 1991; Munro et al., 1992). In addition, the C–O stretching vibrational band appears at between 1940–1950 cm^{-1} in the IR spectrum of P450.CO complexes; this being indicative of non-linearity in the Fe–CO linkage (Jung et al., 1992) and there is also a good correlation between C–O stretching and Fe–CO stretching vibrations both in P450s and hemoproteins in general (Nagai et al., 1991).

Magnetic resonance spectroscopy has been frequently applied to P450 systems, including both NMR and ESR techniques (Weiner, 1986; Lipscomb, 1980; Peisach and Blumberg, 1970; Peisach et al., 1979; Philson et al., 1979). ESR requires the presence of unpaired electrons in the molecule concerned in order to produce a signal and, consequently, the procedure is only applicable to P450 in its ferric form where both spin-states contain free electrons, i.e. five unpaired electrons in the high-spin state and one for the low-spin case (Lewis, 1996a). This feature has enabled ESR spectroscopy to be employed for the study of spin-state equilibria (albeit at low temperatures) in ferric P450, and the field has been reviewed relatively recently (Lewis, 1996a). NMR, on the other hand, can be applied to both heme iron redox states in P450 systems although the paramagnetic shift in the proton NMR signal, due to the proximity of nuclei such as iron, provides a ready means of investigating substrate–heme distances in P450 complexes as evidenced by the work of Roberts and co-workers (Roberts, 1996; Modi et al., 1996a and b). The high-energy regions of the electromagnetic spectrum are involved in Mössbauer (MB) spectroscopy (γ-radiation) and EXAFS (low energy X-ray region), with both techniques being utilized to explore the local heme environments in both P450 (Sharrock et al., 1976) and other hemoproteins (reviewed in Lewis, 1996a). In particular, EXAFS has been employed by Dawson and colleagues (Dawson et al., 1986; Dawson and Sono, 1987; Dawson, 1988; Hawkins and Dawson, 1992) to record Fe–S distances in P450 complexes such that one can investigate the effects of spin-state and redox state changes within the heme locus. Table 6.2 summarizes the experimental timescales involved in a number of different techniques, including spectroscopic procedures.

6.4 Site-directed mutagenesis and residue modification

A number of experimental procedures have been employed for probing the active sites and other key regions of P450 enzymes by changing or modifying certain amino acid residues (reviewed in Lewis, 1996a). These techniques include site-specific mutagenesis of particular amino acids, and covalent adduct formation and chemical modification of targeted amino acid types in the protein sequence of a P450 (Lewis, 1996a, 1998a). Antibody recognition peptide sequences also represent a method for mapping certain functionally or spatially important regions of P450 proteins for immunoquantification, etc. (De Lemos-Chiarandini et al., 1987; Edwards et al., 1998). Consequently, epitope mapping can be useful for indications of how microsomal P450s may become orientated within the endoplasmic reticulum membrane, for example, and which regions of polypeptide are surface-exposed (Kolesanova et al., 1994; Uvarov et al., 1994). However, site-directed mutations of specific amino acids at various regions of P450 sequences are particularly important for determining the relevance of certain residues for substrate binding and recognition, i.e. the substrate recognition sites (SRSs), together with other roles such as the interaction with redox partners, oxygen binding and its activation, in addition to heme binding and structural roles including

Table 6.2 Experimental timescales

Technique	*Timescale (s)*
Ultra-violet (UV) spectroscopy	10^{-15}
Laser flash photolysis	$\geq 10^{-14}$
Infra-red (IR) spectroscopy	10^{-13}
Pulse radiolysis	$\sim 10^{-9}$
Mössbauer (MB) spectroscopy (^{57}Fe)	$10^{-9} - 10^{-6}$
Electron spin resonance spectroscopy (transition metals)	$10^{-9} - 10^{-8}$
Temperature-jump spectrophotometry	$\geq 10^{-8}$
Nuclear magnetic resonance (^{1}H) spectroscopy	$\sim 10^{-5}$
Chemical mixing	$\geq 10^{-3}$

References: Gray and Ellis, 1994; Lippard and Berg, 1994.

Notes:

In comparison with these experimental timescales, it should be recognized that molecular dynamics simulations of proteins tend to be carried out under conditions corresponding to the order of picosends (10^{-12} s) with femtosecond (10^{-15} s) timesteps.

It is generally accepted that functional group vibrations take place within the 10^{-14} to 10^{-13} second timeframe, whereas rotation of methyl groups occur over a period of 10^{-12} to 10^{-11} seconds. However, phenyl ring rotations take place over substantially longer time intervals, being of the order of 10^{-4} to 1 second (Hirst, 1990).

It is also worth noting that an X-ray crystallographic determination is conducted over an even longer timescale, with data being collected in minutes or more, such that the final structure will actually represent a time-averaged conformation for the protein at room temperature in the solid state.

membrane association (see, for example, Lewis, 1998a; Shimizu, 1997; Johnson, 1992; Atkins and Sligar, 1989).

Members of the CYP2 family have been extensively studied (He et al., 1994; Hsu et al., 1993; Ibeanu et al., 1996; Iwasaki et al., 1993; Juvonen et al., 1991; Squires and Negishi, 1988; Lindberg and Negishi, 1989; Waller et al., 1999; Haining et al., 1999) using this technique, but it has also been applied to the CYP1 (Krainev et al., 1992; Hiroya et al., 1992; Furuya et al., 1989a and b) and CYP3 families, together with some of the steroidogenic P450s such as CYP11, CYP17, CYP19 (Chen and Zhou, 1992; Graham-Lorence et al., 1991) and CYP21 (reviewed in Lewis and Lee-Robichaud, 1998). Furthermore, mutagenesis experiments have been carried out on the fungal P450, CYP51, and also on the bacterial enzymes CYP101 (Yasukochi et al., 1994) and CYP102 (Oliver et al., 1997; Noble et al., 1999; Graham-Lorence et al., 1997) as has been reviewed (Lewis, 1998a). Consequently, substantial areas of the P450 chain (especially the SRSs) have been probed using this procedure and some of these are presented in Tables 6.3 to 6.5 which cover the major families of drug-metabolising P450s, and those of the two bacterial template enzymes have been reviewed previously (see Lewis, 1998a). In particular, these compilations include the investigations carried out within the six substrate recognition sites (SRSs) for those enzymes listed, although the original SRS terminology was coined by Gotoh (1992) based on sequence analysis of CYP2 family proteins. It is found that, by and large, the results of site-directed mutagenesis support the view that the general location of SRSs of many P450s tend to match up, especially when a multiple sequence alignment is constructed based on the CYP102 hemoprotein domain as a potential template for homology modelling (Lewis, 1998a). However, SRS3 does not appear to represent part of the substrate-binding site close to the heme, although this could be a surface recognition peptide where a substrate molecule may bind to initially before moving into the vicinity of the heme pocket. The remaining five SRSs can be mapped to specific secondary structural motifs in the CYP102 crystal structure (Lewis, 1998a). In particular, SRS1 forms part of the B′ helix and the following interhelical loop, whereas SRS2 lies within the F helix. As mentioned previously, SRS3 is not located sufficiently close to the heme for it to form part of the active site, this SRS being situated within a loop region between the F and G helices (Lewis, 1998a). However, SRS4 forms the central part of the I helix whereas the remaining two SRSs lie within β-sheet regions (β2 and β4) which, with the other three, make up a 'ring' of residues encircling the heme environment when viewed from above the substrate binding site, as shown in Figure 6.2.

In addition to the important mapping out of putative active site regions of P450s, those areas associated with redox interactions have also been investigated using mutagenesis experiments and also via chemical modification of certain basic amino acids (reviewed in Bernhardt, 1995). In general, it is found that certain surface lysine and arginine residues on P450 are able to enter into electrostatic interactions with complementary acidic amino acid sidechains on typical redox partner proteins, such as cytochrome b_5, reductase and redoxin (Schenkman, 1993). Moreover, those basic residues lying proximal to the heme face and generally equidistant from the central iron atom are primarily associated with redox interactions, as has been shown from both mutagenesis (Stayton and Sligar, 1990; Stayton et al., 1989) and chemical modification studies (Bernhardt et al., 1988; Bernhardt, 1993). Supporting evidence for this form of binding in the P450-reductase interactions has been reported from a recent crystal structure of the FMN domain of CYP102 and the corresponding heme domain (Sevrioukova et al., 1999). It is clear, however, that shorter range forces such as hydrogen bonding also play an important role in such interactions.

Chemical modification of amino acid residues has also been employed for investigating putative active sites of P450s, such as the work on CYP1A1 reported by Henry Strobel's group (Shen and Strobel, 1992; Cvrk et al., 1996; Cvrk and Strobel, 1998). The reactive centres of P450 enzymes may also be probed by another related technique which involves the use of phenyldiazenes and their analogues via the formation of covalent heme adducts (Ortiz de Montellano and Correia, 1995; Correia and Ortiz de Montellano, 1993). The resulting pyrrole derivatives produced from the subsequent breakdown of the heme moiety provide a means for mapping the localized heme environment, because those active site regions hindered by amino acid sidechains give rise to a relatively low ratio of covalent pyrrole adduct formation (Swanson et al., 1991, 1992; Mackman et al., 1996). It is found that there are distinct variations (Mackman et al., 1996) in the

Table 6.3 A summary of site-directed mutagenesis and residue modification in the CYP1A subfamily

Species	*CYP*	*Change*	*CYP102 position*	*Comments*	*SRS*
Rat	1A1	K97[a]	K59	Inhibits reductase interaction	–
Human	1A2	I117T	D80	Affects substrate binding affinity	1
Human	1A2	F226I	L181	Affects substrate binding affinity	2
Rat	1A2	K250L	R203	Enhances alkoxyresorufin O-dealkylation	3
Rat	1A2	R251L	Q204	Enhances alkoxyresorufin O-dealkylation	3
Rat	1A2	K253L	Q206	Enhances alkoxyresorufin O-dealkylation	3
Rat	1A1	K271[a]	A221	Inhibits reductase interaction	–
Rat	1A1	K279[a]	Q229	Inhibits reductase interaction	–
Rat	1A2	L286-S290	M237-K241	Antibody recognition site	–
Rat	1A2	F313Y	L262	Decreases acetanilide hydroxylation	4
Rat	1A2	A315S	A264	Decreases acetanilide hydroxylation	4
Rat	1A2	G316E	G265	Increases 7-ethoxycoumarin O-deethylation	4
Rat	1A2	E318D	E267	Increases 7-ethoxycoumarin O-deethylation	4
Rat	1A2	T319A	T268	Decreases acetanilide hydroxylation	4
Rat	1A2	T322A	G271	Decreases acetanilide hydroxylation	4
Rat	1A2	F325T	S274	Decreases acetanilide hydroxylation	4
Rat	1A1	K407[a]	K349	Inhibits reductase interaction	–
Rat	1A2	E459A	Q403	Decreases acetanilide hydroxylation	–
Rat	1A2	I460S	Q404	Decreases acetanilide hydroxylation	–
Human	1A1	I462V[b]	F405	Increases lung cancer susceptibility	–
Rat	1A1	T501-K503[c]	T438-K640	Peptide fragment interacting with azidocumene	6

References: Lewis et al., 1999c; Parikh et al., 1999; Lainé et al., 1994.

Notes:
SRS = Substrate recognition site (Gotoh, 1992)

a = chemical modification
b = allelic variant
c = photoaffinity labelling

active site topography for different P450s, although similiarities are apparent between several mammalian microsomal P450s and the hemoprotein domain of the unique bacterial enzyme CYP102, thus indicating that the latter could represent a useful template for homology modelling of other Class II P450 enzymes (reviewed in Lewis, 1996a, 1998a). Other studies on P450s include those investigating charge-transfer interactions of the heme locus (McKnight et al., 1993), heme-substrate conformation (Omata et al., 1987), flavoprotein interactions (Sevrioukova et al., 1996, 1999) and co-operative substrate binding kinetics (Shou et al., 1994, 1999).

6.5 Molecular modelling studies

People started to investigate the computer modelling of P450s about twelve years ago, following determination of the first P450 X-ray crystal structure, namely

Table 6.4 A summary* of site-directed mutagenesis experiments in the CYP2 family SRS regions

CYP	*Residue change*	*Corresponding residue in CYP102*	*SRS*	*Comments*
CYP2A subfamily				
2A11	Q104L	A74	1	alters regioselectivity
2A5	V117A	F87	1	alters regioselectivity
2A5	G207P	R179	2	alters regioselectivity
2A5	F209L	L181	2	alters substrate selectivity
2A1	T303S	T268	4	affects catalytic activity
2A5	M365L	T327	5	alters regioselectivity
2A11	R372H	L333	5	affects catalytic activity
2A5	A481V	P441	6	alters regioselectivity
CYP2B subfamily				
2B4	I114F	F87	1	affects steroid 16-hydroxylase stereoselectivity
2B1	I114A	F87	1	affects androgen 15β-hydroxylase activity
2B1	F206L	L181	2	affects androgen regioselectivity
2B1	L209A	A184	2	affects steroid regioselectivity
2B4	S294T	T260	4	affects androgen regioselectivity
2B1	T302S	T268	4	affects androgen regioselectivity
2B1	V363A	A328	5	affects androgen regioselectivity
2B11	I365V	A330	5	affects steroid regioselectivity
2B1	V367A	S332	5	affects androgen regioselectivity
2B1	I477A	T438	6	affects androgen regioselectivity
2B1	G478A	L439	6	affects androgen regioselectivity
2B1	I480A	P441	6	affects androgen regioselectivity
CYP2C subfamily				
2C9	R97A	S72	1	affects catalytic activity
2C19	H99I	A74	1	affects omeprazole metabolism
2C2	G111V	G85	1	affects catalytic activity
2C2	A113V	F87	1	affects catalytic activity
2C3v	V113A	F87	1	affects progesterone regioselectivity
2C2	S115R	S89	1	affects catalytic activity
2C2	T301S	T268	4	affects catalytic activity
2C9	I359L	W325	5	affects warfarin metabolism
2C3	S364T	F331	5	affects progesterone regioselectivity
2C3v	T364S	F331	5	affects progesterone regioselectivity
2C2	H368R	L333	5	affects catalytic activity
2C1	V473S	L437	6	affects catalytic activity
CYP2D subfamily				
2D6	T107I	V78	1	lowers catalytic activity
2D6	E216G	A184	2	affects catalytic activity
2D6	D301G	T260	4	abolishes catalytic activity
2D6	V374M	S332	5	affects metoprolol regioselectivity
2D1	L380F	S332	5	affects bufuralol metabolism
2D6	F481L	T438	6	lowers catalytic activity
CYP2E subfamily				
2E1	T303S	T268	4	affects regioselectivity

References: Lewis, 1998a; Ellis, 2000; Ridderstrom et al., 2000; Richardson & Johnson, 1994.

Notes:
* appended for the sake of brevity and indicating active mutations only
SRS = substrate recognition site (Gotoh, 1992)
SRS3 is not generally regarded as lying close enough to the heme for substrate binding but it may represent a surface recognition/entry site for substrates.

Table 6.5 A summary of CYP3A4 site-directed mutagenesis and interactions

CYP3A4	*CYP102*	*Comments*	*SRS*
V101	L71	possible hydrophobic interaction	1
F102	S72	π-stacking interaction in model	1
T103	Q73	hydrogen bond site in model	1
N104	A74	hydrogen bond site in model	1
M114	D84	possible hydrophobic interaction	1
A117	F87	possible hydrophobic interaction	1
S119	S89	altered binding with progesterone	1
L210	A180	associated with allosteric binding	2
L211	L181	associated with allosteric binding	2
R212	D182	showed little effect on substrate binding	2
F213	E183	π-stacking interaction in model	2
D214	A184	altered binding with testosterone	2
I301	T260	hydrophobic interaction	4
F304	I263	π-stacking interaction in model	4
A305	A264	hydrophobic interaction	4
A370	A328	hydrophobic interaction	5
L373	F331	hydrophobic interaction	5
G479	L437	access for larger substrates	6

References: Szklarz and Halpert, 1998; Harlow and Halpert, 1997; Lewis et al., 1996.

Note:
SRS = substrate recognition site (Gotoh, 1992)

$P450_{cam}$ (CYP101), by Tom Poulos' group (Poulos et al., 1987). At present, there are four crystal structures from bacterial sources (namely CYP101, CYP102, CYP107 and CYP108) and one fungal form (CYP55) which bears close resemblance with the three Class I bacterial P450 crystal structures, but the CYP102 hemoprotein domain exhibits reasonably close homology (usually 20% or more) with many microsomal mammalian P450s as these are all Class II enzymes (Degtyarenko and Archakov, 1993). However, all of the crystallographically resolved P450s share a common structural core, although there are certain differences and these have been described elsewhere (Hasemann et al., 1995). Nevertheless, the tertiary fold appears to be largely conserved in all five crystal structures reported to date, indicating that, as has been shown for the cytochromes c (Creighton, 1993) and for other hemoproteins such as the globins, there are likely to be close analogies across the P450 superfamily. Over 1200 P450 gene sequences have been identified thus far and it is likely that many more will be reported on a regular basis, possibly leading to about 200 P450s per animal species with thousands of individual P450 genes in the entire superfamily (Nelson, 1998). As the rate of crystal structure determination does not keep pace with the frequency of results from gene sequencing, there will always be a role for the modelling of these enzymes by homology with one or more of those which have been structurally characterized. Moreover, as the hemoprotein domain of CYP102 is the only example to date of a Class II P450 with known three-dimensional structure, it is

```
            SRS4  I helix              ><    J helix      > 300   <
cyp102   YQIITFLIAG HETTSGLLSF ALYFLVKNPH VLQKAAEEAA RVLVDP.VPS
cyp2a4   LTTLNLFFAG TETVSTTLRY GFLLLMKYPD IEAKVHEEID RVIGRNRQPK
cyp2a5   LTTLNLFFAG TETVSTTLRY GFLLLMKHPD IEAKVHEEID RVIGRNRQPK
cyp2a6   MTTLNLFIGG TETVSTTLRY GFLLLMKHPE VEAKVHEEID RVIGKNRQPK
cyp2b1   ISLLSLFFAG TETSSTTLRY GFLLMLKYPH VAEKVQKEID QVIGSHRLPT
cyp2b4   LTVLSLFFAG TETTSTTLRY GFLLMLKYPH VTERVQKEIE QVIGSHRPPA
cyp2b6   LNTLSLFFAG TETTSTTLRY GFLLMLKYPH VAERVYREIE QVIGPHRPPE
cyp2c2   TTVSDVFMAG TETTSTTLRY GLLLLMKHPE VIAKVQEEIE RVIGRHRSPC
cyp2c3   TTIWDVFSAG TDTTSNTLKF ALLLLLKHPE ITAKVQEEIE HVIGRHRSPC
cyp2c4   IAVFDLFGAG TETTSTTLRY SLLLLLKHPE VAARVQEEIE RVIGRHRSPC
cyp2c5   IAVSDLFGAG TETTSTTLRY SLLLLLKHPE VAARVQEEIE RVIGRHRSPC
cyp2c8   CTVADLFVAG TETTSTTLRY GLLLLLKHPE VTAKVQEEID HVIGRHRSPC
cyp2c9   NTAVDLFGAG TETTSTTLRY ALLLLLKHPE VTAKVQEEIE RVIGRNRSPC
cyp2c19  ITAADLLGAG TETTSTTLRY ALLLLLKHPE VTAKVQEEIE RVIGRNRSPC
cyp2d1   MVVVDLFTAG MVTTATTLTW ALLLMILYPD VQRRVQQEID EVIGQVRCPE
cyp2d6   IVVADLFSAG MVTTSTTLAW GLLLMILHPD VQRRVQQEID DVIGQVRRPE
cyp2e1r  VTVADMFFAG TETTSTTLRY GLLILLKHPE IEEKLHEEID RVIGPSRMPS
cyp2e1h  VTVADLFFAG TETTSTTLRY GLLILMKYPE IEEKLHEEID RVIGPSRIPA

         < J' ><    K helix    >SRS5330          340       350
cyp102   YKQVKQLKYV GMVLNEALRL WPTAPA.FSL YAKEDTVLGGEYPLEKGDELM
cyp2a4   YEDRMKMPYT EAVIHEIQRF ADLIPMGLAR RVTKDTKFRD.FLLPKGTEVF
cyp2a5   YEDRMKMPYT EAVIHEIQRF ADMIPMGLAR RVTKDTKFRD.FLLPKGTEVF
cyp2a6   FEDRAKMPYM EAVIHEIQRF GDVIPMSLAR RVKKDTKFRD.FFLPKGTEVY
cyp2b1   LDDRSKMPYT DAVIHEIQRF SDLVPIGVPH RVTKDTMFRG.YLLPKNTEVY
cyp2b4   LDDRAKMPYT DAVIHEIQRL GDLIPFGVPH TVTKDTQFRG.YVIPKNTEVF
cyp2b6   LHDRAKMPYT EAVIYEIQRF SDLLPMGVPH IVTQHTSFRG.YIIPKDTEVF
cyp2c2   MQDRSRMPYT DATVHEIQRY INLIPNNVPH TTICNLKFRN.YLIPKGTDVL
cyp2c3   SQDRSRMPYT DAVMHEIQRY VDLVPTSLPH AVTQDIEFNG.YLIPKGTDII
cyp2c4   MQDRSHMPYT DAVIHEIQRF IDLLPTNLPH AVTRDVKFRN.YFIPKGTDII
cyp2c5   MQDRSRMPYT DAVIHEIQRF IDLLPTNLPH AVTRDVRFRN.YFIPKGTDII
cyp2c8   MQDRSHMPYT DAVVHEIQRY SDLVPTGVPH AVTTDTKFRN.YLIPKGTTIM
cyp2c9   MQDRSHMPYT DAVVHEVQRY IDLLPTSLPH AVTCDIKFRN.YLIPKGTTIL
cyp2c19  MQDRGHMPYT DAVVHEVQRY IDLIPTSLPH AVTCDVKFRN.YLIPKGTTIL
cyp2d1   MTDQAHMPYT NAVIHEVQRF GDIAPLNLPR FTSCDIEVQD.FVIPKGTTLI
cyp2d6   MGDQAHMPYT TAVIHEVQRF GDIVPLGVTH MTSRDIEVQG.FRIPKGTTLI
cyp2e1r  VRDRVQMPYM DAVVHEIQRF IDLVPSNLPH EATRDTTFQG.YVIPKGTVVI
cyp2e1h  IKDRQEMPYM DAVVHEIQRF ITLVPSNLPH EATRDTIFRG.YLIPKGTVVV

            <K'>          370   < E1 > < F1>       390       400<
cyp102   VLIPQLHRDKTIWGDDVEEFR PERFENPSAI ..PQHAFKPF GNGQRACIGQ
cyp2a4   PMLGSVLKDPKFF.SNPKDFN PKHFLDDKGQ FKKSDAFVPF SIGKRYCFGE
cyp2a5   PMLGSVLKDPKFF.SNPKDFN PKHFLDDKGQ FKKNDAFVPF SIGKRYCFGE
cyp2a6   PMLGSVLRDPSFF.SNPQDFN PQHFLNEKGQ FKKSDAFVPF SIGKRNCFGE
cyp2b1   PILSSALHDPQYF.DHPDSFN PEHFLDANGA LKKSEAFMPF STGKRICLGE
cyp2b4   PVLSSALHDPRYF.ETPNTFN PGHFLDANGA LKRNEGFMPF SLGKRICLGE
cyp2b6   LILSTALHDPHYF.EKPDAFN PDHFLDANGA LKKTEAFIPF SLGKRICLGE
cyp2c2   TSLSSVLHDDKEF.PNPDRFD PGHFLDASGN FRKSDYFMPF STGKRVCVGE
cyp2c3   PSLTSVLYDDKEF.PNPEKFD PGHFLDESGN FKKSDYFMPF STGKRACVGE
cyp2c4   TSLTSVLHDEKAF.PNPKVFD PGHFLDESGN FKKSDYFMPF SAGKRMCVGE
cyp2c5   TSLTSVLHDEKAF.PNPKVFD PGHFLDEGGN FKKSDYFMPF SAGKRMCVGE
cyp2c8   ALLTSVLHDDKEF.PNPNIFD PGHFLDKNGN FKKSDYFMPF SAGKRICAGE
cyp2c9   ISLTSVLHDNKEF.PNPEMFD PHHFLDEGGN FKKSKYFMPF SAGKRICVGE
cyp2c19  TSLTSVLHDNKEF.PNPEMFD PRHFLDEGGN FKKSNYFMPF SAGKRICVGE
cyp2d1   INLSSVLKDETVW.EKPHRFH PEHFLDAQGN FVKHEAFMPF SAGRRACLGE
cyp2d6   TNLSSVLKDEAVW.EKPFRFH PEHFLDAQGH FVKPEAFLPF SAGRRACLGE
cyp2e1r  PTLDSLLYDKQEF.PDPEKFK PEHFLNEEGK FKYSDYFKPF SAGKRVCVGE
cyp2e1h  PTLDSVLYDNQEF.PDPEKFK PEHFLNENGK FKYSDYFKPF STGKRVCAGE
```

Figure 1.3 (continued).

```
               L helix     >420            430  SRS6  440        450
cyp102   QFALHEATLV LGMMLKHFDF E.DHTNYELD IKETLTLKPE GFVVKAKSKKIPLGG
cyp2a4   GLARMELFLF LTNIMQNFHF KSTQAPQDID VSPRLVGFVT .IPPTYTMSFLSR
cyp2a5   GLARMELFLF LTNIMQNFHF KSTQAPQDID VSPRLVGFAT .IPPTYTMSFLSR
cyp2a6   GLARMELFLF FTTVMQNFRL KSSQSPKDID VSPKHVGFAT .IPRNYTMSFLPR
cyp2b1   GIARNELFLF FTTILQNFSV SSHLAPKDID LTPKESGIGK .IPPTYQICFSAR
cyp2b4   GIARTELFLF FTTILQNFSI ASPVPPEDID LTPRESGVGN .VPPSYQIRFLAR
cyp2b6   GIARAELFLF FTTILQNFSM ASPVAPEDID LTPQECGVGK .IPPTYQIRFLPR
cyp2c2   ALARMELFLF LTAILQNFTP KPLVNPNNVD ENPFSSGIVR .VPPLYRVSFIPV
cyp2c3   GLARMELFLL LTTILQHFTL KPLVDPKDID PTPVENGFVS .VPPSYELCFVPV
cyp2c4   GLARMELFLF LTSILQNFKL QSLVEPKDLD ITAVVNGFVS .VPPSYQLCFIPI
cyp2c5   GLARMELFLF LTSILQNFKL QSLVEPKDLD ITAVVNGFVS .VPPSYQLCFIPI
cyp2c8   GLARMELFLF LTTILQNFNL KSVDDLKNLN TTAVTKGIVS .LPPSYQICFIPV
cyp2c9   ALAGMELFLF LTSILQNFNL KSLVDPKNLD TTPVVNGFAS .VPPFYQLCFIPV
cyp2c19  GLARMELFLF LTFILQNFNL KSLIDPKDLD TTPVVNGFAS .VPPFYQLCFIPV
cyp2d1   PLARMELFLF FTCLLQRFSF .SVPVGQPRP STHGFFAFPV .APLPYQLCAVVREQGL
cyp2d6   PLARMELFLF FTSLLQHFSF .SVPTGQPRP SHHGVFAFLV .SPSPYELCAVPR
cyp2e1r  GLARMELFLL LSAILQHFNL KPLVDPEDID LRNITVGFGR .VPPRYKLCVIPRS
cyp2e1h  GLARMELFLL LCAILQHFNL KPLVDPKDID LSPIHIGFGC .IPPRYKLCVIPRS
```

Figure 1.3 (continued).

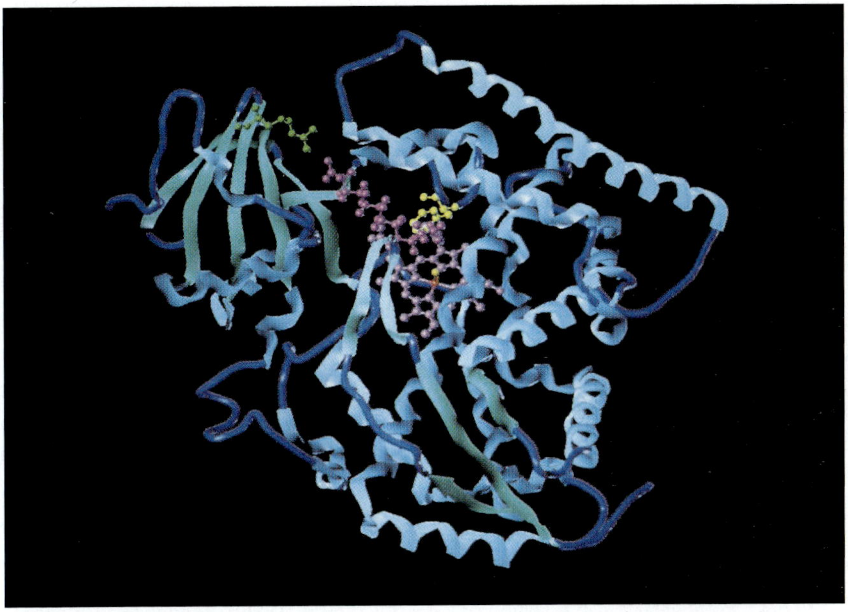

Figure 6.1 Structure of the CYP102 hemoprotein domain showing tertiary and secondary structural elements including α-helices and β-sheets.

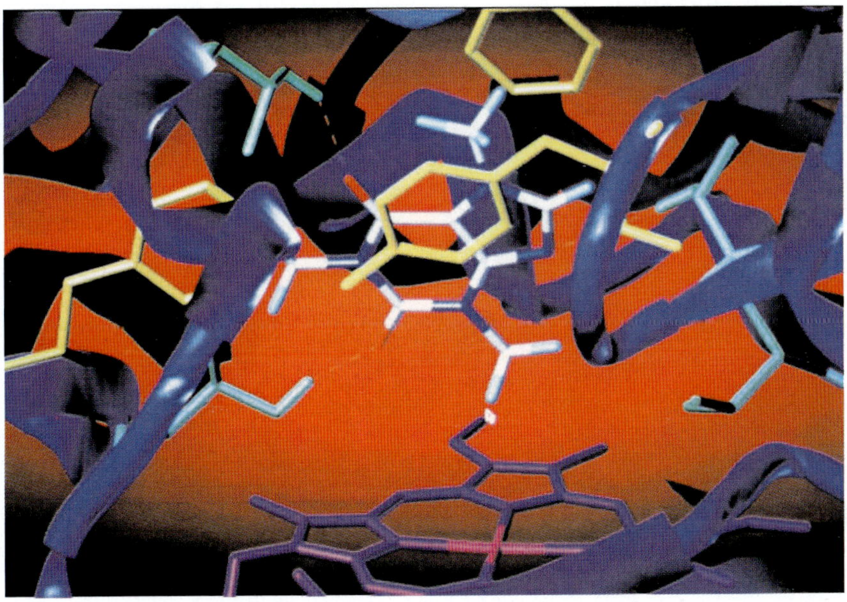

Figure 6.3 The putative active site region of human CYP1A2 containing the substrate caffeine orientated for N-3 demethylation (Lewis et al., 1999c).

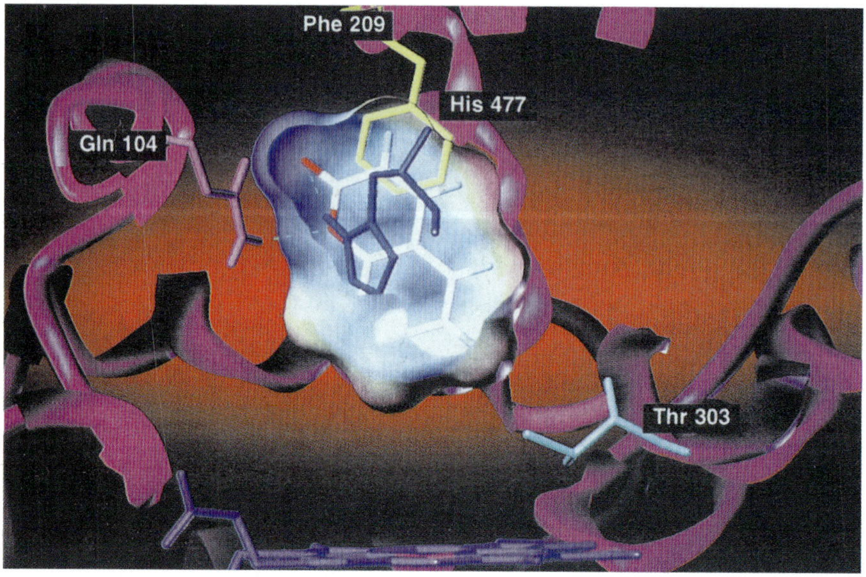

Figure 6.4 Putative active site region of CYP2A6 (Lewis, 1999b) showing the substrate, coumarin, docked close to the heme moiety for 7-hydroxylation.

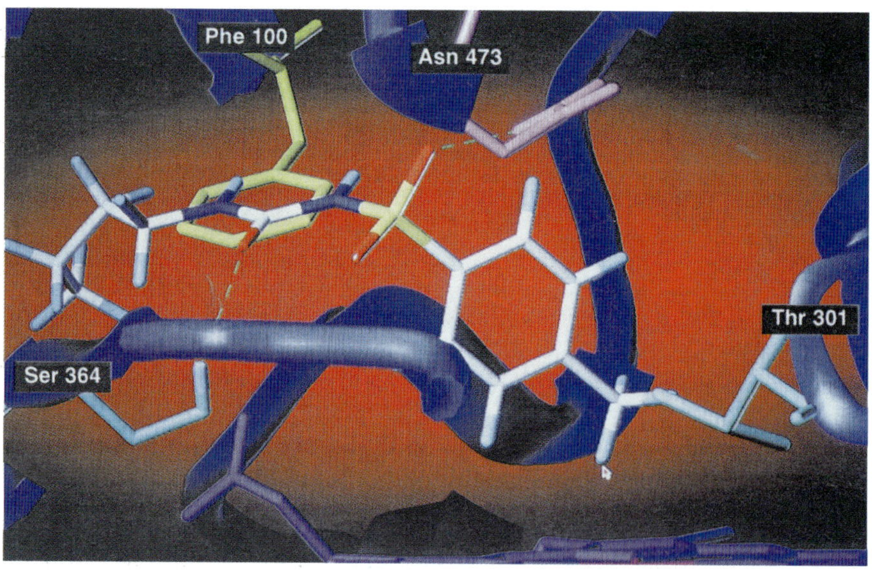

Figure 6.5 Putative active site region of CYP2C9 (Lewis, 1999b) showing the substrate, tolbutamide, docked close to the heme iron for 4-methyl hydroxylation (arrowed).

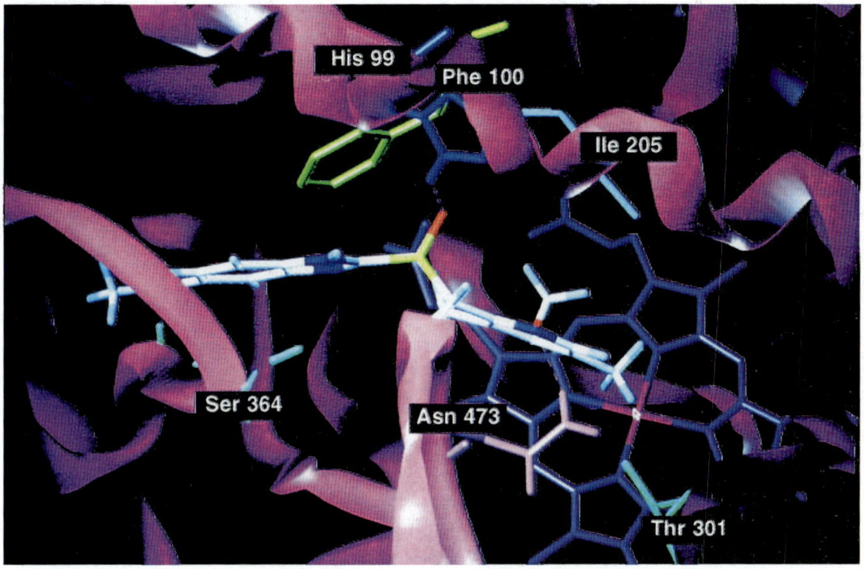

Figure 6.6 Putative active site region of CYP2C19 (Lewis, 1999b) showing the substrate, omeprazole, docked close to the heme group for 5-methyl hydroxylation (arrowed).

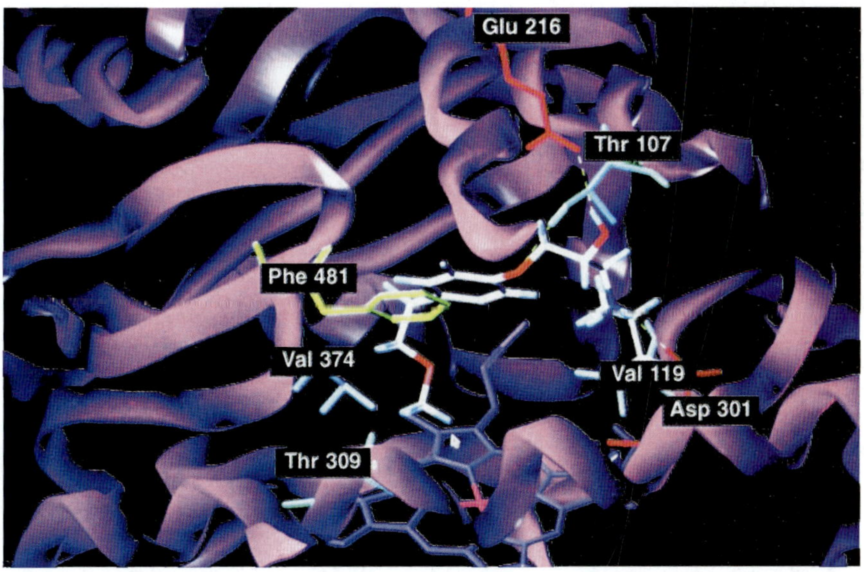

Figure 6.7 Putative active site region of CYP2D6 (Lewis, 1999b) showing the substrate, metoprolol, docked close to the heme group for O-methyl hydroxylation (arrowed).

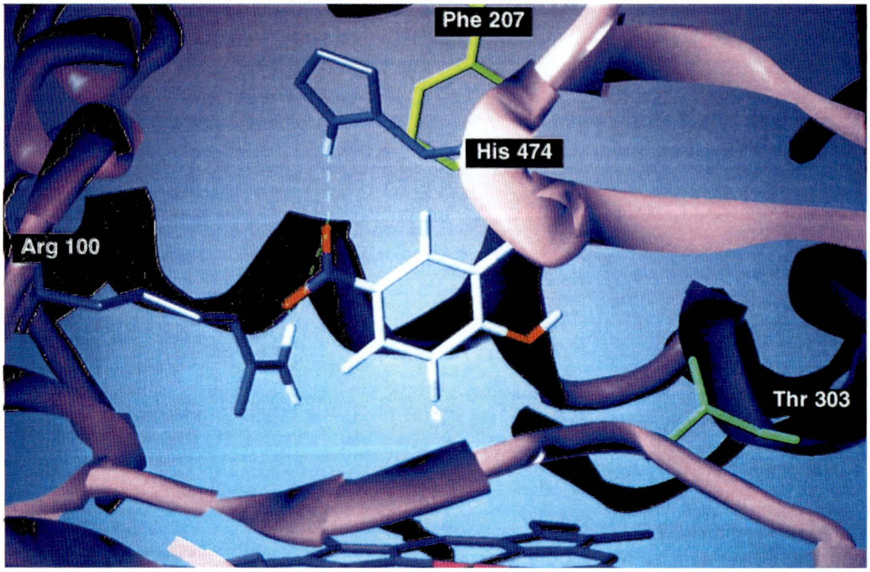

Figure 6.8 Putative active site region of human CYP2E1 (Lewis, 1999b) showing the substrate, *p*-nitrophenol, docked closed to the heme iron for 2-hydroxylation (arrowed).

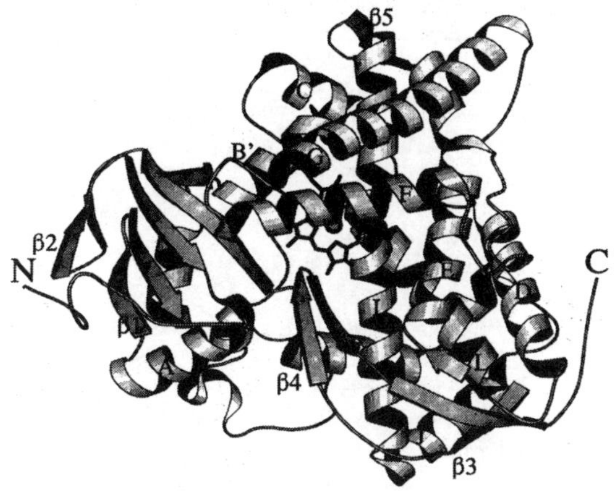

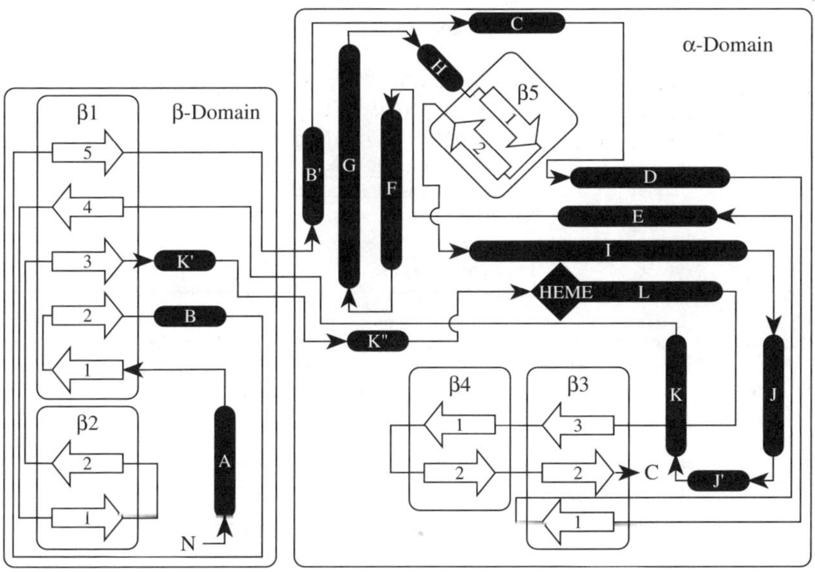

Figure 6.2 Diagrammatic representation of the CYP102 hemoprotein domain showing the location of various structural features (Peterson and Graham-Lorence, 1995). Reproduced from p. 157 of *Cyctochrome P450* (P.R. Ortiz de Montellano, ed.)

clear that this is the preferred molecular template for modelling microsomal P450 enzymes (Lewis, 1996a, 1998a). The relatively low homology of around 20% (although similarity is about 50%) is balanced by the extensive body of information reported from site-specific mutagenesis and other experimental evidence which can be used to generate models of Class II P450s from the CYP102 template,

Table 6.6 Homology models of P450s produced from CYP102

CYP	*Species*	*References*
1A1	rat	Lewis et al., 1994b, 1999c; Lewis and Lake, 1996
1A2	human	Lewis et al., 1999c; Lewis and Lake, 1996; Lozano et al., 1997
1A6	plaice	Lewis et al., 1999c
1B1	human	Lewis et al., 1999c
2A4	mouse	Lewis and Lake, 1995
2A5	mouse	Lewis and Lake, 1995
2A6	human	Lewis and Lake, 1995
2B1	rat	Lewis and Lake, 1997; Lewis, 1995b; Szklarz and Halpert, 1998
2B4	rabbit	Lewis and Lake, 1997; Lewis, 1995b
2B6	human	Lewis et al., 1999e
2C9	human	Lewis et al., 1998c; Payne et al., 1999b; Lewis, 1998a, 1995b
2C18	human	Payne et al., 1999a
2C19	human	Lewis et al., 1998e; Payne et al., 1999a; Lewis, 1998a
2D6	human	Lewis et al., 1997a; Lewis, 1995b, 1998a; de Groot et al., 1996, 1999; Ellis et al., 1995, 1996; Modi et al., 1996a
2E1	rat, mouse and human	Lewis et al., 1997b, 2000a; Tan et al., 1997; Lewis, 1998a
3A4	human	Lewis et al., 1996; Szklarz and Halpert, 1998; Tomlinson et al., 1997
4A1	rat	Lewis and Lake, 1999
4A4	rabbit	Lewis and Lake, 1999; Lewis, 1995b
4A11	human	Lewis and Lake, 1999; Lewis, 1995b
11A1	human	Lewis and Lee-Robichaud, 1998
17A1	human	Lewis and Lee-Robichaud, 1998; Tomlinson et al., 1997
19A1	human	Lewis and Lee-Robichaud, 1998
21A1	human	Lewis and Lee-Robichaud, 1998
51A1	yeast	Lewis et al., 1999a

Note:
Studies other than those of our own group have also employed the CYP102 template in combination with that of CYP101 and CYP108. It should be appreciated also that the final homology models are highly dependent on the nature of the sequence alignment employed.

although portions of other crystallographically resolved P450s may also be included (Jean et al., 1997; de Groot and Vermeulen, 1997; de Groot et al., 1996, 1999). The homology modelling of mammalian and other eukaryotic P450s (see Table 6.6 for further details) from the bacterial crystal structures has been explained previously in some detail (Lewis, 1996a) and will, therefore, only be summarized here in terms of the stages and techniques involved, as follows.

1 *Sequence alignment*
   In this, one should include all available experimental evidence from site-directed mutagenesis and other relevant data in order to produce the best possible alignment between the template sequence (usually that of CYP102) and as many other relevant protein sequences as necessary, including that of the target P450, normally emcompassing all known sequences within the same gene family or subfamily. An example of such alignment is shown in Figure

1.3 and there are several others reported in the literature (Graham-Lorence and Peterson, 1996b; Korzekwa and Jones, 1993; Ouzonis and Melvin, 1991; Lewis, 1996a, 1998a).

2 *Residue replacement*
Using a suitable molecular modelling package, one then systematically modifies the template crystal structure (e.g. substrate-bound CYP102) by changing each amino acid residue required by the alignment to its appropriate counterpart in the target sequence. It is often necessary to delete some residues but also a relatively small number need to be added and this is best achieved via loop-searching of the Protein Databank of known crystal structures for close analogues.

3 *Substrate docking*
Although not essential at this point, it is often helpful to ascertain how well an appropriate substrate (or inhibitor) molecule fits the putative active site of the raw model. Obviously, the most selective substrate available is the preferred choice, and one with a low K_M (or K_D) value usually indicates that the binding interactions are likely to be both specific and of a co-operative nature. The presence of an original bound substrate in the template structure employed (see Table 6.7 for a comparison between active site geometries of substrate-bound and substrate-free CYP102) will facilitate the interactive docking process, and algorithms exist for performing this procedure automatically (De Voss and Ortiz de Montellano, 1996).

4. *Energy minimization*
It is recommended that the raw structure is then energy minimized using an appropriate force field such that any unfavourable contacts are relieved, thus yielding an optimum protein geometry for the new enzyme model. The guide substrate could be removed temporarily for the first few runs of energy minimization via molecular mechanics, and then replaced once the protein has attained a satisfactory minimum energy value. Some investigators employ molecular dynamics (MD) simulations as part of the overall strategy for geometry optimization, and it is possible to use this techinque at various stages in structural refinement of a protein model. Although a discussion of MD is beyond the scope of this account, the interested reader is referred to the excellent work of Gilda Loew and colleagues (Harris and Loew, 1994, 1995; Fruetel et al., 1994; Payne et al., 1999a and b), Ornstein and co-workers (Paulsen et al., 1996; Paulsen and Ornstein, 1992, 1995; Braatz et al., 1994; Filipovic et al., 1992), together with those of others (Dai et al., 1999; Lozano et al., 1997) for further and more detailed information. Dynamic aspects of P450 substrate binding have been observed experimentally for CYP3A4 (Koley et al., 1995, 1996) from CO-binding kinetics and there is strong evidence for allosteric binding in this enzyme in particular (Ueng et al., 1997).

5. *Model analysis*
It is advisable that the energy-minimized model is analysed using ProCheck or ProTable to ensure that the newly derived structure conforms to the generally

Table 6.7 Comparison between heme geometries in CYP102

Property	*Substrate-bound*	*Substrate-free*
Fe–S bond length	2.232 Å	2.114 Å
Fe–N_A bond length	2.002 Å	2.043 Å
Fe–N_B bond length	1.965 Å	1.983 Å
Fe–N_C bond length	1.974 Å	1.979 Å
Fe–N_D bond length	1.956 Å	1.967 Å
Fe–N (average) bond length	1.974 Å	1.993 Å
Fe–OH_2 bond length	–	2.118 Å
Fe–S–C (Cys400) bond angle	123.8°	104.5°
N_D–Fe–S–C bond angle	11.5°	17.9°
Iron state	Fe(III) high-spin	Fe(III) low-spin
Heme symmetry	S_4 'ruffled' (C_{4v})	planar (Oh)
Protein Databank file	1fag	2hpd

References: Ravichandran et al., 1993; Li and Poulos, 1997.

Notes:
The bound substrate molecule employed in the 1fag structure is that of palmitoleic acid which is known to undergo ω-2 hydroxylation by CYP102. NMR measurements indicate that the substrate moves closer to the heme iron following reduction of the enzyme (Modi et al., 1996b).

Similar comparisons between substrate-bound and substrate-free CYP101, together with the carbon monoxide adduct are presented in Lewis, 1996a.

established criteria governing protein geometries. If the new model fails to meet these requirements, it will have to be further refined until the geometry is acceptable. Additionally, models should be consistent with all available experimental evidence from site-directed mutagenesis and from the known substrate selectivity of the enzyme, in that active site contacts orientate substrate molecules for metabolism at the experimentally observed positions (Szklarz et al., 1995, 2000; Halpert et al., 1998; Szklarz and Halpert, 1997, 1998 provide good examples of such an approach). There is now sufficient information available (especially for the human drug-metabolizing enzymes) for it to be standard practice that all new P450 models are thoroughly investigated in this way. Indeed, one should also expect that a molecular template of selective substrates will be able to fit into the putative active site of the new P450 model such that the enzyme's specificity is rationalized in terms of favourable interactions with amino acid sidechains in the heme environment (see, for example, Lewis, 2000b). In any event, some form of testing and at least partial validation of the model is advisable prior to its employment for screening novel compounds.

One such example is presented in Figure 6.3 which shows the putative reactive centre of human CYP1A2 with caffeine docked relative to the heme moiety for N_3-demethylation which is the major route of metabolism catalysed by this enzyme. It can be appreciated from an inspection of this figure that the substrate molecule is positioned for N-demethylation by a combination of three hydrogen bonds with nearby active site threonine residues (Lewis et al., 1999c). There is also a degree of structural complementarity between the caffeine substrate and the CYP1A2 putative binding site, such that the relatively planar substrate molecule is 'sandwiched' between two aromatic amino acid residues (actually, a tyrosine and phenylalanine). Importantly, a number of these active site residues have been the subject of mutagenesis (Parikh et al., 1999) and chemical modification (Cvrk and Strobel, 1998) studies on CYP1A enzymes, where their location within the CYP1A2 heme pocket has been strongly indicated. Furthermore, it can be demonstrated that the caffeine molecule is able to adopt three alternative orientations in the active site such that all known N-demethylations are possible. In each case, the substrate forms three hydrogen bonds with the aforementioned threonines thus explaining the observed metabolite profile for caffeine (Lewis et al., 1999c). Interestingly, other substrates of CYP1A2 are also relatively planar in structure and possess hydrogen bond acceptor atoms in analogous positions to those on caffeine itself and, therefore it is possible to rationalize the selectivity of this enzyme for substrates of that structural type (Lewis et al., 1999c). Figure 6.4 through to Figure 6.8 provide examples of active site regions for enzymes in the CYP2 family, namely CYP2A6 (Figure 6.4), CYP2C9 (Figure 6.5), CYP2C19 (Figure 6.6), CYP2D6 (Figure 6.7) and CYP2E1 (Figure 6.8), which have been described previously (Lewis, 1998b).

Another way of investigating the validity of a newly derived P450 model is through the technique of QSAR analysis (see Rao et al., 2000 for example). In order to do this, one requires a set of biological data to have been generated on either a series of substrates for the same enzyme or for mutagenesis studies to have been conducted on one or more amino acid residues thought to be present in the binding site and, thus, likely to have an effect on catalytic activity (see Szklarz et al., 2000; Szklarz and Halpert, 1998 for examples). It is then necessary to establish which combination of structural parameters for either the substrates or amino acids concerned are able to provide a satisfactory explanation of the biological findings (Lewis, 2000b). In theory, this information should support the model itself and also provide some insight into the molecular mechanisms involved in catalytic activity, together with advancing understanding of the structural roles of key amino acid residues. For Thr252 mutants in $P450_{cam}$ (CYP101) and for the Phe209 mutations in $P450_{coh}$ (CYP2A5) for example, it is found that the relative hydrophobicity of the amino acid residue is an important factor, in addition to its molecular mass (Lewis and Pratt, 1998; Lewis et al., 1999b). These and other QSAR studies on P450 substrates and inhibitors are discussed in the section concerned with structural aspects of P450 substrate selectivity in Chapter 4, and the reader is also referred to several independent QSAR and modelling studies in the P450 field (Fuhr et al., 1993; Jones et al., 1993; Korzekwa and Jones, 1993; Poli-Scaife et al., 1997; Strobl et al., 1993; Waller et al., 1996).

6.6 Molecular orbital calculations

Electronic structure calculations using molecular orbital (MO) and other related quantum-mechanical procedures (Leach, 1996) have been applied to various aspects of the P450 system, including substrates, inhibitors and even to a limited region of the active site in the immediate vicinity of the heme moiety (Segall et al., 1998; de Groot et al., 1998).

The early work by Hanson and colleagues (Hanson et al., 1976, 1977) utilized extended Hückel MO theory to calculate the electronic structure of the localized heme environment of P450, thus demonstrating that the anomalous Soret spectrum of the CO adduct is due to the presence of a heme-thiolate complex. Later, Jung and co-workers investigated the carbonmonoxy-cytochrome P450 system via INDO calculations (Jung, 1985; Rein et al., 1984; Jung et al., 1992) confirming the previous study using another MO procedure and showing that the RR oxidation state marker band frequency (1360–1380 cm^{-1}) correlates with net atomic charge on the heme porphyrin ring. Other MO investigations have also been conducted on the P450 heme moiety by Loew and others (Loew et al., 1977; Loew and Kirchner, 1975; Harris and Loew, 1995; Fruetel et al., 1994; Loew and Harris, 2000) which provide important theoretical evidence for the catalytic properties of P450 enzymes with respect to substrate metabolism, together with explaining further various spectral characteristics of the P450 system. More recently, two independent studies employing different *ab initio* quantum-chemical procedures have afforded profound insights into the electronic status of the P450 heme-thiolate assembly whilst undergoing iron spin-state changes (de Groot et al., 1998; Segall et al., 1998). Such approaches advance significantly our understanding of the bio-inorganic chemistry of the P450 catalytic centre.

The potential ability to predict, or at least rationalize, rates of P450-mediated metabolism in substrates of diverse structure has led to a number of investigations reporting good correlations between electronic parameters such as highest occupied MO energies (Cnubben et al., 1994; Yin et al., 1995) and it is also apparent that the relative ease of hydrogen atom abstraction in substrate molecules gives an indication of the rank order for P450 oxidations (Smith et al., 1998). Indeed, the individual HOMO electron densities may provide a means of establishing likely sites of metabolism, as has been shown for CYP2D6 substrates (Ackland, 1993) whereas electronic structure calculations on porphyrin–substrate complexes indicate that the preferred orientations of substituted anilines for oxidation in various positions are not parallel to the porphyrin ring plane (Zakharieva et al., 1998). The reader is referred to a recent compilation of frontier orbital-generated QSARs (Lewis, 1999a) for further information in this area, and the work of Smith and colleagues on P450 substrate structure–activity relationships (SSARs) is also strongly recommended reading (Smith and Jones, 1992; Smith et al., 1997a and b).

6.7 Conclusions

The structures of P450 enzymes have been investigated by many techniques, both experimental and theoretical, and this has enabled an increasingly clear picture to emerge of the forces underpinning the functionality of these mono-oxygenases. Molecular modelling interfaces closely with the complementary approaches of X-ray crystallography, spectroscopy and molecular biology as they have applied to the P450 system, thus helping to explain both the catalytic and other properties of these hemoproteins at the molecular and even atomic levels. In this way, one is steadily achieving a better comprehension of such biological phenomena as drug metabolism and toxic activation of xenobiotics, where P450s are involved to a major extent, such that future generations can largely enjoy a relatively safer environment in which novel pharmaceuticals, for example, will have been designed and tested specifically for their optimized beneficial effect on human subjects.

Glossary of terms

Active site: The reactive centre of an enzyme where the substrate molecule binds and is then modified via some form of chemical reaction, catalysed by the enzyme, in order to produce a metabolite. For P450s, the active site region is in a hydrophobic pocket lying distal to the heme prosthetic group whereby oxygen is activated by the enzyme.

Allelic variant: A genetic variant of an enzyme which contains a different amino acid sequence and usually gives rise to altered catalytic properties, often lower than the wild-type activity. This is related to polymorphisms which are variations in enzyme activity occurring within a given population due to genetic differences in the relevant enzyme, usually caused by amino acid changes in the protein sequence.

Cytochrome P450: Heme-thiolate enzyme for which the abbreviation is CYP, or P450 (but not CYP450). There is no hyphen between 'P' and '450'. The plural form is cytochromes P450, P450s or CYPs. There is no apostrophe between 'P450' and 's' as P450 is not a possessive noun. Some journals still retain earlier nomenclature. Italics are used when referring to the relevant P450 gene, and mouse P450 gene names are in lower case.

Genotype: The genetic constitution of an organism or individual. In the P450 field, some individuals within a population may possess altered genes encoding for specific P450s which can render 'poor-metabolizer' status in respect of certain drug clearance properties.

Induction: An increase in the expression of a gene in response to a change in the activity of a regulatory protein. In the case of P450s, there are some mechanisms of induction which do not involve a regulatory protein, such as a nuclear receptor.

Iso(en)zymes: Multiple forms of an enzyme which catalyse the same type of reaction but differ from each other in terms of their amino acid sequence, substrate affinity/selectivity and/or regulatory properties. Although the terms 'isozymes' and 'isoforms' are frequently used to describe P450s, it is better to use 'enzymes' as a preferable description because different P450s tend to exhibit altered substrate selectivities and thus catalyse different reactions.

Phenotype: The observable physical characteristics of an organism or individual. For example, 'poor-metabolizers' and 'extensive-metabolizers' of a particular chemical are phenotypes, and this condition may be due to allelic variants of specific P450s.

Polymorphism: A genetic defect in metabolism which is often associated with a mutant form of the normal drug-metabolizing enzyme (DME), such as those occurring in CYP2D6 and CYP2C19 in the case of human hepatic P450s. Individuals showing poor-metabolizer (PM) status, as opposed to the normal (i.e. wild-type) extensive-metabolizer (EM) phenotype, are likely to possess defective DMEs.

Quantitative structure–activity relationship (QSAR): A statistically valid mathematical expression describing the variation in biological potency within a series of chemicals in terms of some combination of their structural properties, including molecular, electronic and physicochemical descriptors.

Selectivity/Specificity: The ability of an enzyme (or receptor) to discriminate between different substrates (or ligands). It has been established that the specificity of P450s towards certain substrates is not quite so exclusive as originally thought. Consequently, it is better to refer to P450 substrate selectivity rather than specificity although, with some enzymes, substrate preferences can reasonably be termed specificity (e.g. steroidogenic P450s).

Substrate: A chemical compound which is metabolized by an enzyme to form a product. In the case of P450, most reactions involve mono-oxygenation of the substrate to form a hydroxylated metabolite, such as an alcohol or phenol. However, many other reaction types are catalysed by P450 enzymes and there are literally thousands of known substrates.

Bibliography

Archakov, A.I. and Bachmanova, G.T., 1990, *Cytochrome P-450 and Active Oxygen*, London: Taylor & Francis.

Gibson, G.G. and Skett, P., 1994, *Introduction to Drug Metabolism*, London: Chapman & Hall.

Guengerich, F.P. (Ed.), 1987, *Mammalian Cytochromes P-450*, Boca Raton, FL: CRC.

Ioannides, C. (Ed.), 1996, *Cytochromes P450: Metabolic and Toxicological Aspects*, Boca Raton, FL: CRC.

Lewis, D.F.V., 1996, *Cytochromes P450: Structure, Function and Mechanism*, London: Taylor & Francis.

Omura, T., Ishimura, Y. and Fujii-Kuriyama, Y. (Eds), 1993, *Cytochrome P-450*, Tokyo: Kodansha.

Ortiz de Montellano, P.R. (Ed.), 1986, *Cytochrome P-450*, New York: Plenum.

Ortiz de Montellano, P.R. (Ed.) 1996, *Cytochrome P450* (2nd edition), New York: Plenum.

Ruckpaul, K. and Rein, H. (Eds), 1984, *Cytochrome P-450*, Berlin: Akademie-Verlag.

Sato, R. and Oura, T., 1976, *Cytochrome P-450*, New York: Academic Press.

Schenkman, J.B. and Griem, H. (Eds), 1993, *Cytochrome P450*, Berlin: Springer-Verlag.

Schenkman, J.B. and Kupfer, D. (Eds), 1982, *Hepatic Cytochrome P-450 Mono-Oxygenase System*, Oxford: Pergamon.

The following multi-volume book series contains a considerable amount of helpful background information on P450 and related subjects:

Frontiers in Biotransformation, edited by K. Ruckpaul and H. Rein, Akademie-Verlag, Berlin.

Volume 1. *Basis and Mechanisms of Regulation of Cytochrome P-450*, 1989.

Volume 2. *Principles, Mechanisms and Biological Consequences of Induction*, 1990.

Volume 3. *Molecular Mechanisms of Adrenal Steroidogenesis and Aspects of Regulation and Application*, 1990.

Volume 4. *Microbial and Plant Cytochromes P-450: Biochemical Characteristics, Genetic Engineering and Practical Implications*, 1991.

Volume 5. *Membrane Organization and Phospholipid Interaction of Cytochrome P-450*, 1991.

Volume 6. *Cytochrome P-450 Dependent Biotransformation of Endogenous Substrates*, 1991.

Volume 7. *Relationships between Structure and Function of Cytochrome P-450: Experiments, Calculations, Models*, 1992.
Volume 8. *Medicinal Implications in Cytochrome P-450 Catalyzed Biotransformations*, 1993
Volume 9. *Regulation and Control of Complex Biological Processes by Biotransformation*, 1994.

An entire volume of Methods in *Enzymology*, Volume 272, 1996 is devoted to recent developments in the P450 field and a special issue of *Comparative Biochemistry and Physiology*, Volume 121C, 1998 is also useful for information on P450s in non-mammalian species.

References

Ackland, M.J. (1993) Correlation between site specificity and electrophilic frontier values in the metabolic hydroxylation of aromatic substrates: a molecular modelling study, *Xenobiotica*, **23**, 1135–1144.

Addison, A.W. and Burman, S. (1985) Ligand-dependent redox chemistry of *Glycera dibranchiata* hemoglobin, *Biochimica et Biophysica Acta*, **828**, 362–368.

Ailhaud, G. (1999) Cell surface receptors, nuclear receptors and ligands that regulate adipose tissue development, *Clinica Chemica Acta*, **286**, 181–190.

Aithal, G.P., Day, C.P., Kesteven, P.J.L. and Daly, A.K. (1999) Association of polymorphisms in the cytochrome P450 CYP2C9 with warfarin dose requirement and risk of bleeding complications, *Lancet*, **353**, 717–719.

Akhtar, M., Njar, V.C.O. and Wright, J.N. (1993) Mechanistic studies on aromatase and related C-C bond cleaving P450 enzymes, *Journal of Steroid Biochemistry and Molecular Biology*, **44**, 375–387.

Akhtar, M., Corina, D., Miller, S., Shyadehi, A.Z. and Wright, J.N. (1994) Mechanism of the acyl-carbon cleavage and related reactions catalyzed by multifunctional P450s: studies on cytochrome $P450_{17\alpha}$, *Biochemistry*, **33**, 4410–4418.

Aldridge, T.C., Tugwood, J.D. and Green, S. (1995) Identification and characterization of DNA elements implicated in the regulation of CYP4A1 transcription, *Biochemical Journal*, **306**, 473–479.

Alexidis, A.N., Rekka, E.A. and Kovrounakis, P.N. (1994) Influence of mercury and cadmium intoxication on hepatic microsomal CYP2E and CYP3A subfamilies, *Research Communications in Molecular Pathology and Pharmacology*, **85**, 67–71.

Alterman, M.A., Chaurasia, C.S., Lu, P., Hardwick, J.P. and Hanzlik, R.P. (1995) Fatty acid discrimination and ω-hydroxylation by cytochrome P450 4A1 and a cytochrome P450 4A1/NADPH-P450 reductase fusion protein, *Archives of Biochemistry and Biophysics*, **320**, 289–296.

Alvares, A.P. and Pratt, W.B. (1990) Pathways of drug metabolism, in: *Principles of Drug Action* (W.B. Pratt and P. Taylor, eds.) Churchill-Livingstone, New York, Chapter 5, 365–422.

Ameer, B. and Weintraub, R.A. (1997) Drug interactions with grapefruit juice, *Clinical Pharmacokinetics*, **33**, 103–121.

Amet, Y., Lucas, D., Zhang-Gouillon, Z.-Q. and French, S.W. (1998) P450-dependent metabolism of lauric acid in alcoholic liver disease: comparison between rat liver and kidney microsomes, *Alcoholism: Clinical and Experimental Research*, **22**, 455–462.

Andersen, J.F., Tatsuta, K., Gunji, H., Ishiyama, T. and Hutchinson, C.R. (1993) Substrate specificity of 6-deoxyerythronolide B hydroxylase, a bacterial cytochrome P450 of erythromycin A biosynthesis, *Biochemistry*, **32**, 1905–1913.

Angus, W.G.R., Larsen, M.C. and Jefcoate, C.R. (1999) Expression of CYP1A1 and CYP1B1 depends on cell-specific factors in human breast cancer cell lines: role of estrogen receptor status, *Carcinogenesis*, **20**, 947–955.

Anzenbacher, P. and Anzenbacherova, E. (2000) Cytochromes P450 and metabolism of xenobiotics, *Cellular and Molecular Life Sciences*, in press.

Archakov, A.I. and Bachmanova, G.I. (1990) *Cytochrome P450 and Active Oxygen*, Taylor & Francis, London.

Arinc, E. (1993) Extrahepatic microsomal forms: lung microsomal cytochrome P450 isoenzymes, in: *Cytochromes P450* (J.B. Schenkman and H. Griem, eds.) Springer-Verlag, Berlin, 373–386.

Astroff, B., Zacharewski, T., Safe, S., Arlotto, M.P., Parkinson, A., Thomas, P. and Levin, W. (1988) 6-Methyl-1,3,8-trichlorodibenzofuran as a 2,3,7,8-tetrachlorodibenzo-p-dioxin antagonist: inhibition of the induction of rat cytochrome P450 isoenzymes and related monooxygenase activities, *Molecular Pharmacology*, **33**, 231–236.

Atkins, W.M. and Sligar, S.G. (1988) The roles of active site hydrogen bonding in cytochrome $P450_{cam}$ as revealed by site-directed mutagenesis, *Journal of Biological Chemistry*, **263**, 18842–18849.

Atkins, W.M. and Sligar, S.G. (1989) Molecular recognition in cytochrome P450: alteration of regioselective alkane hydroxylation via protein engineering, *Journal of the American Chemical Society*, **111**, 2715–2717.

Atkins, W.M. and Sligar, S.G. (1990) Tyrosine-96 as a natural spectroscopic probe of the cytochrome P450cam active site, *Biochemistry*, **29**, 1271–1275.

Averof, M. and Cohen, S.M. (1997) Evolutionary origin of insect wings from ancestral gills, *Nature*, **385**, 627–630.

Bachmann, K.A. (1996) The cytochrome P450 enzymes of hepatic drug metabolism: how are their activities assessed *in vivo*, and what is their clinical relevance?, *American Journal of Therapeutics*, **3**, 150–171.

Backes, W.L. and Canady, W.J. (1981) Methods for the evaluation of hydrophobic substrate binding to cytochrome P450, *Pharmacology and Therapeutics*, **12**, 133–158.

Backes, W.L., Batie, C.J. and Cawley, G.F. (1998) Interactions among P450 enzymes when combined in reconstituted systems: formation of a 2B4–1A2 complex with a high affinity for NADPH-cytochrome P450 reductase, *Biochemistry*, **37**, 12852–12859.

Bäckström, D., Ingelman-Sundberg, M. and Ehrenberg, A. (1983) Oxidation-reduction potential of soluble and membrane-bound rabbit liver microsomal cytochrome P450LM2, *Acta Chemica Scandinavia*, **B37**, 891–894.

Baes, M., Gulick, T., Choi, H.-S., Martinoli, M.G., Simha, D. and Moore, D.D. (1994) A new orphan member of the nuclear hormone receptor superfamily that interacts with a subset of retinoic acid response elements, *Molecular and Cellular Biology*, **14**, 1544–1552.

Baldwin, J.E., Morris, G.M. and Richards, W.G. (1991) Electron transport in cytochromes P450 by covalent switching, *Proceedings of the Royal Society of London, Series B*, **245**, 43–51.

Bast, A. (1986) Is formation of reactive oxygen by cytochrome P450 perilous and predictable?, *Trends in Pharmacological Science*, **7**, 266–270.

Becquemont, L., Le Bot, M.A., Riche, C., Funck-Brentano, C., Jaillon, P. and Beaune, P. (1998) Use of heterologously expressed human cytochrome P450 1A2 to predict tacrine-fluvoxamine drug interaction in man, *Pharmacogenetics*, **8**, 101–108.

Bell, L.C. and Guengerich, F.P. (1997) Oxidation kinetics of ethanol by human cytochrome P450 2E1, *Journal of Biological Chemistry*, **272**, 29643–29651.

Beresford, A.P. (1993) CYP1A1: friend or foe?, *Drug Metabolism Reviews*, **25**, 503–517.

Bernhardt, R. (1993) Chemical probes of cytochrome P450 structure, in: *Cytochrome P450* (J.B. Schenkman and H. Griem, eds.) Springer-Verlag, Berlin, 547–560.

Bernhardt, R. (1995) Cytochrome P450: structure, function and generation of reactive oxygen species, *Reviews of Physiology, Biochemistry and Pharmacology*, **127**, 137–221.

Bernhardt, R., Makower, A., Jänig, G.-R. and Ruckpaul, K. (1984) Selective chemical modification of a functionally linked lysine in cytochrome P450 LM2, *Biochimica et Biophysica Acta*, **785**, 186–190.

Bernhardt, R., Kraft, R., Otto, A. and Ruckpaul, K. (1988) Electrostatic interactions between cytochrome P450 LM2 and NADPH-cytochrome P450 reductase, *Biomedica et Biophysica Acta*, **47**, 581–592.

Bertilsson, G., Heidrich, J., Svensson, K., Åsman, M., Jendeberg, L., Sydow-Bäckman, M., Ohlsson, R., Postlind, H., Blomquist, P. and Berkenstam, A. (1998) Identification of a human nuclear receptor defines a new signaling pathway for CYP3A induction, *Proceedings of the National Academy of Sciences USA*, **95**, 12208–12213.

Black, S.D. (1992) Membrane topology of the mammalian P450 cytochromes, *FASEB Journal*, **6**, 680–685.

Black, S.D. and Coon, M.J. (1982) Structural features of liver microsomal NADPH-cytochrome P450 reductase, *Journal of Biological Chemistry*, **257**, 5929–5938.

Black, S.D. and Coon, M.J. (1986) Comparative structures of P450 cytochromes, in: *Cytochrome P450* (P.R. Ortiz de Montellano, ed.) Plenum, New York, Chapter 6, 161–216.

Blake, R.C. and Coon, M.J. (1989) On the mechanism of action of cytochrome P450, *Journal of Biological Chemistry*, **264**, 3694–3701.

Blanck, J., Rein, H., Sommer, M., Ristau, O., Smettan, G. and Ruckpaul, K. (1983) Correlations between spin equilibrium shift, reduction rate and N-demethylation activity in liver microsomal cytochrome P450 and a series of benzphetamine analogues as substrates, *Biochemical Pharmacology*, **32**, 1683–1688.

Blanck, J., Ristau, O., Zhukov, A.A., Archakov, A.I., Rein, H. and Ruckpaul, K. (1991) Cytochrome P450 spin state and leakiness of the monooxygenase pathway, *Xenobiotica*, **21**, 121–135.

Blumberg, B., Sabbagh, W., Juguila, H., Bolado, J., van Meter, C.M., Ong, E.S. and Evans, R.M. (1998) SXR, a novel steroid and xenobiotic-sensing nuclear receptor, *Genes and Development*, **12**, 3195–3205.

Boddupalli, S.S., Pramanik, B.C., Slaughter, C.A., Estabrook, R.W. and Peterson, J.A. (1992) Fatty acid monooxygenation by $P450_{BM3}$: product identification and proposed mechanisms for the sequential hydroxylation reactions, *Archives of Biochemistry and Biophysics*, **292**, 20–28.

Boobis, A.R., Sesardic, D., Murray, B.P., Edwards, R.J., Singleton, A.M., Rich, K.H., Murray, S., De la Torre, R., Segura, J., Pelkonen, O., Padanen, M., Kobayashi, S., Zhi-Guang, T. and Davies, D.S. (1990) Species variation in the response of the cytochrome

P450-dependent monooxygenase system to inducers and inhibitors, *Xenobiotica*, **20**, 1139–1161.

Bornheim, L.M., Peters, P.G. and Franklin, M.R. (1983) The induction of multiple forms of cytochrome P450 by SKF-525A, *Chemico-Biological Interactions*, **47**, 45–55.

Bourget, W., Ruff, M., Chambon, P., Gronemeyer, H, and Moras, D. (1995) Crystal structure of the ligand binding domain of the human nuclear receptor RXRα, *Nature*, **375**, 377–382.

Bourian, M., Gullsten, H. and Legrum, W. (2000) Genetic polymorphism of CYP2A6 in the German population, *Toxicology*, **144**, 129–137.

Bozak, K.R., Yu, H., Sirevag, R. and Christoffersen, R.E. (1990) Sequence analysis of ripening-related cytochrome P450 cDNAs from avocado fruit, *Proceedings of the National Academy of Sciences USA*, **87**, 3904–3908.

Braatz, J.A., Bass, M.B. and Ornstein, R.L. (1994) An evaluation of molecular models of the cytochrome P450 *Streptomyces griseolus* enzymes P450SU1 and P450SU2, *Journal of Computer-Aided Molecular Design*, **8**, 607–622.

Branden, C. and Tooze, J. (1991) *Introduction to Protein Structure*, Garland, New York.

Brasier, M.D. and McIlroy, D. (1998) *Neonereites uniserialis* from c. 600 Ma year old rocks in western Scotland and the emergence of animals, *Journal of the Geological Society, London*, **155**, 5–12.

Brown, C.A. and Black, S.D. (1989) Membrane topology of mammalian cytochromes P450 from liver endoplasmic reticulum, *Journal of Biological Chemistry*, **264**, 4442–4449.

Brown, D. (2000) P450: enzymes with the answers on drug risks, *The Washington Post*, April 10, p.A9.

Brzozowski, A.M., Pike, A.C.W., Dauter, Z., Hubbard, R.E., Bonn, R., Engstrom, O., Ohman, L., Greene, G.L., Gustafsson, J.-A. and Carlquist, M. (1997) Molecular basis of agonism and antagonism in the oestrogen receptor, *Nature*, **389**, 753–758.

Burbach, K.M., Poland, A. and Bradfield, C.A. (1992) Cloning of the Ah receptor cDNA reveals a distinctive ligand-activated transcription factor, *Proceedings of the National Academy of Sciences USA*, **89**, 8185–8189.

Burke, M.D. and Mayer, R.T. (1983) Differential effects of phenobarbitone and 3-methylcholanthrene induction in the hepatic microsomal metabolism and cytochrome P450 binding of phenoxasone and a homologous series of its n-alkyl ethers (alkoxyresorufins), *Chemico-Biological Interactions*, **45**, 243–258.

Byfield, M.P., Hamza, M.S.A. and Pratt, J.M. (1993) Hemes and hemoproteins. Part 8. Coordination of amines and amino acids by the iron (III) porphyrin microperoxidase-8, *Journal of the Chemical Society*, Dalton Transactions, 1641–1645.

Campbell, B. (1998) The use of kinetic and metabolic predictions in drug development, in: *Drug Metabolism Towards the Next Millenium* (N. Gooderham, ed.) IOS Press, Amsterdam, 144–158.

Canfield, D.E. and Teske, A. (1996) Late Proterozoic rise in atmospheric oxygen concentration inferred from phylogenetic and sulphur-isotope studies, *Nature*, **382**, 127–132.

Capdevila, J.H., Falck, J.R. and Estabrook, R.W. (1992) Cytochrome P450 and the arachidonate cascade, *FASEB Journal*, **6**, 731–736.

Capdevila, J.H., Zeldin, D., Makita, K., Karara, A. and Falck, J.R. (1995) Cytochrome P450 and the metabolism of arachidonic acid and oxygenated eicosanoids, in:

Cytochrome P450 (P.R. Ortiz de Montellano, ed.) Plenum, New York, Chapter 13, 443–471.

Causevic, M., Wolf, C.R. and Palmer, C.N.A. (1999) Substitution of a conserved amino acid residue alters the ligand binding properties of peroxisome proliferator activated receptors, *FEBS Letters*, **463**, 205–210.

Champion, P.M. (1989) Elementary electronic excitations and the mechanism of cytochrome P450, *Journal of the American Chemical Society*, **111**, 3433–3434.

Champion, P.M., Stallard, B.R., Wagner, G.C. and Gunsalus, I.C. (1982) Resonance Raman detection of an Fe-S bond in cytochrome $P450_{cam}$, *Journal of the American Chemical Society*, **104**, 5469–5473.

Chang, T.K.H. and Waxman, D.J. (1996) The CYP2A Subfamily in: *Cytochromes P450: metabolic and toxicological aspects* (C. Ioannides, ed.) CRC Press, Boca Raton, FL, Chapter 5, 99–134.

Chapple, C. (1998) Molecular genetic analysis of plant cytochrome P450-dependent monooxygenases, *Annual Review of Plant Physiology and Plant Molecular Biology*, **49**, 311–343.

Chen, Q., Galleano, M. and Cederbaum, A.I. (1997) Cytotoxicity and apoptosis produced by arachidonic acid in HepG2 cells overexpressing human cytochrome P4502E1, *Journal of Biological Chemistry*, **272**, 14532–14541.

Chen, S. and Zhou, D. (1992) Functional domains of aromatase cytochrome P450 inferred from comparative analysis of amino acid sequences and substantiated by site-directed mutagenesis experiments, *Journal of Biological Chemistry*, **267**, 22587–22594.

Chen, S., Besman, M.-J., Shively, J.E., Yanagibashi, K. and Hall, P.F. (1989) Human aromatase, *Drug Metabolism Reviews*, **20**, 511–517.

Chin, D.-H., Chiericato, G., Nanni, E.J. and Sawyer, D.T. (1982) Proton-induced disproportionation of superoxide ion in aprotic media, *Journal of the American Chemical Society*, **104**, 1296–1299.

Cho, A.K. and Miwa, G.T. (1974) The role of ionization in the N-demethylation of some N,N-dimethyl amines, *Drug Metabolism and Disposition*, **2**, 477–483.

Choi, H.-S., Chung, M., Tzameli, I., Simha, D., Lee, Y.-K., Seol, W. and Moore, D.D. (1997) Differential transactivation by two isoforms of the orphan nuclear hormone receptor CAR, *Journal of Biological Chemistry*, **272**, 23565–23571.

Cholerton, S., Daly, A.K. and Idle, J.R. (1992) The role of individual human cytochromes P450 in drug metabolism and clinical response, *Trends in Pharmacological Science*, **17**, 434–439.

Ciolino, H.P., Daschner, P.J. and Yeh, G.C. (1999) Dietary flavonols quercetin and kaempferol are ligands of the aryl hydrocarbon receptor that affect CYP1A1 transcription differentially, *Biochemical Journal*, **340**, 715–722.

Clarke, S.E. (1998) *In vitro* assessment of human cytochrome P450, *Xenobiotica*, **28**, 1255–1273.

Cleaves, H.J. and Miller, S.L. (1998) Oceanic protection of prebiotic organic compounds from UV radiation, *Proceedings of the National Academy of Sciences USA*, **95**, 7260–7263.

Cloud, P. (1976) Beginnings of biospheric evolution and their biogeographical consequences, *Paleobiology*, **2**, 351–387.

Cnubben, N.H.P., Peelen, S., Borst, J.-W., Vervoort, J., Veeger, C. and Rietjens, I.M.C.M. (1994) Molecular orbital-based quantitative structure-activity relationship for the cytochrome P450-catalyzed 4-hydroxylation of halogenated anilines, *Chemical Research in Toxicology*, **7**, 590–598.

Cohen, M.B. and Feyereisen, R. (1995) A cluster of cytochrome P450 genes of the CYP6 family in the house fly, *DNA and Cell Biology*, **14**, 73–82.

Cole, P.A. and Robinson, C.H. (1991) Mechanistic studies on a placental aromatase model reaction, *Journal of the American Chemical Society*, **113**, 8130–8137.

Conney, A.M., Chang, R.L., Jerina, D.M. and Wei, S.-J.C. (1994) Studies on the metabolism of benzo[a]pyrene and dose-dependent differences in the mutagenic profile of its ultimate carcinogenic metabolite, *Drug Metabolism Reviews*, **26**, 125–163.

Coon, M.J. and White, R.E. (1980) Cytochrome P450: a versatile catalyst in monooxygenation reactions, in: *Metal Ion Activation of Dioxygen* (T.G. Spiro, ed.) Wiley, New York, 73–123.

Coon, M.J., Vaz, A.D.N. and Bestervelt, L.L. (1996) Peroxidative reactions of diversozymes, *FASEB Journal*, **10**, 428–434.

Coon, M.J., Vaz, A.D.N., McGinnity, D.F. and Peng, H.-M. (1998) Multiple activated oxygen species in P450 catalysis, *Drug Metabolism and Disposition*, **26**, 1190–1193.

Correia, M.A. and Ortiz de Montellano, P.R. (1993) Inhibitors of cytochrome P450 and possibilities for their therapeutic application, *Frontiers in Biotransformation*, **8**, 75–146.

Corton, J.C., Lapinskas, P.J. and Gonzalez, F.J. (2000) Central role of PPARα in the mechanism of action of hepatocarcinogenic peroxisome proliferators, *Mutation Research*, **448**, 139–151.

Coulson, C.J., King, D.J. and Wiseman, A. (1984) Chemotherapeutic and agrochemical applications of cytochrome P450 ligands, *Trends in Biochemical Sciences*, **9**, 446–449.

Creighton, T.E. (1993) *Proteins: Structure and Molecular Properties*, Freeman, New York.

Crespi, C.L., Penman, B.W., Gelboin, H.V. and Gonzalez, F.J. (1991) A tobacco smoke-derived nitrosamine, 4-(methylnitrosamino)-1-(3-pyridyl)-1-butanone is activated by multiple human cytochrome P450s including the polymorphic human cytochrome P450 2D6, *Carcinogenesis*, **12**, 1197–1201.

Crofts, F., Cosma, G.N., Currie, D., Taioli, E., Toniolo, P. and Garte, S.J. (1993) A novel CYP1A1 gene polymorphism in African-Americans, *Carcinogenesis*, **14**, 1729–1731.

Cupp-Vickery, J.R. and Poulos, T.L. (1995) Structure of cytochrome $P450_{eryF}$ involved in erythromycin biosynthesis, *Structural Biology*, **2**, 144–153.

Cvrk, T. and Strobel, H.W. (1998) Photoaffinity labeling of cytochrome P4501A1 with azidocumene: identification of cumene hydroperoxide binding region, *Archives of Biochemistry and Biophysics*, **349**, 95–104.

Cvrk, T., Hodek, P. and Strobel, H.W. (1996) Identification and characterization of cytochrome P4501A1 amino acid residues interacting a radiolabeled photoaffinity diazido-benzphetamine analogue, *Archives of Biochemistry and Biophysics*, **330**, 142–152.

Daff, S.N., Chapman, S.K., Turner, K.L., Holt, R.A., Govindaraj, S., Poulos, T.L. and Munro, A.W. (1997) Redox control of the catalytic cycle of flavocytochrome $P450_{BM3}$, *Biochemistry*, **36**, 13816–13823.

Dai, R., Zhai, S., Wei, X., Pincus, M.R., Vestal, R.E. and Friedman, F.K. (1999) Inhibition of human cytochrome P450 1A2 by flavones: a molecular modeling study, *Journal of Protein Chemistry*, **17**, 643–650.

Dailey, H.W. and Strittmatter, P. (1979) Modification and identification of cytchromes b_5 carboxyl groups involved in protein-protein interaction with cytochrome b_5 reductase, *Journal of Biological Chemistry*, **254**, 5388–5396.

Daly, A.K., Brockmöller, J., Broly, F., Eichelbaum, M., Evans, W.E., Gonzalez, F.J., Huang, J.-D., Idle, J.R., Ingelman-Sundberg, M., Ishizaki, T., Jacqz-Aigrain, E., Meyer, V.A., Nebert, D.W., Steen, V.M., Wolf, C.R. and Zanger, V.M. (1996) Nomenclature for human CYP2D6 alleles, *Pharmacogenetics*, **6**, 193–201.

Davies, M.D., Qin, L., Beck, J.L., Suslick, K.S., Koga, H., Horiuchi, T. and Sligar, S.G. (1990) Putiredoxin reduction of cytochrome $P450_{cam}$. Dependence of electron transfer on the identity of putidaredoxin's C-terminal amino acid, *Journal of the American Chemical Society*, **112**, 7396–7398.

Davis, S.C., Sui, Z., Peterson, J.A. and Ortiz de Montellano, P.R. (1996) Oxidation of ω-oxo fatty acids by cytochrome $P450_{BM3}$ (CYP102), *Archives of Biochemistry and Biophysics*, **328**, 35–42.

Dawson, J.H. (1988) Probing structure-function relations in heme-containing oxygenases and peroxidases, *Science*, **240**, 433–439.

Dawson, J.H. and Sono, M. (1987) Cytochrome P450 and chloroperoxidase: thiolate-ligated heme enzymes. Spectroscopic determination of their active site structures and mechanistic implications of thiolate ligation, *Chemical Reviews*, **87**, 1255–1276.

Dawson, J.H., Andersson, L.A. and Sono, M. (1982) Spectroscopic investigations of ferric cytochrome $P450_{cam}$ ligand complexes, *Journal of Biological Chemistry*, **257**, 3606–3617.

Dawson, J.H., Kau, L.-S., Penner-Hahn, J.E., Sono, M., Eble, K.S., Bruce, G.S., Hager, L.P. and Hodgson, K.O. (1986) Oxygenated cytochrome $P450_{cam}$ and chloroperoxidase: direct evidence for sulfur donor ligation trans to dioxygen and structural characterization using EXAFS spectroscopy, *Journal of the American Chemical Society*, **108**, 8114–8116.

de Groot, M.J. and Vermeulen, N.P.E. (1997) Modeling the active sites of cytochrome P450s and glutathione S-transferases, two of the most important biotransformation enzymes, *Drug Metabolism Reviews*, **29**, 747–799.

de Groot, M.J., Vermeulen, N.P.E., Kramer, J.D., van Acker, F.A.A. and Donné-Opden Kelder, G.M. (1996) A three-dimensional protein model for human cytochrome P450 2D6 based on the crystal structures of P450 101, P450 102 and P450 108, *Chemical Research in Toxicology*, **9**, 1079–1091.

de Groot, M.J., Havenith, R.W.A., Vinkers, H.M., Zwaans, R., Vermeulen, N.P.E. and van Lenthe, J.H. (1998) Ab initio calculations on iron-porphyrin model systems for intermediates in the oxidative cycle of cytochrome P450s, *Journal of Computer-Aided Molecular Design*, **12**, 183–193.

de Groot, M.J., Ackland, M.J., Horne, V.A., Alex, A.A. and Jones, B.C. (1999) Novel approach to predicting P450-mediated drug metabolism: development of a combined protein and pharmacophore model for CYP2D6, *Journal of Medicinal Chemistry*, **42**, 1515–1524.

De Lemos-Chiarandini, C., Frey, A.B., Sabatinini, D.D. and Kreibich, G. (1987) Determination of the membrance topology of phenobarbital-inducible rat liver cytochrome P450 isoenzyme PB-4 using site-specific antibodies, *Journal of Cell Biology*, **104**, 209–219.

De Matteis, F., Dawson, S.J., Boobis, A.R. and Comoglio, A. (1991) Inducible bilirubin-degrading system of rat liver microsomes: role of cytochrome P4501A1, *Molecular Pharmacology*, **40**, 686–691.

De Voss, J.J. and Ortiz de Montellano, P.R. (1996) Substrate docking algorithms and the prediction of substrate specificity, *Methods in Enzymology*, **272**, 336–347.

Degawa, M., Arai, H., Miura, S.-I. and Hashimoto, Y. (1993) Preferential inhibitions of hepatic P450IA2 expression and induction by lead nitrate, *Carcinogenesis*, **14**, 1091–1094.

Degtyarenko, K.N. (1995) Structural domains of P450-containing mono-oxygenase systems, *Protein Engineering*, **8**, 737–747.

Degtyarenko, K.N. and Archakov, A.I. (1993) Molecular evolution of P450 superfamily and P450-containing monooxygenase systems, *FEBS Letters*, **332**, 1–8.

Deprez, E., Gerber, N.C., Di Primo, C., Douzou, P., Sligar, S.G. and Hui Bon Hoa, G. (1994) Electrostatic control of the substrate access channel in cytochrome $P450_{cam}$, *Biochemistry*, **33**, 14464–14468.

Dey, A. and Nebert, D.W. (1998) Markedly increased constitutive CYP1A1 mRNA levels in the fertilized ovum of the mouse, *Biochemical and Biophysical Research Communications*, **251**, 657–661.

Di Primo, C., Sligar, S.G., Hui Bon Hoa, G. and Douzon, P. (1992) A critical role of protein-bound water in the the catalytic cycle of cytochrome $P450_{cam}$, *FEBS Letters*, **312**, 252–254.

Ding, X. and Coon, M.J. (1993) Extrahepatic microsomal forms: olfactory cytochrome P450, in: *Cytochromes P450* (J.B. Schenkman and H. Griem, eds.) Springer-Verlag, Berlin, 351–361.

Doehmer, J. and Griem, H. (1993) Cytochromes P450 in genetically engineered cell cultures: The gene technological approach, in: *Cytochrome P450* (J.B. Schenkman and H. Griem, eds.) Springer-Verlag, Berlin, 415–429.

Doolittle, R.F., Feng, D.-F., Tsang, S., Cho, G. and Little, E. (1996) Determining divergence times of the major kingdoms of living organisms with a protein clock, *Science*, **271**, 470–477.

Doostdar, H., Burke, M.D. and Mayer, R.T. (2000) Bioflavonoids: selective substrates and inhibitors for cytochrome P450 CYP1A and CYP1B1, *Toxicology*, **144**, 31–38.

Duport, C., Spagnoli, R., Degryse, E. and Pompon, D. (1998) Self-sufficient biosynthesis of pregnenolone and progesterone in engineered yeast, *Nature Biotechnology*, **16**, 186–189.

Durst, F. (1991) Biochemistry and physiology of plant cytochrome P450, *Frontiers in Biotransformation*, **4**, 191–232.

Durst, F. and Benveniste, I. (1993) Cytochrome P450 in plants, in: *Cytochrome P450* (J.B. Schenkman and H. Griem, eds.) Springer-Verlag, Berlin, Chapter 19, 293–310.

Durst, F. and Nelson, D.R. (1995) Diversity and evolution of plant P450 and P450 reductases, *Drug Metabolism and Drug Interactions*, **12**, 189–206.

Durst, F. and O'Keefe, D.P. (1995) Plant cytochromes P450: an overview, *Drug Metabolism and Drug Interactions*, **12**, 171–187.

Durst, F., Benveniste, I., Salaun, J.-P. and Werck, D. (1994) Function and diversity of plant cytochrome P450, in: *Cytochrome P450 Biochemistry, Biophysics and Molecular Biology* (M.C. Lechner, ed.) Libbey, Paris, 23–30.

Eddershaw, P.J. and Dickins, M. (1999) Advances in *in vitro* drug metabolism screening, *Pharmaceutical Science and Technology Today*, **2**, 13–19.

Edwards, P.A. and Ericsson, J. (1999) Sterols and isoprenoids: Signaling molecules derived from the cholesterol biosynthetic pathway, *Annual Review of Biochemistry*, **68**, 157–185.

Edwards, R.J., Singleton, A.M., Murray, B.P., Davies, D.S. and Boobis, A.R. (1995) Short synthetic peptides exploited for reliable and specific targeting of antibodies to the C-termini of cytochrome P450 enzymes, *Biochemical Pharmacology*, **49**, 39–47.

Edwards, R.J., Adams, D.A., Watts, P.S., Davies, D.S. and Boobis, A.R. (1998) Development of a comprehensive panel of antibodies against the major xenobiotic metabolising forms of cytochromes P450 in humans, *Biochemical Pharmacology*, **56**, 377–387.

Egawa, T., Ogura, T., Makino, R., Ishimura, Y. and Kitagawa, T. (1991) Observation of the O-O stretching Raman band for cytochrome $P450_{cam}$ under catalytic conditions, *Journal of Biological Chemistry*, **266**, 10246–10248.

Eichelbaum, M. and Gross, A.S. (1990) The genetic polymorphism of debrisoquine/sparteine metabolism – clinical aspects, *Pharmacology and Therapeutics*, **46**, 377–394.

Ekins, S. and Wrighton, S.A. (1999) The role of CYP2B6 in human xenobiotic metabolism, *Drug Metabolism Reviews*, **31**, 719–754.

Ekins, S., Bravi, G., Wikel, J.H. and Wrighton, S.A. (1999a) Three-dimensional-quantitative structure activity relationship analysis of cytochrome P450 3A4 substrates, *Journal of Pharmacology and Experimental Therapeutics*, **291**, 424–433.

Ekins, S., Bravi, G., Ring, B.J., Gillespie, T.A., Gillespie, J.S., Vandenbranden, M., Wrighton, S.A. and Wikel, J.H. (1999b) Three-dimensional quantitative structure activity relationship analyses of substrates for CYP2B6, *Journal of Pharmacology and Experimental Therapeutics*, **288**, 21–29.

Ekins, S., Bravi, G., Binkley, S., Gillespie, J.S., Ring, B.J., Wikel, J.H. and Wrighton, S.A. (1999c) Three- and four-dimensional quantitative structure activity relationship (3D/4D-QSAR) analyses of CYP2C9 inhibitors, *Drug Metabolism and Disposition*, **28**, 994–1002.

Elfarra, A.A. (1993) Aliphatic halogenated hydrocarbons, in: *Toxicology of the Kidney* (J.B. Hook and R.S. Goldstein, eds.) Raven, New York, 387–413.

Ellis, S.W. (2000) Active site residues of CYP2D6 – fact or fiction, *Proceedings of the Australian Society of Clinical Experimental Pharmacology and Toxicology*, **7**, 12.

Ellis, S.W., Ching, M.S., Watson, P.F., Henderson, C.J., Simula, A.P., Lennard, M.S., Tucker, G.T. and Woods, H.F. (1992) Catalytic activities of human debrisoquine 4-hydroxylase cytochrome P450 (CYP2D6) expressed in yeast, *Biochemical Pharmacology*, **44**, 617–620.

Ellis, S.W., Hayhurst, G.P., Smith, G., Lightfoot, T., Wang, M.M.S., Simula, A.P., Ackland, M.J., Sternberg, M.J.E., Lennard, M.S., Tucker, G.T. and Wolf, C.R. (1995) Evidence that aspartic acid 301 is a critical substrate-contact residue in the active site of cytochrome P450 2D6, *Journal of Biological Chemistry*, **270**, 29055–29058.

Ellis, S.W., Rowland, K., Ackland, M.J., Rekka, E., Simula, A.P., Lennard, M.S., Wolf, C.R. and Tucker, G.T. (1996) Influence of amino acid residue 374 of cytochrome P450 2D6 (CYP2D6) on the regio- and enantioselective metabolism of metoprolol, *Biochemical Journal*, **316**, 647–654.

England, P.A., Harford-Cross, C.F., Stevenson, J.-A., Rouch, D.A. and Wong, L.-L. (1998) The oxidation of naphthalene and pyrene by cytochrome $P450_{cam}$, *FEBS Letters*, **424**, 271–274.

English, N., Hughes, V. and Wolf, C.R. (1994) Common pathways of cytochrome P450 gene regulation by peroxisome proliferators and barbiturates in *Bacillus megaterium* ATCC 14581, *Journal of Biological Chemistry*, **269**, 26837–26841.

English, N., Hughes, V. and Wolf, C.R. (1996) Induction of cytochrome $P450_{BM3}$ (CYP102) by non-steroidal anti-inflammatory drugs in *Bacillus megaterium*, *Biochemical Journal*, **316**, 279–283.

Epstein, P.M., Curti, M., Jansson, I., Huang, C.-H. and Schenkman, J.B. (1989) Phosphorylation of cytochrome P450: regulation by cytochrome b_5, *Archives of Biochemistry and Biophysics*, **271**, 424–432.

Escher, P. and Wahli, W. (2000) Peroxisome proliferator-activated receptors: insight into multiple cellular functions, *Mutation Research*, **448**, 121–138.

Estabrook, R.W. (1996) The remarkable P450s: a historical overview of these versatile hemoprotein catalysts, *FASEB Journal*, **10**, 202–204.

Estabrook, R.W. (1998) Impact of the human genome project on drug metabolism and chemical toxicity, British Toxicology Society Meeting, Guildford, UK, April 19–22.

Estabrook, R.W., Cooper, D.Y. and Rosenthal, O. (1963) The light reversible carbon monoxide inhibition of the steroid C21-hydroxylase system in the adrenal cortex, *Biochemische Zeitschrift*, **338**, 741–755.

Estabrook, R.W., Mason, J.I., Simpson, E.R., Peterson, J.A. and Waterman, M.R. (1991) The heterologous expression of the cytochromes P450: a new approach for the study of enzyme activities and regulation, *Advances in Enzyme Regulation*, **31**, 365–383.

Faulkner, K.M., Shet, S.S., Fisher, C.W. and Estabrook, R.W. (1995) Electrocatalytically driven ω-hydroxylation of fatty acids using cytochrome P450 4A1, *Proceedings of the National Academy of Sciences USA*, **92**, 7705–7709.

Faustman, E.M., Silbernagel, S.M., Fenske, R.A., Burbacher, T.M. and Ponce, R.A. (2000) Mechanisms underlying children's susceptibility to environmental toxicants, *Environmental Health Perspectives*, **108**, Supplement 1, 13–21.

Filipovic, D., Paulsen, M.D., Loida, P.J., Sligar, S.G. and Ornstein, R.L. (1992) Ethylbenzene hydroxylation by cytochrome $P450_{cam}$, *Biochemical and Biophysical Research Communications*, **189**, 488–495.

Finch, S.A.E. and Stier, A. (1991) Rotational diffusion of homo- and hetero-oligomers of cytochrome P450, the functional significance of cooperativity and the membrane structure, *Frontiers in Biotransformation*, **5**, 34–70.

Fisher, M.T. and Sligar, S.G. (1985) Control of heme protein redox potential and reduction rate: linear free energy relation between potential and ferric spin state equilibrium, *Journal of the American Chemical Society*, **107**, 5018–5019.

Fisslthaler, B., Popp, R., Kiss, L., Potente, M., Harder, D.R., Fleming, I. and Busse, R. (1999) Cytochrome P450 2C is an EDHF synthase in coronary arteries, *Nature*, **401**, 493–497.

Fleming, I. (1980) *Frontier Orbitals and Organic Chemical Reactions*, Wiley, New York.

Forman, B.M., Ruan, B., Chen, J., Schroepfer, G.J. and Evans, R.M. (1997) The orphan nuclear receptor LXRα is positively and negatively regulated by distinct products of mevalonate metabolism, *Proceedings of the National Academy of Sciences USA*, **94**, 10588–10593.

Forman, B.M., Tzameli, I., Choi, H.-S., Chen, J., Simha, D., Seol, W., Evans, R.M. and Moore, D.D. (1998) Androstane metabolites bind to and deactivate the nuclear receptor CAR-β, *Nature*, **395**, 612–615.

Forrester, L.M., Henderson, C.J., Glancy, M.J., Black, D.M., Park, B.K., Ball, S.E., Kitteringham, N.R., McLaren, A.W., Miles, J.S., Skett, P. and Wolf, C.R. (1992) Relative expression of cytochrome P450 isoenzymes in human liver and association with the metabolism of drugs and xenobiotics, *Biochemical Journal*, **281**, 359–368.

Fowler, S.M., England, P.A., Westlake, A.C.G., Rouch, D.R., Nickerson, D.P., Blunt, C., Braybrook, D., West, S., Wong, L.-L. and Flitsch, S. (1994) Cytochrome $P450_{cam}$ monooxygenase can be redesigned to catalyse the regioselective aromatic hydroxylation of diphenylmethane, *Chemical Communications*, 2761–2762.

Frausto da Silva, J.J.R. and Williams, R.J.P. (1991) *The Biological Chemistry of the Elements*, Clarendon Press, Oxford.

Fruetel, J., Chang, Y.-T., Collins, J., Loew, G.H. and Ortiz de Montellano, P.R. (1994) Thioanisole sulfoxidation by cytochrome $P450_{cam}$ (CYP101): experimental and calculated absolute stereochemistries, *Journal of the American Chemical Society*, **116**, 11643–11648.

Fuhr, U., Strobl, G., Manaut, F., Anders, E.-M., Sargel, F., Lopez-de-Brinas, E., Chu, D.T.W., Pernet, A.G., Mahr, G., Sanz, F. and Staib, A.H. (1993) Quinoline antibacterial agents: relationship between structure and *in vitro* inhibition of the human cytochrome P450 isoform CYP1A2, *Molecular Pharmacology*, **43**, 191–199.

Fujii-Kuriyama, Y., Imataka, H., Sogawa, K., Yasumoto, K.-I. and Kikuchi, Y. (1992) Regulation of CYP1A1 expression, *FASEB Journal*, **6**, 706–710.

Fulco, A.J. (1991) $P450_{BM-3}$ and other inducible bacterial P450 cytochromes: biochemistry and regulation, *Annual Review of Pharmacology and Toxicology*, **31**, 177–203.

Fulco, A.J. and Ruettinger, R.T. (1987) Occurrence of a barbiturate-inducible catalytically self-sufficient 119,000 dalton cytochrome P450 monooxygenase in bacilli, *Life Sciences*, **40**, 1769–1775.

Funae, Y. and Imaoka, S. (1993) Cytochrome P450 in rodents, in: *Cytochrome P450* (J.B. Schenkman and H. Griem, eds.) Springer-Verlag, Berlin, 221–238.

Furuya, H., Shimizu, T., Hirano, K., Hatano, M., Fujii-Kuriyama, Y., Raag, R. and Poulos, T.L. (1989a) Site-directed mutagenesis of rat liver cytochrome $P450_d$: catalytic activities toward benzphetamine and 7-ethoxy coumarin, *Biochemistry*, **28**, 6848–6857.

Furuya, H., Shimizu, T., Hatano, M. and Fujii-Kuriyama, Y. (1989b) Mutations at the distal and proximal sites of cytochrome P450d changed regio-specificity of acetanilide hydroxylations, *Biochemical and Biophysical Research Communications*, **160**, 669–676.

Gao, H. and Hansch, (1996) QSAR of P450 oxidation: on the value of comparing k_{cat} and K_m with k_{cat}/K_m, *Drug Metabolism Reviews*, **28**, 513–526.

Garfinkel, D. (1958) Studies on pig liver microsomes. I. Enzymic and pigment composition of different microsomal fractions, *Archives of Biochemistry and Biophysics*, **77**, 493–509.

George, J. and Farrell, G.C. (1991) Role of human hepatic cytochromes P450 in drug metabolism and toxicity, *Australian and New Zealand Journal of Medicine*, **21**, 356–362.

Gerber, N.C. and Sligar, S.G. (1992) Catalytic mechanism of cytochrome P450: evidence for a distal charge relay, *Journal of the American Chemical Society*, **114**, 8742–8743.

Gerber, N.C. and Sligar, S.G. (1994) A role for Asp-251 in cytochrome $P450_{cam}$ oxygen activation, *Journal of Biological Chemistry*, **269**, 4260–4266.

Geren, L.M., O'Brien, P., Stonehuerner, J. and Millett, F. (1984) Identification of specific carboxylate groups in adrenodoxin that are involved in the interaction with adrenodoxin reductase, *Journal of Biological Chemistry*, **259**, 2155–2160.

Gibson, G.G. (1985) Cytochrome P450 spin state: regulation and functional significance in: *Microsomes and Drug Oxidations* (A.R. Boobis, J. Caldwell, F. De Matteis and C.R. Elcombe, eds.) Taylor & Francis, London, 33–41.

Gibson, G.G. (1992) Co-induction of cytochrome P450 4A1 and peroxisome proliferation: a causal or casual relationship?, *Xenobiotica*, **22**, 1101–1109.

Gibson, G.G. and Skett, P. (1994) *Introduction to Drug Metabolism*, Chapman & Hall, London.

Gibson, G.G. and Tamburini, P.P. (1984) Cytochrome P450 spin state: inorganic biochemistry of haem iron ligation and functional significance, *Xenobiotica*, **14**, 27–47.

Gillam, E.M.J., Guo, Z. and Guengerich, F.P. (1994) Expression of modified human cytochrome P450 2E1 in *Escherichia coli*, purification, spectral and catalytic properties, *Archives of Biochemistry and Biophysics*, **312**, 59–66.

Goldstein, J.A. and de Morais, S.M.F. (1994) Biochemistry and molecular biology of the human CYP2C subfamily, *Pharmacogenetics*, **4**, 285–299.

Gonder, J.C., Proctor, R.A. and Will, J.A. (1985) Genetic differences in oxygen toxicity are correlated with cytochrome P450 inducibility, *Proceedings of the National Academy of Sciences USA*, **82**, 6315–6319.

Gonzalez, F.J. (1989) The molecular biology of cytochromes P450, *Pharmacological Reviews*, **40**, 243–248.

Gonzalez, F.J. (1992) Human cytochromes P450: problems and prospects, *Trends in Pharmacological Science*, **13**, 346–352.

Gonzalez, F.J. (1996) The CYP2D subfamily, in: *Cytochrome P450: metabolic and toxicological aspects* (C. Ioannides, ed.) CRC Press, Boca Raton, FL, Chapter 8, 183–210.

Gonzalez, F.J. and Fernandez-Salguero, P. (1998) The aryl hydrocarbon receptor: studies using the AhR-null mice, *Drug Metabolism and Disposition*, **26**, 1194–1198.

Gonzalez F.J. and Gelboin, H.V. (1991) Human cytochromes P450: evolution, catalytic activities and inter-individual variations in expression, *Progress in Clinical and Biological Research*, **372**, 11 –20.

Gonzalez, F.J. and Gelboin, H.V. (1992) Human cytochromes P450: evolution and cDNA-directed expression, *Environmental Health Perspectives*, **98**, 81–85.

Gonzalez, F.J. and Gelboin, H.V. (1994) Role of human cytochromes P450 in the metabolic activation of chemical carcinogens and toxins, *Drug Metabolism Reviews*, **26**, 165–183.

Gonzalez, F.J. and Idle, J.R. (1994) Pharmacogenetic phenotyping and genotyping, *Clinical Pharmacokinetics*, **26**, 59–70.

Gonzalez, F.J. and Korzekwa, K.R. (1995) Cytochromes P450 expression systems, *Annual Review of Pharmacology and Toxicology*, **35**, 369–390.

Gonzalez, F.J. and Lee, Y.-H. (1996) Constitutive expression of hepatic cytochrome P450 genes, *FASEB Journal*, **10**, 1112–1117.

Gonzalez, F.J. and Nebert, D.W. (1990) Evolution of the P450 gene superfamily, *Trends in Genetics*, **6**, 182–186.

Gonzalez, F.J., Matsunaga, T. and Nagata, K. (1989) Structure and regulation of P450s in the rat P450IIA gene subfamily, *Drug Metabolism Reviews*, **20**, 827–837.

Gonzalez, F.J., Crespi, C.L. and Gelboin, H.V. (1991) cDNA-expressed human cytochrome P450s: a new age of molecular toxicology and human risk assessment, *Mutation Research*, **247**, 113–127.

Gorman, N., Walton, H.S., Sinclair, J.F. and Sinclair, P.R. (1998) CYP1A-catalyzed uroporphyrinogen oxidation in hepatic microsomes from non-mammalian vertebrates (chick and duck embryos, scup and alligator), *Comparative Biochemistry and Physiology*, **121C**, 405–412.

Gotoh, O. (1992) Substrate recognition sites in cytochrome P450 family 2 (CYP2) proteins inferred from comparative analyses of amino acid and coding nucleotide sequences, *Journal of Biological Chemistry*, **267**, 83–90.

Gotoh, O. and Fujii-Kuriyama, Y. (1989) Evolution, structure and gene regulation of cytochrome P450, *Frontiers in Biotransformation*, **1**, 195–243.

Gotoh, O., Tagashira, Y., Morohashi, K. and Fujii-Kuriyama, Y. (1985) Possible steroid binding site common to adrenal cytochrome $P450_{scc}$ and prostatic steroid binding protein, *FEBS Letters*, **188**, 8–10.

Graham, S.E. and Peterson, J.A. (1999) How similar are P450s and what can their differences teach us?, *Archives of Biochemistry and Biophysics*, **369**, 24–29.

Graham-Lorence, S. and Peterson, J.A. (1996a) P450s: Structural similarities and functional differences, *FASEB Journal*, **10**, 206–214.

Graham-Lorence, S.E. and Peterson, J.A. (1996b) Structural alignments of P450s and extrapolations to the unknown, *Methods in Enzymology*, **272**, 315–326.

Graham-Lorence, S., Khalil, M.W., Lorence, M.C., Mendelson, C.R. and Simpson, E.R. (1991) Structure-function relationships of human aromatase cytochrome P450 using molecular modeling and site-directed mutagenesis, *Journal of Biological Chemistry*, **266**, 11939–11946.

Graham-Lorence, S., Truan, G., Peterson, J.A., Falck, J.R., Wei, S., Helvig, C. and Capdevila, J. (1997) An active site substitution, F87V, converts cytochrome $P450_{BM3}$ into a regio- and stereoselective (14S, 15R)-arachidonic acid epoxygenase, *Journal of Biological Chemistry*, **272**, 1127–1135.

Gray, H.B. and Ellis, W.R. (1994) Electron transfer, in: *BioInorganic Chemistry* (I. Bertini, H.B. Gray, S.J. Lippard and J.S. Valentine, eds.) University Sciences Books, Mill Valley, CA, 315–363.

Gray, R.D. (1992) The molecular basis of electron transfer in cytochrome P450 enzyme systems, *Frontiers in Biotransformation*, **7**, 321–350.

Greschner, S., Sharonov, Y.A. and Jung, C. (1993) Substrate induced changes of the active site electronic states in reduced cytochrome $P450_{cam}$ and the photolysis product of its CO complex, *FEBS Letters*, **315**, 153–158.

Griffin, B.W. and Peterson, J.A. (1972) Camphor binding by *Pseudomonas putida* cytochrome P450. Kinetics and thermodynamics of the reaction, *Biochemistry*, **11**, 4740–4746.

Griffin, B.W. and Peterson, J.A. (1975) *Pseudomonas putida* cytochrome P450: the effect of the ferric hemoprotein on the relaxation of solvent water protons, *Journal of Biological Chemistry*, **250**, 6445–6451.

Groves, J.T. (1997) The importance of being selective, *Nature*, **389**, 329–330.

Groves, J.T. and Watanabe, Y. (1988) Reactive iron porphyrin derivatives related to the catalytic cycles of cytochrome P450 and peroxidase. Studies on the mechanism of oxygen activation, *Journal of the American Chemical Society*, **110**, 8443–8452.

Guengerich, F.P. (1983) Oxidation-reduction properties of rat liver cytochromes P450 and NADPH-cytochrome P450 reductase related to catalysis in reconstituted systems, *Biochemistry*, **22**, 2811–2820.

Guengerich, F.P. (1987) *Mammalian Cytochromes P450*, CRC Press, Boca Raton, FL.

Guengerich, F.P. (1988) Roles of cytochrome P450 enzymes in chemical carcinogenesis and cancer chemotherapy, *Cancer Research*, **48**, 2946–2954.

Guengerich, F.P. (1989a) Characterization of human microsomal cytochrome P450 enzymes, *Annual Review of Pharmacology and Toxicology*, **29**, 241–264.

Guengerich, F.P. (1989b) Structure and function of cytochrome P450, *Frontiers in Biotransformation*, **1**, 101–150.

Guengerich, F.P. (1990) Mechanism-based inactivation of human liver microsomal cytochrome P450 IIIA4 by gestodene, *Chemical Research in Toxicology*, **3**, 363–371.

Guengerich, F.P. (1991a) Molecular advances for the cytochrome P450 superfamily, *Trends in Pharmacological Science*, **12**, 281–283.

Guengerich, F.P. (1991b) Reactions and significance of cytochrome P450 enzymes, *Journal of Biological Chemistry*, **266**, 10019–10022.

Guengerich, F.P. (1992a) Characterization of human cytochrome P450 enzymes, *FASEB Journal*, **6**, 745–748.

Guengerich, F.P. (1992b) Human cytochrome P450 enzymes, *Life Sciences*, **50,** 1471–1478.

Guengerich, F.P. (1992c) Metabolic activation of carcinogens, *Pharmacology and Therapeutics*, **54**, 17–61.

Guengerich, F.P. (1992d) Cytochrome P450: advances and prospects, *FASEB Journal*, **6**, 667–678.

Guengerich, F.P. (1993) Cytochrome P450 enzymes, *American Scientist*, **81**, 440–447.

Guengerich, F.P. (1994) Catalytic selectivity of human cytochrome P450 enzymes: relevance to drug metabolism and toxicity, *Toxicology Letters*, **70**, 133–138.

Guengerich, F.P. (1995a) Human cytochrome P450 enzymes, in: *Cytochrome P450* (P.R. Ortiz de Montellano, ed.) Plenum, New York, Chapter 14, 473–535.

Guengerich, F.P. (1995b) Cytochrome P450 proteins and potential utilization in biodegradation, *Environmental Health Perspectives*, **103**, Supplement 5, 25–28.

Guengerich, F.P. (1997) Comparisons of catalytic selectivity of cytochrome P450 subfamily enzymes from different species, *Chemico-Biological Interactions*, **106**, 161–182.

Guengerich, F.P. (1999) Cytochrome P450 3A4: regulation and role in drug metabolism, *Annual Review of Pharmacology and Toxicology*, **39**, 1–17.

Guengerich, F.P. and Johnson, W.W. (1997) Kinetics of ferric cytochrome P450 reduction by NADPH-cytochrome P450 reductase: rapid reduction in the absence of substrate and variations among cytochrome P450 systems, *Biochemistry*, **36**, 14741–14750.

Guengerich, F.P., Ballou, D.P. and Coon, M.J. (1976) Spectral intermediates in the reaction of oxygen with purified liver microsomal cytochrome P450, *Biochemical and Biophysical Research Communications*, **70**, 951–956.

Guengerich, F.P., Shimada, T., Raney, K.D., Yun, C.-H., Meyer, D.J., Ketterer, B., Harris, T.M., Groopman, J.D. and Kadlubar, F.F. (1992) Elucidation of catalytic specificities of human cytochrome P450 and glutathione S-transferase enzymes and relevance to molecular epidemiology, *Environmental Health Perspectives*, **98**, 75–80.

Guengerich, F.P., Hosea, N.A., Parikh, A., Bell-Parikh, L.C., Johnson, W.W., Gillam, E.M.J. and Shimada, T. (1998) Twenty years of biochemistry of human P450s, *Drug Metabolism and Disposition*, **26**, 1175–1178.

Hahn, M.E. (1998) The aryl hydrocarbon receptor: a comparative perspective, *Comparative Biochemistry and Physiology*, Part C, **121**, 23–53.

Haining, R.L., Jones, J.P., Henne, K.R., Fisher, M.B. and Koop, D.R. (1999) Enzymatic determinants of the substrate specificity of CYP2C9: role of B-C loop residues in providing the pi-stacking anchor site for warfarin binding, *Biochemistry*, **38**, 3825–3292.

Hakkola, J., Pasanen, M., Hukkanen, J., Pelkonen, O., Maenpaa, J., Edwards, R.J., Boobis, A.R. and Raunio, H. (1996) Expression of xenobiotic-metabolizing cytochrome P450 forms in human full-term placenta, *Biochemical Pharmacology*, **51**, 403–411.

Hakkola, J., Pasanen, M., Pelkonen, O., Hukkanen, J., Evisalmi, S., Anttila, S., Rane, A., Mantyla, M., Purkunen, R., Saarikoski, S., Tooming, M. and Raunio, H. (1997) Expression of CYP1B1 in human adult and fetal tissues and differential inducibility of CYP1B1 and CYP1A1 by Ah receptor ligands in human placenta and cultured cells, *Carcinogenesis*, **18**, 391–397.

Hakkola, J., Pelkonen, O., Pasanen, M. and Raunio, H. (1998) Xenobiotic-metabolizing cytochrome P450 enzymes in the human feto-placental unit: role in intrauterine toxicity, *Critical Reviews in Toxicology*, **28**, 35–72.

Hallahan, D.L. and West, J.M. (1995) Cytochrome P450 in plant/insect interactions: geraniol 10-hydroxylase and the biosyntheses of iridoid monoterpenoids, *Drug Metabolism and Drug Interactions*, **12**, 369–382.

Hallahan, D.L., Cheriton, A.K., Hyde, R., Clark, I. and Forde, B.G. (1993) Plant cytochrome P450 and agricultural biotechnology, *Biochemical Society Transactions*, **21**, 1068–1073.

Halpert, J.R. (1995) Structural basis of selective cytochrome P450 inhibition, *Annual Review of Pharmacology and Toxicology*, **35**, 29–53.

Halpert, J.R., Jaw, J.-Y. and Johnson, E.F. (1989) Design of specific mechanism-based inactivators of hepatic and adrenal microsomal cytochromes P450 responsible for progesterone 21-hydroxylation, *Drug Metabolism Reviews*, **20**, 645–655.

Halpert, J.R., Domanski, T.L., Adali, O., Biagini, C.P., Cosme, J., Dierks, E.A., Johnson, E.F., Jones, J.P., Ortiz de Montellano, P.R., Philpot, R.M., Sibbesen, O., Wyatt, W.K. and Zheng, Z. (1998) Structure-function of cytochromes P450 and flavin-containing monooxygenases, *Drug Metabolism and Disposition*, **26**, 1223–1231.

Hankinson, O. (1995) The aryl hydrocarbon receptor complex, *Annual Review of Pharmacology and Toxicology*, **35**, 307–340.

Hansch, C. and Zhang, L. (1993) Quantitative structure-activity relationships of cytochrome P450, *Drug Metabolism Reviews*, **25**, 1–48.

Hanson, L.K., Eaton, W.A., Sligar, S.G., Gunsalus, I.C., Gouterman, M. and Connell, C.R. (1976) Origin of the anomalous Soret Spectra of carboxy cytochrome P450, *Journal of the American Chemical Society*, **98**, 2672–2674.

Hanson, L.K., Sligar, S.G. and Gunsalus, I.C. (1977) Electronic structure of cytochrome P450, *Croatica Chemica Acta*, **49**, 237–250.

Hanukoglu, I. (1992) Steroidogenic enzymes: structure, function and role in regulation of steroid hormone biosynthesis, *Journal of Steroid Biochemistry and Molecular Biology*, **43**, 779–804.

Harborne, J.B. (1993) *Introduction to Ecological Biochemistry*, Academic Press, New York.

Harcourt, R.D. (1977) Valence formulas for the $Fe(II)O_2$ linkage of oxyhemoglobin and cytochrome P450-dependent mono-oxygenases, *International Journal of Quantum Chemistry*, Quantum Biology Symposium, **4**, 143–153.

Hargreaves, M.B., Jones, B.C., Smith, D.A. and Gescher, A. (1994) Inhibition of p-nitrophenol hydroxylase in rat liver microsomes by small aromatic and heterocyclic molecules, *Drug Metabolism and Disposition*, **22**, 806–810.

Harland, W.B., Armstrong, R.L., Craig, L.E., Smith, A.G. and Smith, D.G. (1989) *A Geological Time Scale*, Cambridge University Press, Cambridge.

Harlow, G.R. and Halpert, J.R. (1997) Alanine-scanning mutagenesis of a putative substrate recognition site in human cytochrome P450 3A4, *Journal of Biological Chemistry*, **272**, 5396–5402.

Harris, D. and Loew, G.H. (1994) A role for Thr 252 in cytochrome $P450_{cam}$ oxygen activation, *Journal of the American Chemical Society*, **116**, 11671–11674.

Harris, D. and Loew, G.H. (1995) Prediction of regiospecific hydroxylation of camphor analogs by cytochrome $P450_{cam}$, *Journal of the American Chemical Society*, **117**, 2738–2746.

Hasemann, C.A., Ravichandran, K.G., Peterson, J.A. and Deisenhofer, J. (1994) Crystal structure and refinement of cytochrome $P450_{terp}$ at 2.3Å resolution, *Journal of Molecular Biology*, **236**, 1169–1185.

Hasemann, C.A., Kurumbail, R.G., Boddupalli, S.S., Peterson, J.A. and Deisenhofer, J. (1995) Structure and function of cytochromes P450: a comparative analysis of three crystal structures, *Structure*, **3**, 41–62.

Hasinoff, B.B. (1985) Quantitative structure-activity relationships for the reaction of hydrated electrons with heme proteins, *Biochimica et Biophysica Acta*, **829**, 1–5.

Hattori, K., Krouse, H.-R. and Campbell, F.A. (1983) The start of sulfur oxidation in continental environments: about 2×10^9 years ago, *Science*, **221**, 549–551.

Hawkins, B.K. and Dawson, J.H. (1992) Oxygen activation by heme-containing mono-oxygenases: cytochrome P450 and secondary amine mono-oxygenase. Active site structure and mechanisms of action, *Frontiers in Biotransformation*, **7**, 216–278.

Hazzard, J.T., Govindaraj, S., Poulos, T.L. and Tollin, G. (1997) Electron transfer between the FMN and heme domains of cytochrome $P450_{BM3}$, *Journal of Biological Chemistry*, **272**, 7922–7926.

He, Y., Luo, Z., Klekotka, P.A., Burnett, V.L. and Halpert, J.R. (1994) Structural determinants of cytochrome P4502B1 specificity: evidence of five substrate recognition sites, *Biochemistry*, **33**, 4419–4424.

He, M., Kunze, K.L. and Trager, W.F. (1995) Inhibition of (S)-warfarin metabolism by sulfinpyrazone and its metabolites, *Drug Metabolism and Disposition*, **23**, 659–663.

Hedges, S.B., Parker, P.H., Sibley, C.G. and Kumar, S. (1996) Continental breakup and the ordinal diversification of birds and mammals, *Nature*, **381**, 226–229.

Heim, M.H. and Meyer, U.A. (1991) Genetic polymorphism of debrisoquine oxidation, *Methods in Enzymology*, **206**, 173–183.

Hellmold, H., Rylander, T., Magnusson, M., Reihner, E., Warner, M. and Gustafsson, J.-A. (1998) Characterization of cytochrome P450 enzymes in human breast tissue for reduction mammaplasties, *Journal of Clinical Endocrinology and Metabolism*, **83**, 886–895.

Helms, V., Deprez, E., Gill, E., Barret, C., Hui Bon Hoa, G. and Wade, R.C. (1996) Improved binding of cytochrome $P450_{cam}$ substrate analogues designed to fill extra space in the substrate binding pocket, *Biochemistry*, **35**, 1485–1499.

Hildebrandt, P. (1992) Resonance Raman spectroscopy of cytochrome P450, *Frontiers in Biotransformation*, **7**, 166–215.

Hill, H.A.O., Röder, A. and Williams, R.J.P. (1970) The chemical nature and reactivity of cytochrome P450, *Structure and Bonding*, **8**, 123–151.

Hines, R.N., Piechocki, M.P. and Boncher, P.D. (1994) Molecular mechanisms controlling CYP1A gene expression, *Frontiers in Biotransformation*, **9**, 85–110.

Hintz, M.J., Mock, D.M., Peterson, L.L., Tuttle, K. and Peterson, J.A. (1982) Equilibrium and kinetic studies of the interaction of cytochrome $P450_{cam}$ and putidaredoxin, *Journal of Biological Chemistry*, **257**, 14324–14332.

Hiroya, K., Ishigooka, M., Shimizu, T. and Hatano, M. (1992) Role of Glu 318 and Thr 319 in the catalytic function of cytochrome $P450_d$ (P4501A2): effects of mutations on the methanol hydroxylation, *FASEB Journal*, **6**, 749–751.

Hirst, D.M. (1990) *A Computational Approach to Chemistry*, Blackwell, Oxford.

Hollebone, B.R. (1986) Categorization of lipophilic xenobiotics by the enthalpic structure-function response of hepatic mixed-function oxidase, *Drug Metabolism Reviews*, **17**, 93–143.

Hollebone, B.R. and Brownlee, L.J. (1995) A thermodynamic QSAR analysis of the polysubstrate monooxygenase responses to xenobiotic chemicals, *Toxicology Letters*, **79**, 157–168.

Holton, T.A., Brugliera, F., Lester, D.R., Tanaka, Y., Hyland, C.D., Menting, J.G.T., Lu, C.-Y., Farey, E., Stevenson, T.W. and Cornish, E.C. (1993) Cloning and expression of cytochrome P450 genes controlling flower colour, *Nature*, **366**, 276–279.

Honkakoski, P. and Negishi, M. (1997) The structure, function and regulation of cytochrome P450 2A enzymes, *Drug Metabolism Reviews*, **29**, 977–996.

Honkakoski, P. and Negishi, M. (2000) Regulation of cytochrome P450 (CYP) genes by nuclear receptors, *Biochemical Journal*, **347**, 321–337.

Honkakoski, P., Zelko, I., Sueyoshi, T. and Negishi, M. (1998) The nuclear orphan receptor CAR-retinoid X receptor heterodimer activates the phenobarbital-responsive enhancer module of the CYP2B gene, *Molecular and Cellular Biology*, **18**, 5652–5658.

Horn, E.P., Tucker, M.A., Lambert, G., Silverman, D., Zametkin, D., Sinha, R., Hartge, T., Landi, M.T. and Caporaso, N.E. (1995) A study of gender-based cytochrome P450IA2 variability: a possible mechanism for the male excess of bladder cancer, *Cancer Epidemiology, Biomarkers and Prevention*, **4**, 529–533.

Houston, J.B. and Carlile, D.J. (1997) Prediction of hepatic clearance from microsomes, hepatocytes and liver slices, *Drug Metabolism Reviews*, **29**, 891–922.

Hsu, M.-H., Griffin, K.J., Wang, Y., Kemper, B. and Johnson, E.F. (1993) A single amino acid substitution confers progesterone 6β-hydroxylase activity to rabbit cytochrome P4502C3, *Journal of Biological Chemistry*, **268**, 6939–6944.

Huff, J., Lucier, G. and Tritscher, A. (1994) Carcinogenicity of TCDD: Experimental, mechanistic and epidemiological evidence, *Annual Review of Pharmacology and Toxicology*, **34**, 343–372.

Ibeanu, G.S., Ghanayem, B.I., Linko, P., Li, L., Pedersen, L.G. and Goldstein, J.A. (1996) Identification of residues 99, 220 and 221 of human cytochrome P450 2C19 as key determinants of omeprazole hydroxylase activity, *Journal of Biological Chemistry*, **271**, 12496–12501.

Imai, Y., Sato, R. and Iyanagi, T. (1977) Rate-limiting step in the reconstituted microsomal drug hydroxylase system, *Journal of Biochemistry*, **62**, 239–249.

Imai, Y., Shimada, H., Watanabe, Y., Matsushima-Hibiya, Y., Makino, R., Koga, H., Horiuchi, T. and Ishimura, Y. (1989) Uncoupling of the cytochrome $P450_{cam}$ monooxygenase reaction by a single mutation, threonine-252 to alanine or valine: a possible role of the hydroxy amino acid in oxygen activation, *Proceedings of the National Academy of Sciences USA*, **86**, 7823–7827.

Imaoka, S., Yamada, T., Hiroi, T., Hayashi, K., Sakaki, T., Yabusaki, Y. and Funae, Y. (1996) Multiple forms of human P450 expressed in *Saccharomyces cerevisiae*, systemic characterization and comparison with those in rat, *Biochemical Pharmacology*, **51**, 1041–1050.

Ingelman-Sundberg, M., Oscarson, M., Persson, I., Masimirembwa, S., Dahl, M.-L., Bertilsson, L., Sjoqvist, F. and Johansson, I. (1995) Genetic polymorphism of human drug metabolizing enzymes. Recent aspects on polymorphic forms of cytochrome P450,

COST B1 Conference on Variability and Specificity in Drug Metabolism, European Commission, Brussels, pp. 93–110.

Inouye, K. and Coon, M.J. (1985) Properties of the tryptophan residue in rabbit liver microsomal cytochrome P450 isozyme 2 as determined by fluorescence, *Biochemical and Biophysical Research Communications*, **128,** 676–682.

Ioannides, C. (1996) *Cytochromes P450: Metabolic and Toxicological Aspects*, CRC Press, Boca Raton, FL.

Ioannides, C. (1999) Effect of diet and nutrition on the expression of cytochromes P450, *Xenobiotica*, **29**, 109–154.

Ioannides, C. and Lewis, D.F.V. (1995) *Drugs, Diet and Disease, Volume 1: Mechanistic Approaches to Cancer*, Ellis Horwood, Chichester.

Ioannides, C. and Parke, D.V. (1990) The cytochrome P4501 gene family of microsomal proteins and their role in the metabolic activation of chemicals, *Drug Metabolism Reviews*, **22**, 1–85.

Ioannides, C., and Parke, D.V. (1993) Induction of cytochrome P4501 as an indicator of potential chemical carcinogenesis, *Drug Metabolism Reviews*, **25**, 485–501.

Issemann, I. and Green, S. (1990) Activation of a member of the steroid hormone receptor superfamily by peroxisome proliferators, *Nature*, **347**, 645–650.

Ito, K., Iwatsubo, T., Kanamitsu, S., Ueda, K., Suzuki, H. and Sugiyama, Y. (1998) Prediction of pharmacokinetic alterations caused by drug-drug interactions: metabolic interaction in the liver, *Pharmacological Reviews*, **50**, 387–411.

Ivanov, Y.D., Kanaeva, I.P., Eldarov, M.A., Skryabin, K.G., Lehnerer, M., Schulze, J., Hlavica, P. and Archakov, A.I. (1997) An optical biosensor study of the interaction parameters and role of hydrophobic tails of cytochrome P450 2B4, b_5 and NADPH-flavoprotein in complex formation, *Biochemistry and Molecular Biology International*, **42**, 731–737.

Iwasaki, M., Lindberg, R.L.P., Juvonen, R.O. and Negishi, M. (1993) Site-directed mutagenesis of mouse steroid 7α-hydroxylase (cytochrome $P450_{7\alpha}$): role of residue-209 in determining steroid-cytochrome P450 interaction, *Biochemical Journal*, **291**, 569–573.

Iwata, H. and Stegeman, J.J. (2000) In situ RT-PCR detection of CYP1A mRNA in pharyngeal epithelium and chondroid cells from chemically untreated fish, *Biochemical and Biophysical Research Communications*, **271**, 130–137.

Iyanagi, T. (1987) On the mechanisms of one- and two-electron transfer by flavin enzymes, *Chemica Scripta*, **27A**, 31–36.

Jacobs, M.N. and Lewis, D.F.V. (2000a) Homology modelling of the pregnane-X receptor (PXR) from the human oestrogen receptor alpha crystal structure, *Toxicology Letters*, **116** (Supplement 1) 58.

Jacobs, M.N. and Lewis, D.F.V. (2000b) Homology modelling of the aryl hydrocarbon receptor (AhR) ligand binding domain from the human oestrogen receptor alpha crystal structure, *Toxicology Letters*, **116** (Supplement 1), 59.

Janowski, B.A., Willy, P.J., Devi, T.R., Falck, J.R. and Mangelsdorf, D.J. (1996) An oxysterol signalling pathway mediated by the nuclear receptor LXRα, *Nature*, **383**, 728–731.

Jansson, I. (1993) Post-translational modification of cytochrome P450, in: *Cytochrome P450* (J.B. Schenkman and H. Griem, eds.) Springer-Verlag, Berlin, 561–580.

Jansson, I., Curti, M., Epstein, P.M., Peterson, J.A. and Schenkman, J.B. (1990) Relationship between phosphorylation and cytochrome P450 destruction, *Archives of Biochemistry and Biophysics*, **283**, 285–292.

Jean, P., Pothier, J., Dansette, P.M., Mansuy, D. and Viari, A. (1997) Automated multiple analysis of protein structures: application to homology modeling of cytochromes P450, *Proteins: Structure, Function and Genetics*, **28**, 388–404.

Johnson, E.F. (1992) Mapping determinants of the substrate selectivities of P450 enzymes by site-directed mutagenesis, *Trends in Pharmacological Science*, **13**, 122–126.

Johnson, E.F., Kronbach, T. and Hsu, M.-H. (1992) Analysis of the catalytic specificity of cytochrome P450 enzymes through site-directed mutagenesis, *FASEB Journal*, **6**, 700–705.

Johnson, E.F., Palmer, C.N.A., Griffin, K.J. and Hsu, M.-H. (1996) Role of the peroxisome proliferator-activated receptor in cytochrome P450 4A gene regulation, *FASEB Journal*, **10**, 1241–1248.

Jones, B.C., Hawksworth, G., Horne, V., Newlands, A., Tute, M. and Smith, D.A. (1993) Putative active site model for CYP2C9 (tolbutamide hydroxylase), *British Journal of Clinical Pharmacology*, **36**, 143P-144P.

Jones, J.P. and Korzekwa, K.R. (1996) Predicting the rates and regioselectivity of reactions mediated by the P450 superfamily, *Methods in Enzymology*, **272**, 326–335.

Jones, S.A., Moore, L.B., Shenk, J.L., Wisely, G.B., Hamilton, G.A., McKee, D.D., Tomkinson, N.C.O., LeCluyse, E.L., Lambert, M.H., Willson, T.M., Kliewer, S.A. and Moore, J.T. (2000) The pregnane X receptor: a promiscuous xenobiotic receptor that has diverged during evolution, *Molecular Endocrinology*, **14**, 27–39.

Joo, H., Lin, Z. and Arnold, F.H. (1999) Laboratory evolution of peroxide-mediated cytochrome P450 hydroxylation, *Nature*, **399**, 670–673.

Juchau, M.R. (1990) Substrate specificities and functions of the P450 cytochromes, *Life Sciences*, **47**, 2385–2394.

Jung, C. (1985) Quantum chemical explanation of the 'hyper' spectrum of the carbon monoxide complex of cytochrome P450, *Chemical Physics Letters*, **113,** 589–596.

Jung, C., Hui Bon Hoa, G., Schröder, K.-L., Simon, M. and Doucet, J.P. (1992) Substrate analogue induced changes of the CO-stretching mode in cytochrome $P450_{cam}$-carbon monoxide complex, *Biochemistry*, **31**, 12855–12862.

Juvonen, R.O., Iwasaki, M. and Negishi, M. (1991) Structural function of residue-209 in coumarin 7-hydroxylase ($P450_{coh}$), *Journal of Biological Chemistry*, **266**, 16431–16435.

Kagawa, N. and Waterman, M.R. (1995) Regulation of steroidogenic and related P450s, in: *Cytochrome P450* (P.R. Ortiz de Montellano, ed.), Plenum, New York, Chapter 12, 419–442.

Kappus, H. (1993) Metabolic reactions: role of cytochrome P450 in the formation of reactive oxygen species, in: *Cytochrome P450* (J.B. Schenkman and H. Griem, eds.) Springer-Verlag, Berlin, 145–154.

Karki, S.B., Dinnocenzo, J.P., Jones, J.P. and Korzekwa, K.R. (1995) Mechanism of oxidative amine dealkylation of substituted N,N-dimethylanilines by cytochrome P450: application of isotope effect profiles, *Journal of the American Chemical Society*, **117**, 3657–3664.

Kassner, R.J. (1973) A theoretical model for the effects of local nonpolar heme environments on the redox potentials in cytochromes, *Journal of the American Chemical Society*, **95**, 2674–2677.

Kato, R. and Yamazoe, Y. (1993) Hormonal regulation of cytochrome P450 in rat liver, in: *Cytochrome P450* (J.B. Schenkman and H. Griem, eds.) Springer-Verlag, Berlin, 447–459.

Kawajiri, K. and Hayashi, S.-I. (1996) The CYP1 family, in: *Cytochromes P450: Metabolic and Toxicological Aspects* (C. Ioannides, ed.) CRC Press, Boca Raton, FL, Chapter 4, 77–97.

Kawajiri, K., Watanabe, J. and Hayashi, S.-I. (1992) Roles in genetic polymorphisms of drug metabolizing enzymes in humans, *Journal of Basic Clinical Physiology and Pharmacology*, **3**, 76–77.

Kedzie, K.M., Philpot, R.M. and Halpert, J.R. (1991) Functional expression of mammalian cytochromes P450IIB in the yeast *Saccharomyces cerevisiae*, *Archives of Biochemistry and Biophysics*, **291**, 176–186.

Kehrer, J.P. (1993) Free radicals as mediators of tissue injury and disease, *Critical Reviews in Toxicology*, **23**, 21–48.

Kellner, D.G., Maves, S.A. and Sligar, S.G. (1997) Engineering cytochrome P450s for bioremediation, *Current Opinion in Biotechnology*, **8**, 274–278.

Kemper, B. (1993) Mammalian cytochrome P450 genes, *Frontiers in Biotransformation*, **8**, 1–58.

Kikuchi, Y., Yasukochi, Y., Nagata, Y., Fukuda, M. and Takagi, M. (1994) Nucleotide sequence and functional analysis of the meta-cleavage pathway involved in biphenyl and polychlorinated biphenyl degradation in *Pseudomonas* sp. strain KKS102, *Journal of Bacteriology*, **176**, 4269–4276.

Klein, M.L. and Fulco, A.J. (1993) Critical residues involved in FMN binding and catalytic activity in cytochrome $P450_{BM3}$, *Journal of Biological Chemistry*, **268**, 7553–7561.

Kliewer, S.A. and Willson, T.M. (1998) The nuclear receptor PPARγ – bigger than fat, *Current Opinion in Genetics and Development*, **8**, 576–581.

Kliewer, S.A., Lehmann, J.M. and Willson, T.M. (1998a) Orphan nuclear receptors: shifting endocrinology into reverse, *Science*, **284**, 757–760.

Kliewer, S.A., Moore, J.T., Wade, L., Staudinger, J.L., Watson, M.A., Jones, S.A., McKee, D.D., Oliver, B.B., Willson, T.M., Zetterstrom, R.H., Pearlmann, T. and Lehmann, J.M. (1998b) An orphan nuclear receptor activated by pregnanes defines a novel steroid signaling pathway, *Cell*, **92**, 73–82.

Klingenberg, M. (1958) Pigments of rat liver microsomes, *Archives of Biochemistry and Biophysics*, **75**, 376–386.

Knoll, A.M. (1992) The early evolution of eukaryotes: A geological perspective, *Science*, **256**, 622–627.

Knowles, R.G. and Moncada, S. (1994) Nitric oxide synthases in mammals, *Biochemical Journal*, **298**, 249–258.

Kobayashi, K., Iwamoto, T. and Honda, K. (1994) Spectral intermediate in the reaction of ferrous cytochrome $P450_{cam}$ with superoxide anion, *Biochemical and Biophysical Research Communications*, **201**, 1348–1355.

Koga, H., Sagara, Y., Yaoi, T., Tsajumura, M., Nakamura, K., Sekimizu, K., Makino, R., Shimada, H., Ishimura, Y., Yura, K., Co, M., Ikeguchi, M. and Horiuchi, T. (1993) Essential role of the Arg112 residue of cytochrome $P450_{cam}$ for electron transfer from reduced putidaredoxin, *FEBS Letters*, **331**, 109–113.

Kolesanova, E.F., Kozin, S.A., Lemeshko, A.O. and Archakov, A.I. (1994) Epitope mapping of cytochrome P450 2B4 by peptide scanning, *Biochemistry and Molecular Biology International*, **32**, 465–473.

Koley, A.P., Buters, J.T.M., Robinson, R.C., Markowitz, A. and Friedman, F.K. (1995) CO binding kinetics of human cytochrome P450 3A4, *Journal of Biological Chemistry*, **270**, 5014–5018.

Koley, A.P., Robinson, R.C. and Friedman, F.K. (1996) Cytochrome P450 conformation and substrate interactions as probed by CO binding kinetics, *Biochimie*, **78**, 706–713.

Koop, D.R. (1990) Induction of ethanol-inducible cytochrome P450IIE1 by 3-amino-1,2,4-triazole, *Chemical Research in Toxicology*, **3**, 377–383.

Korzekwa, K.R. and Jones, J.P. (1993) Predicting the cytochrome P450 mediated metabolism of xenobiotics, *Pharmacogenetics*, **3**, 1–18.

Korzekwa, K.R., Jones, J.P. and Gillette, J.R. (1990) Theoretical studies on cytochrome P450 mediated hydroxylation: a predictive model for hydrogen atom abstractions, *Journal of the American Chemical Society*, **112**, 7042–7046.

Koskela, S., Hakkola, J., Hukkanen, J., Pelkonen, O., Sorri, M., Saranen, A., Anttila, S., Fernandez-Salguero, P., Gonzalez, F.J. and Raunio, H. (1999) Expression of CYP2A genes in human liver and extrahepatic tissues, *Biochemical Pharmacology*, **57**, 1407–1413.

Koymans, L., Donné-Op den Kelder, G.M., Koppele Te, J.M. and Vermeulen, N.P.E. (1993) Cytochromes P450: their active site structure and mechanism of oxidation, *Drug Metabolism Reviews*, **25**, 325–387.

Krainev, A.G., Shimizu, T., Hiroya, K. and Hatano, M. (1992) Effects of mutations at Lys250, Arg251 and Lys253 of cytochrome P450 1A2 on the catalytic activities and the bindings of bifunctional axial ligands, *Archives of Biochemistry and Biophysics*, **298**, 198–203.

Krust, A., Green, S., Argos, P., Kumar, V., Walter, P., Bornert, J.-M. and Chambon, P. (1986) The chicken oestrogen receptor sequence: homology with v-*erbA* and the human oestrogen and glucocorticoid receptors, *EMBO Journal*, **5**, 891–897.

Kukielka, E. and Cederbaum, A.I. (1994) DNA strand cleavage as a sensitive assay for the production of hydroxyl radicals by microsomes: role of cytochrome P450 2E1 in the increased activity after ethanol treatment, *Biochemical Journal*, **302**, 773–779.

Kulisch, G.P. and Vilker, V.L. (1991) Application of *Pseudomonas putida* PpG 786 containing P450 cytochrome monooxygenase for removal of trace naphthalene concentrations, *Biotechnology Progress*, **7**, 93–98.

Kumaki, K., Sato, M., Kon, H. and Nebert, D.W. (1978) Correlation of type I, type II and reverse type I difference spectra with absolute changes in spin state of hepatic microsomal cytochrome P450 iron from five mammalian species, *Journal of Biological Chemistry*, **253**, 1048–1058.

Kupfer, D. (1980) Endogenous substrates of mono-oxygenases: fatty acids and prostaglandins, *Pharmacology and Therapeutics*, **11**, 469–496.

Lainé, R., Urban, P. and Pompon, D. (1994) Segment directed mutagenesis of human cytochrome P450 1A2 by PCR and effect on the catalytic activity and substrate selectivity, in: *Cytochrome P450* – 8th International Conference (M.C. Lechner, ed.) John Libby, Paris, 451–453.

Lake, B.G. (1995) Mechanisms of carcinogenicity of peroxisome-proliferating drugs and chemicals, *Annual Review of Pharmacology and Toxicology*, **35**, 483–507.

Lake, B.G. (1999) Coumarin metabolism, toxicity and carcinogenicity: relevance for human risk assessment, *Food and Chemical Toxicology*, **37**, 423–453.

Lake, B.G. and Lewis, D.F.V. (1996) The CYP4 family, in: *Cytochrome P450: Metabolic and Toxicological Aspects* (C. Ioannides, ed.) CRC Press, Boca Raton, FL, Chapter 11, 271–297.

Lake, B.G., Renwick, A.B., Cunningham, M.E., Price, R.J., Surry, D. and Evans, D.C. (1998) Comparison of the effects of some CYP3A and other enzyme inducers on replicative DNA synthesis and cytochrome P450 isoforms in rat liver, *Toxicology*, **131**, 9–20.

Lambeth, J.D. (1990) Enzymology of mitochondrial side-chain cleavage by cytochrome $P450_{scc}$, *Frontiers in Biotransformation*, **3**, 58–100.

Lambeth, J.D., Green, L.M. and Millett, F. (1984) Adrenodoxin interaction with adrenodoxin reductase and cytochrome $P450_{scc}$, *Journal of Biological Chemistry*, **259**, 10025–10029.

Landers, J.P. and Bunce, N.J. (1991) The Ah receptor and the mechanism of dioxin toxicity, *Biochemical Journal*, **276**, 273–287.

Langenbach, R., Smith, P.B. and Crespi, C. (1992) Recombinant DNA approaches for the development of metabolic systems used in *in vitro* toxicology, *Mutation Research*, **277**, 251–275.

Larsen, M.C., Angus, W.G.R., Brake, P.B., Elton, S.E., Sukow, K.A. and Jefcoate, C.R. (1998) Characterization of CYP1B1 and CYP1A1 expression in human mammary epithelial cells: role of the aryl hydrocarbon receptor in polycyclic aromatic hydrocarbon metabolism, *Cancer Research*, **58**, 2366–2374.

Lau, S.S. and Monks, T.J. (1993) Nephrotoxicity of bromobenzene: the role of quinone-thioethers, in: *Toxicology of the Kidney* (J.B. Hook and R.S. Goldstein, eds.) Raven, New York, 415–436.

Laudet, V., Auwerx, J., Gustafsson, J.-A. and Wahli, W. (1999) A unified nomenclature system for the nuclear receptor superfamily, *Cell*, **97**, 161–163.

Leach, A.R. (1996) *Molecular Modelling: Principles and Applications*, Longman, Harlow.

Lecoeur, S., Bonierbale, E., Challine, D., Gautier, J.-C., Valadon, P., Dansette, P.M., Catinot, R., Ballet, F., Mansuy, D. and Beaunne, P.H. (1994) Specificity of *in vitro* covalent binding of tienilic acid metabolites to human liver microsomes in relationship to the type of hepatotoxicity: comparison with two directly hepatotoxic drugs, *Chemical Research in Toxicology*, **7**, 434–442.

Leeder, J.S., Gaedigk, A., Lu, X. and Cook, V.A. (1996) Epitope mapping studies with human anti-cytochrome P450 3A antibodies, *Molecular Pharmacology*, **49**, 234–243.

Lee-Robichaud, P., Wright, J.N., Akhtar, M.E., and Akhtar, M. (1995) Modulation of the activity of human 17α-hydroxylase-17,20-lyase (CYP17) by cytochrome b_5: endocrinological and mechanistic implications, *Biochemical Journal*, **308**, 901–908.

Lee-Robichaud, P., Akhtar, M.E. and Akhtar, M. (1998) Control of androgen biosynthesis in the human through the interaction of Arg^{347} and Arg^{358} of CYP17 with cytochrome b_5, *Biochemical Journal*, **332**, 293–296.

Lee-Robichaud, P., Akhtar, M.E. and Akhtar, M. (1999) Lysine mutagenesis identifies cationic charges of human CYP17 that interact with cytochrome b_5 to promote male sex-hormone biosynthesis, *Biochemical Journal*, **342**, 309–312.

Lehmann, J.M., Kliewer, S.A., Moore, L.B., Smith-Oliver, T.A., Oliver, B.B., Su, J-L., Sundseth, S.S., Winegar, D.A., Blanchard, D.E., Spencer, T.A. and Willson, T.M. (1997) Activation of the nuclear receptor LXR by oxysterols defines a new hormone response pathway, *Journal of Biological Chemistry*, **272**, 3137–3140.

Lehnerer, M., Schulze, J., Bernhardt, R. and Hlavica, P. (1999) Some properties of mitochondrial adrenodoxin associated with its nonconventional electron donor function toward rabbit liver microsomal cytochrome P450 2B4, *Biochemical and Biophysical Research Communications*, **254**, 83–87.

Lehnerer, M., Schulze, J., Achterhold, K., Lewis, D.F.V. and Hlavica, P. (2000) Identification of key residues in rabbit liver microsomal cytochrome P450 2B4: importance in interactions with NADPH-cytochrome P450 reductase, *Journal of Biochemistry*, **127**, 163–169.

Lehninger, A.L., Nelson, D.L. and Cox, M.M. (1993) *Principles of Biochemistry*, Worth, New York.

Lewis, D.F.V. (1986) Physical methods in the study of the active site geometry of cytochrome P450, *Drug Metabolism Reviews*, **17**, 1–66.

Lewis, D.F.V. (1992a) Computer modelling of cytochromes P450 and their substrates: a rational approach to the prediction of carcinogenicity, *Frontiers in Biotransformation*, **7**, 90–136.

Lewis, D.F.V. (1992b) Computer-assisted methods in the evaluation of chemical toxicity, *Reviews in Computational Chemistry*, **3**, 173–222.

Lewis, D.F.V. (1995a) COMPACT and the importance of frontier orbitals in toxicity mediated by the cytochrome P450 mono-oxygenase system, *Toxicology Modelling*, **1**, 85–97.

Lewis, D.F.V. (1995b) Three-dimensional models of human and other mammalian microsomal P450s constructed from an alignment with P450102 ($P450_{BM3}$), *Xenobiotica*, **25**, 333–366.

Lewis, D.F.V. (1996a) *Cytochromes P450: Structure, Function and Mechanism*, Taylor & Francis, London.

Lewis, D.F.V. (1996b) Molecular modeling of mammalian cytochromes P450, in: *Cytochrome P450: Metabolic and Toxicological Aspects* (C. Ioannides, ed.) CRC Press, Boca Raton, FL, Chapter 14, 355–398.

Lewis, D.F.V. (1997a) Sex and drugs and P450, *Chemistry and Industry*, No.20, October 20, 381–384.

Lewis, D.F.V. (1997b) Quantitative structure-activity relationships in substrates, inducers and inhibitors of cytochrome P4501 (CYP1), *Drug Metabolism Reviews*, **29**, 589–650.

Lewis, D.F.V. (1998a) The CYP2 family: models, mutants and interactions, *Xenobiotica*, **28**, 617–661.

Lewis, D.F.V. (1998b) Molecular modelling in drug metabolism: the cytochrome P450, in: *Drug Metabolism: Towards the Next Millennium* (N. Gooderham, ed.) IOS Press, Amsterdam, 1–12.

Lewis, D.F.V. (1998c) Models to predict drug metabolism, *Manufacturing Chemist*, **69** (May issue), 15–19.

Lewis, D.F.V. (1999a) Frontier orbitals in chemical and biological activity: quantitative relationships and mechanistic implications, *Drug Metabolism Reviews*, **31**, 755–816.

Lewis, D.F.V. (1999b) Homology modelling of human cytochromes P450 involved in xenobiotic metabolism and rationalization of substrate selectivity, *Experimental and Toxicologic Pathology*, **51**, 369–374.

Lewis, D.F.V. (2000a) Structural characteristics of human P450s involved in drug metabolism: QSARs and lipophilicity profiles, *Toxicology*, **144**, 197–203.

Lewis, D.F.V. (2000b) On the recognition of mammalian microsomal cytochrome P450 substrates and their characteristics, *Biochemical Pharmacology*, **60**, 293–306.

Lewis, D.F.V. (2000c) Modelling human cytochromes P450 for evaluating drug metabolism, *Drug Metabolism and Drug Interactions*, **16**, 307–324.

Lewis, D.F.V. (2001a) COMPACT: A structural approach for the modelling of the biological activity of xenobiotics, *Journal of Chemical Technology and Biotechnology*, **76**, 1–8.

Lewis, D.F.V. (2001b) Gender variations in mammalian P450 regulation and substrate metabolism, in: *Gender and Pesticides* (M.N. Jacobs, ed.) Pesticides Trust, London, in press.

Lewis, D.F.V. and Dickins, M. (2001) Structure-activity relationships for human cytochrome P450 substrates, inhibitors and inducers, *Drug Discovery Today*, submitted.

Lewis, D.F.V. and Hlavica, P. (2000) Interactions between redox partners in various cytochrome P450 systems: functional and structural aspects, *Biochimica et Biophysica Acta*, **1460**, 353–374.

Lewis, D.F.V. and Lake, B.G. (1995) Molecular modelling of members of the P4502A subfamily: applicaton to studies of enzyme specificity, *Xenobiotica*, **25**, 585–598.

Lewis, D.F.V. and Lake, B.G. (1996) Molecular modelling of CYP1A subfamily members based on an alignment with CYP102: rationalization of CYP1A substrate specificity in terms of active site amino acid residues, *Xenobiotica*, **26**, 723–753.

Lewis, D.F.V. and Lake, B.G. (1997) Molecular modelling of mammalian CYP2B isoforms and their interaction with substrates, inhibitors and redox partners, *Xenobiotica*, **27**, 443–478.

Lewis, D.F.V. and Lake, B.G. (1998a) Molecular modelling and quantitative structure-activity relationships (QSAR) studies on the interaction of omeprazole with cytochrome P450 isozymes, *Toxicology*, **125**, 31–44.

Lewis, D.F.V. and Lake, B.G. (1998b) Molecular modelling of the rat peroxisome proliferator-activated receptor α (rPPARα) by homology with the human retinoic acid X receptor α (hRARα) and investigation of peroxisome proliferator binding interactions: QSARs, *Toxicology in Vitro*, **12**, 619–632.

Lewis, D.F.V. and Lake, B.G. (1999) Molecular modelling of CYP4A subfamily members based on sequence homology with CYP102, *Xenobiotica*, **29**, 763–781.

Lewis, D.F.V. and Lee-Robichaud, P. (1998) Molecular modelling of steroidogenic cytochromes P450 from families CYP11, CYP17, CYP19 and CYP21 based on the CYP102 crystal structure, *Journal of Steroid Biochemistry and Molecular Biology*, **66**, 217–233.

Lewis, D.F.V. and Pratt, J.M. (1998) The cytochrome P450 catalytic cycle and mechanism of oxygenation, *Drug Metabolism Reviews*, **30**, 739–786.

Lewis, D.F.V. and Sheridan, G. (2001) Cytochromes P450, oxygen and evolution, *The Scientific World*, **1**, 151–167.

Lewis, D.F.V., Moereels, H., Lake, B.G., Ioannides, C. and Parke, D.V. (1994a) Molecular modeling of enzymes and receptors involved in carcinogenesis: QSARs and COMPACT-3D, *Drug Metabolism Reviews*, **26**, 262–285.

Lewis, D.F.V., Ioannides, C. and Parke, D.V. (1994b) Molecular modelling of cytochrome P450 CYP1A1: A putative access channel explains differences in induction potency between the isomers benzo(a)pyrene and benzo(e)pyrene and 2- and 4-acetylamino-fluorene, *Toxicology Letters*, **71**, 235–243.

Lewis, D.F.V., Ioannides, C. and Parke, D.V. (1995a) A quantitative structure-activity relationship study on a series of 10 para-substituted toluenes binding to cytochrome P450 2B4 (CYP2B4) and also their hydroxylation rates, *Biochemical Pharmacology*, **50**, 619–625.

Lewis, D.F.V., Ioannides, C. and Parke, D.V. (1995b) Molecular orbital-generated QSARs in an homologous series of alkoxy resorufins and studies of their interactive docking with cytochromes P450, *Xenobiotica*, **25**, 1355–1369.

Lewis, D.F.V., Eddershaw, P.J., Goldfarb, P.S. and Tarbit, M.H. (1996) Molecular modelling of CYP3A4 from an alignment with CYP102: identification of key interactions between putative active site residues and CYP3A-specific chemicals, *Xenobiotica*, **26**, 1067–1086.

Lewis, D.F.V., Eddershaw, P.J., Goldfarb, P.S. and Tarbit, M.H. (1997a) Molecular modelling of cytochrome P4502D6 (CYP2D6) based on an alignment with CYP102: structural studies on specific CYP2D6 substrate metabolism, *Xenobiotica*, **27**, 319–340.

Lewis, D.F.V, Bird, M.G. and Parke, D.V. (1997b) Molecular modelling of CYP2E1 enzymes from rat, mouse and man: an explanation for species differences in butadiene metabolism and potential carcinogenicity, and rationalization of CYP2E substrate specificity, *Toxicology*, **118**, 93–113.

Lewis, D.F.V., Ioannides, C. and Parke, D.V. (1998a) Cytochrome P450 and species differences in xenobiotic metabolism and activation of carcinogens, *Environmental Health Perspectives*, **106**, 633–641.

Lewis, D.F.V., Watson, E. and Lake, B.G. (1998b) Evolution of the cytochrome P450 superfamily: sequence alignments and pharmacogenetics, *Mutation Research*, **410**, 245–270.

Lewis, D.F.V., Dickins, M., Weaver, R.J., Eddershaw, P.J., Goldfarb, P.S. and Tarbit, M.H. (1998c) Molecular modelling of human CYP2C subfamily enzymes CYP2C9 and CYP2C19: rationalization of substrate specificity and site-directed mutagenesis experiments in the CYP2C subfamily, *Xenobiotica*, **28**, 235–268.

Lewis, D.F.V., Eddershaw, P.J., Dickins, M., Tarbit, M.H. and Goldfarb, P.S. (1998d) Structural determinants of P450 substrate specificity, binding affinity and catalytic rate, *Chemico-Biological Interactions*, **115**, 175–199.

Lewis, D.F.V., Ioannides, C. and Parke, D.V. (1998e) An improved and updated version of the COMPACT procedure for the evaluation of P450-mediated chemical activation, *Drug Metabolism Reviews*, **30**, 709–737.

Lewis, D.F.V., Ioannides, C. and Parke, D.V. (1998f) Further validation of the COMPACT approach for the prospective safety evaluation of chemicals. Re-evaluation of 200 miscellaneous chemicals by comparison with rodent carcinogenicity data from the U.S. NCI/NTP, *Mutation Research*, **412**, 41–54.

Lewis, D.F.V., Wiseman, A. and Tarbit, M.H. (1999a) Molecular modelling of lanosterol 14α-demethylase (CYP51) from *Saccharomyces cerevisiae* via homology with CYP102, a unique bacterial cytochrome P450 isoform: quantitative structure-activity relationships (QSARs), *Journal of Enzyme Inhibition*, **14**, 175–192.

Lewis, D.F.V., Dickins, M., Eddershaw, P.J., Tarbit, M.H. and Goldfarb, P.S. (1999b) Cytochrome P450 substrate specificities, substrate structural templates and enzyme active site geometries, *Drug Metabolism and Drug Interactions*, **15**, 1–49.

Lewis, D.F.V., Lake, B.G., George, S., Dickins, M., Beresford, A.P., Eddershaw, P.J., Tarbit, M.H., Goldfarb, P.S. and Guengerich, F.P. (1999c) Molecular modelling of CYP1 family isoforms CYP1A1, CYP1A2, CYP1A6 and CYP1B1 based on sequence homology with CYP102, *Toxicology*, **139**, 53–79.

Lewis, D.F.V., Dickins, M., Lake, B.G., Eddershaw, P.J., Tarbit, M.M. and Goldfarb, P.S. (1999d) Molecular modelling of the human cytochrome P450 isoform CYP2A6 and investigations of CYP2A substrate selectivity, *Toxicology*, **133**, 1–33.

Lewis, D.F.V., Lake, B.G., Dickins, M., Eddershaw, P.J., Tarbit, M.H. and Goldfarb, P.S. (1999e) Molecular modelling of the phenobarbital-inducible P450 isoforms: CYP2B1,

CYP2B4 and CYP2B6 by homology with the substrate-bound CYP102 crystal structure, and evaluation of CYP2B substrate binding affinity, *Xenobiotica*, **29**, 361–393.

Lewis, D.F.V., Dobrota, M., Taylor, M. and Parke, D.V. (1999f) Metal toxicity and redox potential: a QSAR evaluation, *Environmental Toxicology and Chemistry*, **18**, 2199–2204.

Lewis, D.F.V., Lake, B.G., Bird, M.G., Dickins, M., Eddershaw, P.J., Tarbit, M.H. and Goldfarb, P.S. (2000a) Molecular modelling of human CYP2E1 by homology with the CYP102 hemoprotein domain: investigation of the interactions of substrates and inhibitors within the putative active site of the CYP2E1 isoform, *Xenobiotica*, **50**, 1–25.

Lewis, D.F.V., Ioannides, C., Schulte-Hermann and Parke, D.V. (2000b) Quantitative structure-activity relationships in a series of endogenous and synthetic steroids exhibiting induction of CYP3A activity and hepatomegaly associated with increased DNA sythesis, *Journal of Steroid Biochemistry and Molecular Biology*, **74**, 179–185.

Lewis, D.F.V., Modi, S. and Dickins, M. (2001) Quantitative structure-activity relationships (QSARs) within substrates of human cytochromes P450 involved in drug metabolism, *Drug Metabolism Reviews*, submitted.

Li, A.P., Kaminski, D.L. and Rasmussen, A. (1995) Substrates of human hepatic cytochrome P450 3A4, *Toxicology*, **104**, 1–8.

Li, H. and Poulos, T.L. (1994) Structural variation in heme enzymes: a comparative analysis of peroxidase and P450 crystal structures, *Structure*, **2**, 461–464.

Li, H. and Poulos, T.L. (1995) Modeling protein-substrate interactions in the heme domain of cytochrome $P450_{BM3}$, *Acta Crystallographica*, **D51**, 21–32.

Li, H. and Poulos, T.L. (1997) The structure of the cytochrome P450BM-3 haem domain complexed with the fatty acid substrate, palmitoleic acid, *Nature Structural Biology*, **4**, 140–146.

Li, Q.-S., Schwaneberg, U., Fischer, P. and Schmid, R.D. (2000) Directed evolution of the fatty acid hydroxylase P450 BM3 into an indole-hydroxylating catalyst, *Chemistry: A European Journal*, **6**, 1531–1536.

Lightfoot, T., Ellis, S.W., Mahling, J., Ackland, M.J., Blaney, F.E., Bijloo, G.J., de Groot, M.J., Vermeulen, N.P.E., Blackburn, G.M., Lennard, M.S. and Tucker, G.T. (2000) Regioselective hydroxylation of debrisoquine by cytochrome P4502D6: implications for active site modelling, *Xenobiotica*, **30**, 219–233.

Lijinsky, W. (1993) Life-span and cancer: the induction time of tumours in diverse animal species treated with nitrosodiethylamine, *Carcinogenesis*, **14,** 2373–2375.

Lin, F.H., Stohs, S.J., Birnbaum, L.S., Clark, G., Lucier, G.W. and Goldstein, J.A. (1991) The effects of 2,3,7,8-tetrachlorodibenzo-p-dioxin (TCDD) on the hepatic estrogen and glucocorticoid receptors in congeneric strains of Ah responsive and Ah nonresponsive C57BL/6J mice, *Toxicology and Applied Pharmacology*, **108**, 129–139.

Lin, J.H. (1998) Applications and limitations of interspecies scaling and *in vitro* extrapolation in pharmacokinetics, *Drug Metabolism and Disposition*, **26**, 1202–1212.

Lindberg, R.L.P. and Negishi, M. (1989) Alteration of mouse cytochrome $P450_{coh}$ substrate specificity by mutation of a single amino-acid residue, *Nature*, **339**, 632–634.

Lippard, S.J. and Berg, J.M. (1994) *Principles of Bioinorganic Chemistry*, University Science Books, Mill Valley, CA.

Lipscomb, J.D. (1980) Electron paramagnetic resonance detectable states of cytochrome $P450_{cam}$, *Biochemistry*, **19**, 3590–3599.

Lipscomb, J.D., Sligar, S.G., Namtvedt, M.J. and Gunsalus, I.C. (1976) Autooxidation and hydroxylation reactions of oxygenated cytochrome $P450_{cam}$, *Journal of Biological Chemistry*, **251**, 1116–1124.

Liu, P.T., Ioannides, C., Shavila, J., Symons, A.M. and Parke, D.V. (1993) Effects of ether anaesthesia and fasting on various cytochrome P450 of rat liver and kidney, *Biochemical Pharmacology*, **45**, 871–877.

Loew, G.H. and Harris, D.L. (2000) Role of the heme active site and protein environment in structure, spectra and function of the cytochrome P450s, *Chemical Reviews*, **100**, 407–419.

Loew, G.H. and Kirchner, R.F. (1975) Electronic structure and electric field gradients in oxyhemoglobin and cytochrome P450 model compounds, *Journal of the American Chemical Society*, **97**, 7388–7390.

Loew, G.H., Kert, C.J., Hjelmeland, L.M. and Kirchner, R.F. (1977) Active site models of horseradish peroxidase compound 1 and a cytochrome P450 analogue: electronic structure and electric field gradients, *Journal of the American Chemical Society*, **99**, 3534–3536.

Lorenz, J., Glatt, H.R., Fleischmann, R., Ferlinz, R. and Oesch, F. (1984) Drug metabolism in man and its relationship to that in three rodent species: monooxygenase, epoxide hydrolase, and glutathione S-transferase activities in subcellular fractions of lung and liver, *Biochemical Medicine*, **32**, 43–56.

Lovelock, J.E. (1988) *The Ages of Gaia*, Oxford University Press, Oxford.

Lozano, J.J., Lopez-de-Brinas, E., Centeno, N.B., Guigo, R. and Sanz, F. (1997) Three-dimensional modelling of human cytochrome P4501A2 and its interaction with caffeine and MeIQ, *Journal of Computer-Aided Molecular Design*, **11**, 395–408.

Lu, A.Y.H. (1998a) A journey in cytochrome P450 and drug metabolism research, *Drug Metabolism and Disposition*, **26**, 1168–1173.

Lu, A.Y.H. (1998b) Drug metabolism research challenges in the new millenium: individual variability in drug therapy and drug safety, *Drug Metabolism and Disposition*, **26**, 1217–1222.

Mackenzie, P.I. (1990) Structure and regulation of UDP glucuronosyl transferases, *Frontiers in Biotransformation*, **2**, 211–243.

Mackman, R., Guo, Z., Guengerich, F.P. and Ortiz de Montellano, P.R. (1996) Active site topology of human cytochrome P450 2E1, *Chemical Research in Toxicology*, **9**, 223–226.

Makita, K., Falck, J.R. and Capdevila, J.H. (1996) Cytochrome P450, the arachidonic acid cascade, and hypertension: new vistas for an old enzyme system, *FASEB Journal*, **10**, 1456–1463.

Mancy, A., Antignac, M., Minoletti, C., Dijols, S., Mouries, V., Ha Duong, N.-T., Battioni, P., Dansette, P.M. and Mansuy, D. (1999) Diclofenac and its derivatives as tools for studying human cytochromes P450 active sites: particular efficiency and regioselectivity of P450 2Cs, *Biochemistry*, **38**,14264–14270.

Mansuy, D. (1998) The great diversity of reactions catalyzed by cytochromes P450, *Comparative Biochemistry and Physiology*, Part C, **121**, 5–14.

Mansuy, D and Renaud, J.-P. (1995) Heme-thiolate proteins different from the cytochromes P450 catalyzing monooxygenations, in: *Cytochrome P450* (P.R. Ortiz de Montellano, ed.) Plenum, New York, Chapter 15, 537–574.

Marcus, R.A. and Sutin, N. (1985) Electron transfers in chemistry and biology, *Biochimica et Biophysica Acta*, **811**, 265–322.

Margulis, L. and Sagan, D. (1995) *What is Life?*, Weidenfeld & Nicolson, London.

Marnett, L.J. and Kennedy, T.A. (1995) Comparison of the peroxidase activity of hemoproteins and cytochrome P450, in: *Cytochrome P450* (P.R. Ortiz de Montellano, ed.) Plenum, New York, Chapter 2, 49–80.

Martin, A.P. and Palumbi, S.R. (1993) Body size, metabolic rate, generation time, and the molecular clock, *Proceedings of the National Academy of Sciences USA*, **90**, 4087–4091.

Martinis, S.A., Atkins, W.M., Stayton, P.S. and Sligar, S.G. (1989) A conserved residue of cytochrome P450 is involved in heme-oxygen stability and activation, *Journal of the American Chemical Society*, **111**, 9252–9253.

Martinis, S.A., Ropp, J.D., Sligar, S.G. and Gunsalus, I.C. (1991) Molecular recognition by cytochrome $P450_{cam}$: substrate specificity, catalysis and electron transfer, *Frontiers in Biotransformation*, **4**, 54–86.

Martucci, C.P. and Fishman, J. (1993) P450 enzymes of estrogen metabolism, *Pharmacology and Therapeutics*, **57**, 237–257.

Masters, B.S.S., McMillan, K., Sheta, E.A., Nishimura, J.S., Roman, L.J. and Martasek, P. (1996) Neuronal nitric oxide synthase, a modular enzyme formed by convergent evolution: structure, studies of a cysteine thiolate-liganded heme protein that hydroxylates L-arginine to produce NO as a cellular signal, *FASEB Journal*, **10**, 552–558.

Mathews, F.S., Argos, P. and Levine, M. (1972) The structure of cytochrome b_5 at 2.0Å resolution, *Cold Spring Harbor Symposium on Quantum Biology*, **36**, 387–395.

Maurel, P. (1996) The CYP3 family, in: *Cytochromes P450 – Metabolic and Toxicological Consequences* (C. Ioannides, ed.) CRC Press, Boca Raton, FL, Chapter 10, 241–270.

McKay, J.A., Murray, G.I., Ah-See, A.K., Greenlee, W.F., Marcus, C.B., Burke, M.D. and Melvin, W.T. (1996) Differential expression of CYP1A1 and CYP1B1 in human breast cancer, *Biochemical Society Transactions*, **24**, 3275.

McKenna, N.J., Xu, J., Nawaz, Z., Tsai, S.Y., Tsai, M.-J. and O'Malley, B.W. (1999) Nuclear receptor cofactors: multiple enzymes, multiple complexes, multiple functions, *Journal of Steroid Biochemistry and Molecular Biology*, **69**, 3–12.

McKinnon, R.A. and Nebert, D.W. (1998) Cytochrome P450 knockout mice: new toxicological models, *Clinical and Experimental Pharmacology and Physiology*, **25**, 783–787.

McKnight, J., Cheeseman, M.R., Thomson, A.J., Miles, J.S. and Munro, A.W. (1993) Identification of charge-transfer transitions in the optical spectrum of low-spin ferric cytochrome P450 *Bacillus megaterium*, *European Journal of Biochemistry*, **213**, 683–687.

Michalik, L. and Wahli, W. (1999) Peroxisome proliferator-activated receptors: three isotypes for a multitude of functions, *Current Opinion in Biotechnology*, **10**, 564–570.

Micka, J., Milatovich, A., Menon, A., Grabowski, G.A., Puga, A. and Nebert, D.W. (1997) Human Ah receptor (AhR) gene: localization to 7p15 and suggestive correlation of polymorphism with CYP1A1 inducibility, *Pharmacogenetics*, **7**, 95–101.

Miles, J.S. and Wolf, C.R. (1991) Developments and perspectives on the role of cytochrome P450s in chemical carcinogenesis, *Carcinogenesis*, **12**, 2195–2199.

Miles, J.S., Munro, A.W., Rospendowski, B.N., Smith, W.E., McKnight, J. and Thomson, A.J. (1992) Domains of the catalytically self-sufficient cytochrome $P450_{BM3}$, *Biochemical Journal*, **288**, 503–509.

Miners, J.O. and Birkett, D.J. (1998) Cytochrome P4502C9: an enzyme of major importance in human drug metabolism, *British Journal of Clinical Pharmacology*, **45**, 525–538.

Miura, Y., Hisaki, H., Oda, S. and Nagai, T. (1989) Oxygenation of fatty acids by *suncus* kidney microsomes, in: *Cytochrome P450 – Biochemistry and Biophysics* (I. Schuster, ed.) Taylor & Francis, London, 125–128.

Modi, S., Primrose, W.V., Lian, L.-Y. and Roberts, G.C.K. (1995) Effect of replacement of ferriprotoporphyrin IX in the haem domain of cytochrome $P450_{BM3}$ on substrate binding and catalytic activity, *Biochemical Journal*, **310**, 939–943.

Modi, S., Paine, M.J., Sutcliffe, M.J., Lian, L.Y., Primrose, W.U., Wolf, C.R. and Roberts, G.C.K. (1996a) A model for human cytochrome P450 2D6 based on homology modelling and NMR studies of substrate binding, *Biochemistry*, **35**, 4540–4550.

Modi, S., Sutcliffe, M.J., Primrose, W.U., Lian, L.-Y. and Roberts, G.C.K. (1996b) The catalytic mechanism of cytochrome $P450_{BM3}$ involves a 6Å movement of the bound substrate on reduction, *Nature Structural Biology*, **3**, 414–417.

Mojzsis, S.J., Arrhenius, G., McKeegan, K.D., Harrison, T.M., Nutman, A.P. and Friend, C.R.L. (1996) Evidence for life on Earth before 3,800 million years ago, *Nature*, **384**, 55–59.

Moore, L.B., Parks, D.J., Jones, S.A., Bledsoe, R.K., Cousler, T.G., Stimmel, J.B., Goodwin, B., Liddle, C., Blanchard, S.G., Willson, T.M., Collins, J.L. and Kliewer, S.A. (2000a) Orphan nuclear receptors constitutive antrostane receptor and pregnane X receptor share xenobiotic and steroid ligands, *Journal of Biological Chemistry*, **275**, 15122–15127.

Moore, L.B., Goodwin, B., Jones, S.A., Wisely, G.B., Serabjit-Singh, C.J., Willson, T.M., Collins, J.L. and Kliewer, S.A. (2000b) St. John's Wort induces hepatic drug metabolism through activation of the pregnane X receptor, *Proceedings of the New York Academy of Sciences USA*, **97**, 7500–7502.

Morgan, E.T. (1997) Regulation of cytochrome P450 during inflammation and infection, *Drug Metabolism Reviews*, **29**, 1129–1188.

Morgan, E.T., Sewer, M.B., Iber, H., Gonzalez, F.J., Lee, Y.-H., Tukey, R.H., Okino, S., Vu, T., Chen, Y.-H., Sidhu, J.S. and Omiecinski, C.J. (1998) Physiological and pathophysiological regulation of cytochrome P450, *Drug Metabolism and Disposition*, **26**, 1232–1240.

Morohashi, K.-I. and Omura, T. (1996) Ad4BP/SF-1, a transcription factor essential for the transcription of steroidogenic cytochrome P450 genes and for the establishment of the reproductive function, *FASEB Journal*, **10**, 1569–1577.

Morrison, H.G., Oleksiak, M.F., Cornell, N.W., Sogin, M.L. and Stegeman, J.J. (1995) Identification of cytochrome P450 1A (CYP1A) genes from two teleost fish, toadfish (*Opsanus tau*) and scup (*Stenotomus chrysops*) and phylogenetic analysis of CYP1A genes, *Biochemical Journal*, **308**, 97–104.

Moser, C.C. and Dutton, P.L. (1992) Engineering protein structure for electron transfer function in photosynthetic reaction centers, *Biochimica et Biophysica Acta*, **1101**, 171–176.

Mueller, A., Mueller, J.J., Uhlmann, H., Bernhardt, R. and Heinemann, U. (1998) New aspects of electron transfer revealed by the crystal structure of a truncated bovine adrenodoxin, Adx (4–108), *Structure*, **6**, 269–280.

Mueller, E.J., Loida, P.J. and Sligar, S.G. (1995) Twenty-five years of $P450_{cam}$ research, in: *Cytochrome P450* (P.R. Ortiz de Montellano, ed.) Plenum, New York, Chapter 3, 83–124.

Mukhtar, H. and Khan, W.A. (1989) Cutaneous cytochrome P450, *Drug Metabolism Reviews*, **20**, 657–673.

Müller, H.-G., Schunk, W.-H. and Kargel, E. (1991) Cytochromes P450 in alkane-assimilating yeasts, *Frontiers in Biotransformation*, **4**, 87–126.

Munro, A. (1994) P450 Biotechnology, *The Biochemist*, Apr/May issue, 50–52.

Munro, A.W. and Lindsay, J.G. (1996) Bacterial cytochromes P450, *Molecular Microbiology*, **20**, 1115–1125.

Munro, A.W., Malarkey, K. and Miles, J.S. (1992) Investigating the function of cytochrome P450 BM-3: a role for the phylogenetically conserved tryptophan residue, *Biochemical Society Transactions*, **21**, 66(S).

Munro, A.W., Malarkey, K., McKnight, J., Thomson, A.J., Kelly, S.M., Price, N.C., Lindsay, J.G., Coggins, J.R. and Miles, J.S. (1994a) The role of tryptophan 97 of cytochrome $P450_{BM3}$ from *Bacillus megaterium* in catalytic function, *Biochemical Journal*, **303**, 423–428.

Munro, A.W., Lindsay, J.G., Coggins, J.R., MacDonald, I., Smith, W.E. and Rospendowski, B.N. (1994b) Resonance Raman spectroscopic studies on intact cytochrome P450 BM3, *Biochemical Society Transactions*, **22**, 54S.

Munro, A.W., Daff, S., Chapman, S.K., Cook, R.M., Lindsay, J.G. and Coggins, J.R. (1996a) Kinetic analysis of P450 BM3 from *Bacillus megaterium*, in: *Flavins and Flavoproteins*, University of Calgary Press, Calgary, 467–470.

Munro, A.W., Daff, S., Coggins, J.R., Lindsay, J.G. and Chapman, S.K. (1996b) Probing electron transfer in flavocytochrome P450 BM3 and its component domains, *European Journal of Biochemistry*, **239**, 403–409.

Munro, A.W., Noble, M.A., Miles, C.S., Daff, S.N., Green, A.J., Quaroni, L., Rivers, S., Ost, T.W.B., Reid, G.A. and Chapman, S.K. (1999) Flavocytochrome $P450_{BM3}$: a paradigm for the analysis of electron transfer and its control in the P450s, *Biochemical Society Transactions*, **27**, 190–196.

Murray, G.I., Foster, C.O., Barnes, T.S., Weaver, R.J., Ewen, S.W.B., Melvin, W.T. and Burke, M.D. (1991) Expression of cytochrome P4501A in breast cancer, *British Journal of Cancer*, **63**, 1021–1025.

Murray, M. (1992) P450 enzymes: inhibition mechanisms, genetic regulation and effects of liver disease, *Clinical Pharmacokinetic Concepts*, **23**, 132–146.

Murray, M. and Reidy, G.F. (1990) Selectivity in the inhibition of mammalian cytochromes P450 by chemical agents, *Pharmacological Reviews*, **42**, 85–101.

Nagai, M., Yoneyama, Y. and Kitagawa, T. (1991) Unusual CO bonding geometry in abnormal subunits of hemoglobin M Boston and hemoglobin M Saskatoon, *Biochemistry*, **30**, 6495–6503.

Nakahara, K., Shoun, H., Adachi, S.-I. and Shiro, Y. (1994) Crystallization and preliminary X-ray diffraction studies of nitric oxide reductase cytochrome P450nor from *Fusarium oxysporum*, *Journal of Molecular Biology*, **239**, 158–159.

Narhi, L.O. and Fulco, A.J. (1986) Characterization of a catalytically self-sufficient 119,000-Dalton cytochrome P450 monooxygenase induced by barbiturates in *Bacillus megaterium*, *Journal of Biological Chemistry*, **261**, 7160–7169.

Nebert, D.W. (1997a) Polymorphisms in drug-metabolizing enzymes: what is their clinical relevance and why do they exist?, *American Journal of Human Genetics*, **60**, 265–271.

Nebert, D.W. (1997b) Pharmacogenetics: 65 candles on the cake, *Pharmacogenetics*, **7**, 435–440.

Nebert, D.W. and Carvan, M.J. (1997) Ecogenetics: from ecology to health, *Toxicology and Industrial Health*, **13**, 163–192.

Nebert, D.W. and Duffy, J.J. (1997) How knockout mouse lines will be used to study the role of drug-metabolizing enzymes and their receptors during reproduction and development, and in environmental toxicity, cancer and oxidative stress, *Biochemical Pharmacology*, **53**, 249–254.

Nebert, D.W. and Gonzalez, F.J. (1987) P450 genes: structure, evolution and regulation, *Annual Review of Biochemistry*, **56**, 945–993.

Nebert, D.W., Nelson, D.R. and Feyereisen, R. (1989a) Evolution of the cytochrome P450 genes, *Xenobiotica*, **19**, 1149–1160.

Nebert, D.W., Nelson, D.R., Adesnik, M., Coon, M.J., Estabrook, R.W., Gonzalez, F.J., Guengerich, F.P., Gunsalus, I.C., Johnson, E.F., Kemper, B., Levin, W., Phillips, I.R., Sato, R. and Waterman, M.R. (1989b) The P450 superfamily: updated listing of all genes and recommended nomenclature for the chromosomal loci, *DNA*, **8**, 1–13.

Nebert, D.W., Petersen, D.D. and Fornace, A.J. (1990) Cellular responses to oxidative stress: the [Ah] gene battery as a paradigm, *Environmental Health Perspectives*, **88**, 13–25.

Nebert, D.W., Nelson, D.R., Coon, M.J., Estabrook, R.W., Feyereisen, R., Fujii-Kuriyama, Y., Gonzalez, F.J., Guengerich, F.P., Gunsalus, I.C., Johnson, E.F., Loper, J.C., Sato, R., Waterman, M.R. and Waxman, D.J. (1991a) The P450 superfamily: update on new sequences, gene mapping and recommended nomenclature, *DNA and Cell Biology*, **10**, 1–14.

Nebert, D.W., Petersen, D.D. and Puga, A. (1991b) Human Ah locus polymorphism and cancer: inducibility of CYP1A1 and other genes by combustion products and dioxin, *Pharmacogenetics*, **1**, 68–78.

Nebert, D.W., McKinnon, R.A. and Puga, A. (1996) Human drug-metabolizing enzyme polymorphisms: effect on risk of toxicity and cancer, *DNA and Cell Biology*, **15**, 273–280.

Nedelcheva, V. and Gut, I. (1994) P450 in the rat and man: methods of investigation, substrate specificities and relevance to cancer, *Xenobiotica*, **24**, 1151–1175.

Negishi, M., Iwasaki, M., Juvonen, R.O., Sueyoshi, T., Darden, T.A. and Pedersen, L.G. (1996a) Structural flexibility and functional versatility of cytochrome P450 and rapid evolution, *Mutation Research*, **350**, 43–50.

Negishi, M., Uno, T., Darden, T.A., Sueyoshi, T. and Pedersen, L.G. (1996b) Structural flexibility and functional versatility of mammalian P450 enzymes, *FASEB Journal*, **10**, 683–689.

Nelson, D.R. (1995) Cytochrome P450 nomenclature and alignment of selected sequences, in: *Cytochrome P450* (P.R. Ortiz de Montellano, ed.) Plenum, New York, Appendix A, 575–606.

Nelson, D.R. (1998) Metazoan cytochrome P450 evolution, *Comparative Biochemistry and Physiology*, Part C, **121**, 15–22.

Nelson, D.R. (1999) Cytochrome P450 and the individuality of species, *Archives of Biochemistry and Biophysics*, **369**, 1–10.

Nelson, D.R. and Strobel, H.W. (1987) Evolution of cytochrome P450 proteins, *Molecular Biology and Evolution*, **4**, 572–593.

Nelson, D.R., Kamataki, T., Waxman, D.J., Guengerich, F.P., Estabrook, R.W., Feyereisen, R., Gonzalez, F.J., Coon, M.J., Gunsalus, I.C., Gotoh, O., Okuda, K. and Nebert, D.W. (1993) The P450 superfamily: update on new sequences, gene mapping, accession numbers, early trivial names of enzymes and nomenclature, *DNA and Cell Biology*, **12**, 1–51.

Nelson, D.R., Koymans, L., Kamataki, T., Stegeman, J.J., Feyereisen, R., Waxman, D.J., Waterman, M.R., Gotoh, O., Coon, M.J., Estabrook, R.W., Gunsalus, I.C. and Nebert, D.W. (1996) P450 superfamily: update on new sequences, gene mapping, accession numbers and nomenclature, *Pharmacogenetics*, **6**, 1–42.

Nelson, S.D. (1982) Metabolic activation and drug toxicity, *Journal of Medicinal Chemistry*, **25**, 753–765.

Neubert, D. (1997) Vulnerability of the endocrine system to xenobiotic influence, *Regulatory Toxicology and Pharmacology*, **26**, 9–29.

Newcomb, M., Le Tadic, M.-H., Putt, D.A. and Hollenberg, P.F. (1995) An incredibly fast apparent oxygen rebound rate constant for hydrocarbon hydroxylation by cytochrome P450 enzymes, *Journal of the American Chemical Society*, **117**, 3312–3313.

Nims, R.W. and Lubet, R.A. (1996) The CYP2B subfamily, in: *Cytochromes P450 – Metabolic and Toxicological Consequences* (C. Ioannides, ed.) CRC Press, Boca Raton, FL, Chapter 6, 135–160.

Nishida, H., Inaka, K., Yamanaka, M., Kaida, S., Kobayashi, K. and Miki, K. (1995) Crystal structure of NADH-cytochrome b_5 reductase from pig liver at 2.4Å resolution, *Biochemistry*, **34**, 2763–2767.

Noble, M.A., Miles, C.S., Chapman, S.K., Lysek, D.A., Mackay, A.C., Reid, G.A., Hanzlik, R.P. and Munro, A.W. (1999) Roles of key active-site residues in flavocytochrome P450 BM3, *Biochemical Journal*, **339**, 371–379.

Nolte, R.T., Wisely, G.B., Westin, S., Cobb, J.E., Lambert, M.H., Kurokawa, R., Rosenfeld, M.G., Willson, T.M., Glass, C.K. and Milburn, M.V. (1998) Ligand binding and coactivator assembly of the peroxisome proliferator-activated receptor-γ, *Nature*, **395**, 137–143.

Nyarko, A.K., Kellner-Weibel, G.L. and Harvison, P.J. (1997) Cytochrome P450-mediated metabolism and nephrotoxicity of N-(3,5-dichlorophenyl)succinimide in Fischer 344 rats, *Fundamental and Applied Toxicology*, **37**, 117–124.

Oberley, L.W. and Oberley, T.D. (1995) Reactive oxygen species in the aetiology of cancer, in: *Drugs, Diet and Disease, Volume 1: Mechanistic Approaches to Cancer* (C. Ioannides and D.F.V. Lewis, eds.) Ellis Horwood, Chichester, 47–63.

Ogg, M.S. (1998) *In vitro assessment of the regulation of the human CYP3A4 gene*, PhD Thesis, School of Biological Sciences, University of Surrey, UK.

Okazaki, O. and Guengerich, F.P. (1993) Evidence for specific base catalysis in N-dealkylation reactions catalyzed by cytochrome P450 and chloroperoxidase, *Journal of Biological Chemistry*, **268**, 1546–1552.

Okey, A.B. (1990) Enzyme induction in the cytochrome P450 system, *Pharmacology and Therapeutics*, **45**, 241–298.

Okey, A.B. and Vella, L.M. (1982) Binding of 3-methylcholanthrene and 2,3,7,8-tetra-chlorodibenzo-p-dioxin to a common Ah receptor site in mouse and rat hepatic cytosols, *European Journal of Biochemistry*, **127**, 39–47.

Okita, R.T. and Masters, B.S.S. (1992) Biotransformations: the cytochromes P450, in: *Textbook of Biochemistry* (T.M. Devlin, ed.) Wiley-Liss, New York, 981–999.

Okuda, K., Ogishima, T. and Noshiro, M. (1993) Cholesterol 7α-hydroxylase and 12α-hydroxylase, in: *Cytochrome P450* (J.B. Schenkman and H. Griem, eds.) Springer-Verlag, Berlin, 601–610.

Oliver, C.F., Modi, S., Sutcliffe, M.J., Primrose, W.U., Lian, L.-Y. and Roberts, G.C.K. (1997) A single mutation in cytochrome P450 BM3 changes substrate orientation in a catalytic intermediate and the regiospecificity of hydroxylation, *Biochemistry*, **36**, 1567–1572.

Omata, Y., Aibara, K. and Ueno, Y. (1987) Conformation between the substrate-binding site and heme of cytochrome P450 studied by excitation energy transfer, *Biochimica et Biophysica Acta*, **912**, 115–123.

Omura, T. and Sato, R. (1962) A new cytochrome in liver microsomes, *Journal of Biological Chemistry*, **237**, 1375–1376.

Omura, T. and Sato, R. (1964) The carbon monoxide-binding pigment of liver microsomes, *Journal of Biological Chemistry*, **239**, 2370–2385.

Omura, T., Ishimura, Y. and Fujii-Kuriyama, Y. (1993) *Cytochrome P450*, 2nd edition, Kodansha, Tokyo.

Oprea, T., Hummer, G. and Garcia, A.E. (1997) Identification of a functional water channel in cytochrome P450 enzymes, *Proceedings of the National Academy of Sciences USA*, **94**, 2133–2138.

Orr, W.C, and Sohal, R.S. (1994) Extension of life-span by overexpression of superoxide dismutase and catalase in *drosophila melanogaster*, *Nature*, **263**, 1128–1130.

Ortiz de Montellano, P.R. (1986) *Cytochrome P450*, Plenum, New York.

Ortiz de Montellano, P.R. (1987) Control of the catalytic activity of prosthetic heme by the structure of hemoproteins, *Accounts of Chemical Research*, **20**, 354–359.

Ortiz de Montellano, P.R. (1989) Cytochrome P450 catalysis: radical intermediates and dehydrogenation reactions, *Trends in Pharmacological Science*, **10**, 354–359.

Ortiz de Montellano, P.R. (1995) *Cytochrome P450*, 2nd edition, Plenum, New York.

Ortiz de Montellano, P.R. (1998) Heme oxygenase mechanism: evidence for an electrophilic, ferric peroxide species, *Accounts of Chemical Research*, **31**, 543–549.

Ortiz de Montellano, P.R. and Correia, M.A. (1995) Inhibition of cytochrome P450 enzymes, in: *Cytochrome P450* (P.R. Ortiz de Montellano, ed.) Plenum, New York, Chapter 9, 305–364.

Ortiz de Montellano, P.R. and Graham-Lorence, S.E. (1993) Structure of cytochrome P450: heme-binding and heme reactivity, in: *Cytochrome P450* (J.B. Schenkman and H. Griem, eds.) Springer-Verlag, Berlin, 169–181.

Ortiz de Montellano, P.R., Nishida, C., Rodriguez-Crespo, I. and Gerber, N. (1998) Nitric oxide synthase structure and electron transfer, *Drug Metabolism and Disposition*, **26**, 1185–1189.

Otsuka, J. (1970) One interpretation of the thermal equilibrium between high-spin and low-spin states in ferrihemoproteins, *Biochimica et Biophysica Acta*, **214**, 233–235.

Ouzonis, C.A. and Melvin, W.T. (1991) Primary and secondary structural patterns in eukaryotic cytochrome P450 families correspond to structures of the helix-rich domain of *Pseudomonas putida* cytochrome $P450_{cam}$, *European Journal of Biochemistry*, **198**, 307–315.

Paine, A.J. (1995) Heterogeneity of cytochrome P450 and its toxicological significance, *Human and Experimental Toxicology*, **14**, 1–7.

Parikh, A., Josephy, P.D. and Guengerich, F.P. (1999) Selection and characterization of human cytochrome P450 1A2 mutants with altered catalytic properties, *Biochemistry*, **38**, 5283–5289.

Park, S.-Y., Shimizu, H., Adachi, S.-I., Nakagawa, A., Tanaka, I., Nakahara, K., Shoun, H., Obayashi, E., Nakamura, H., Iizuka, T. and Shiro, Y. (1997) Crystal structure of nitric oxide reductase from denitrifying fungus *Fusarium oxysporum*, *Nature Structural Biology*, **4**, 827–832.

Parke, D.V. (1984) The cytochromes P450 and mechanisms of chemical carcinogenesis, *Environmental Health Perspectives*, **102**, 852–853.

Parke, D.V. (1987) Activation mechanisms to chemical toxicity, *Archives of Toxicology*, **60**, 5–15.

Parke, D.V. (1994) The cytochromes P450 and mechanisms of chemical carcinogenesis, *Environmental Health Perspectives*, **102**, 852–853.

Parke, D.V. and Ioannides, C. (1994) The effect of nutrition on chemical toxicity, *Drug Metabolism Reviews*, **26**, 739–765.

Parkinson, A. (1996) An overview of current cytochrome P450 technology for assessing the safety and efficacy of new materials, *Toxicologic Pathology*, **24**, 45–57.

Parodi, A.J. (2000) Role of N-oligosaccharide endoplasmic reticulum processing reactions in glycoprotein folding and degradation, *Biochemical Journal*, **348**, 1–13.

Parton, R.F., Vankelecom, I.F.J., Casselman, M.J.A., Bezoukhanova, C.P., Uytterhoeven, J.B. and Jacobs, P.A. (1994) An efficient mimic of cytochrome P450 from a zeolite-encaged iron complex in a polymer membrane, *Nature*, **370**, 541–544.

Pascussi, J.-M., Drocourt, L., Fabre, J.-M., Maurel, P. and Vilarem, M.-J. (2000) Dexamethasone induces pregnane X receptor and retinoid X receptor-α expression in human hepatocytes: synergistic increase of CYP3A4 induction by pregnane X receptor activators, *Molecular Pharmacology*, **58**, 361–372.

Patzelt, H. and Woggon, W.-D. (1992) O-insertion into nonactivated C-H bonds: the first observation of the O_2 cleavage by a P450 enzyme model in the presence of a thiolate ligand, *Helvetica Chimica Acta*, **75**, 523–530.

Paulsen, M.D. and Ornstein, R.L. (1992) Predicting the product specificity and coupling of cytochrome $P450_{cam}$, *Journal of Computer-Aided Molecular Design*, **6**, 449–460.

Paulsen, M.D. and Ornstein, R.L. (1995) Dramatic differences in the motions of the mouth of open and closed cytochrome $P450_{BM-3}$ by molecular dynamics simulations, *Proteins: Structure, Function and Genetics*, **21**, 237–243.

Paulsen, M.D., Manchester, J.I. and Ornstein, R.L. (1996) Using molecular modeling and molecular dynamics simulation to predict P450 oxidation products, *Methods in Enzymology*, **272**, 347–357.

Payne, V.A., Chang, Y.-T. and Loew, G.H. (1999a) Homology modeling and substrate binding study of human CYP2C9 enzyme, *Proteins: Structure, Function and Genetics*, **37**, 176–190.

Payne, V.A., Chang, Y.-T. and Loew, G.H. (1999b) Homology modeling and substrate binding study of human CYP2C18 and CYP2C19 enzymes, *Proteins: Structure, Function and Genetics*, **37**, 204–217.

Peet, D.J., Turley, S.D., Ma, W., Janowski, B.A., Lobaccaro, J.-M.A., Hammer, R.E. and Mangelsdorf, D.J. (1998) Cholesterol and bile acid metabolism are impaired in mice lacking the nuclear oxysterol receptor LXRα, *Cell*, **93**, 693–704.

Peisach, J. and Blumberg, W.E. (1970) Electron paramagnetic resonance of the high- and low-spin forms of cytochrome P450 in liver and in liver microsomes from a methylcholanthrene-treated rabbit, *Proceedings of the National Academy of Sciences USA*, **67**, 171–179.

Peisach, J., Mims, W.B. and Davis, J.L. (1979) Studies of the electron-nuclear coupling between Fe(III) and ^{14}N in cytochrome P450 and in a series of low-spin heme compounds, *Journal of Biological Chemistry*, **254**, 12379–12389.

Pelkonen, O., Maenpaa, J., Taavitsainen, P., Rautio, A. and Raunio, H. (1998) Inhibition and induction of human cytochrome P450 (CYP) enzymes, *Xenobiotica*, **28**, 1203–1253.

Pelkonen, O., Rautio, A., Raunio, H. and Pasanen, M. (2000) CYP2A6: a human coumarin 7-hydroxylase, *Toxicology*, **144**, 139–147.

Pelletier, H. and Kraut, J. (1992) Crystal structure of a complex between electron transfer partners, cytochrome c peroxidase and cytochrome c, *Science*, **258**, 1748–1755.

Peterson, J.A. and Graham-Lorence, S.E. (1995) Bacterial P450s: structural similarities and functional differences, in: *Cytochrome P450* (P.R. Ortiz de Montellano, ed.) Plenum, New York, Chapter 5, 151–180.

Peterson, J.A. and Mock, D.M. (1979) Cytochrome $P450_{cam}$ and putidaredoxin interaction during electron transfer, *Acta Biologica et Medica Germanica*, **38**, 153–162.

Peterson, J.A., Ullrich, V. and Hildebrandt, A.G. (1971) Metyrapone interaction with *Pseudomonas putida* cytochrome P450, *Archives of Biochemistry and Biophysics*, **145**, 531–542.

Peterson, J.A., Sevrioukova, I., Truan, G. and Graham-Lorence, S.E. (1997) $P450_{BM3}$: a tale of two domains – or is it three?, *Steroids*, **62**, 117–123.

Petzold, D.R., Rein, H., Schwarz, D., Sommer, S. and Ruckpaul, K. (1985) Relation between the structure of benzphetamine analogues and their binding properties to cytochrome $P450_{LM2}$, *Biochimica et Biophysica Acta*, **829**, 253–261.

Phillips, I.R. and Shephard, E.A. (eds.) (1998) *Cytochrome P450 Protocols*, Humana Press, Totowa, NJ.

Philson, S.B., Debrunner, P.G., Schmidt, P.G. and Gunsalus, I.C. (1979) The effect of cytochrome $P450_{cam}$ on the NMR relaxation rate of water protons, *Journal of Biological Chemistry*, **254**, 10173–10179.

Pinot, F., Benveniste, I., Salaun, J.-P., Loreau, O., Noel, J.-P., Schreiber, L. and Durst, F. (1999) Production *in vitro* by the cytochrome P450 CYP94A1 of major cutin monomers and potential messengers in plant-pathogen interactions: enantioselectivity studies, *Biochemical Journal*, **342**, 27–32.

Pliska, V., Testa, B. and van de Waterbeemd, H. (eds.) (1996) *Lipophilicity in Drug Action and Toxicology*, Verlag Helvetica Chimica, Weinheim.

Pochapsky, T.C., Ye, X.M., Ratnaswamy, G. and Lyons, T.A. (1994) An NMR-derived model for the solution structure of oxidized putidaredoxin, a 2-Fe, 2-S ferredoxin from *pseudomonas*, *Biochemistry*, **33**, 6424–6432.

Poland, A. and Knutson, J.C. (1982) 2,3,7,8-Tetrachlorodibenzo-p-dioxin and related halogenated aromatic hydrocarbons: examination of the mechanism of toxicity, *Annual Review of Pharmacology and Toxicology*, **22**, 517–554.

Poli-Scaife, S., Attais, R., Dansette, P.M. and Mansuy, D. (1997) The substrate binding site of human liver cytochrome P450 2C9: an NMR study, *Biochemistry*, **36**, 12672–12682.

Porter, T.D. and Coon, M.J. (1991) Cytochrome P450: multiplicity of isoforms, substrates, and catalytic and regulatory mechanisms, *Journal of Biological Chemistry*, **266**, 13469–13472.

Porter, T.D. and Kasper, C.B. (1986) NADPH-cytochrome P450 oxidoreductase: Flavin mononucleotide and flavin adenine dinucleotide domains evolved from different flavoproteins, *Biochemistry*, **25**, 1682–1687.

Poulos, T.L. (1986) The crystal structure of cytochrome $P450_{cam}$, in: *Cytochrome P450* (P.R. Ortiz de Montellano, ed.) Plenum, New York, Chapter 13, 505–523.

Poulos, T.L. (1988) Cytochrome P450: molecular architecture, mechanism and prospects for rational inhibitor design, *Pharmaceutical Research*, **5**, 67–75.

Poulos, T.L. (1991) Modelling of mammalian P450s on the basis of the $P450_{cam}$ X-ray crystal structure, *Methods in Enzymology*, **206**, 11–30.

Poulos, T.L. (1996a) Approaches to crystallizing P450s, *Methods in Enzymology*, **272**, 358–368.

Poulos, T.L. (1996b) Ligands and electrons and haem proteins, *Nature Structural Biology*, **3**, 401–403.

Poulos, T.L. and Howard, A.J. (1987) Crystal structures of metyrapone- and phenyl imidazole-inhibited complexes of cytochrome $P450_{cam}$, *Biochemistry*, **26**, 8165–8174.

Poulos, T.L. and Raag, R. (1992) Cytochrome $P450_{cam}$: crystallography, oxygen activation and electron transfer, *FASEB Journal*, **6**, 674–679.

Poulos, T.L., Finzel, B.C. and Howard, A.J. (1986) Crystal structure of substrate-free *Pseudomonas putida* cytochrome P450, *Biochemistry*, **25**, 5314–5322.

Poulos, T.L., Finzel, B.C. and Howard, A.J. (1987) High-resolution crystal structure of cytochrome $P450_{cam,}$ *Journal of Molecular Biology*, **195**, 687–700.

Poulos, T.L., Cupp-Vickery, J. and Li, H. (1995) Structural studies on prokaryotic cytochromes P450, in: *Cytochrome P450* (P.R. Ortiz de Montellano, ed.) Plenum, New York, Chapter 4, 125–150.

Powell, P.K., Wolf, I. and Lasker, J.M. (1996) Identification of CYP4A11 as the major lauric acid ω-hydroxylase in human liver microsomes, *Archives of Biochemistry and Biophysics*, **335**, 219–226.

Pratt, J.M., Ridd, T.I. and King, L.J. (1995) Activation of H_2O_2 by P450: evidence that the hydroxylating intermediate is iron(III)-coordinated H_2O_2 and not the ferryl FeO^{3+} complex, *Chemical Communications*, 2297–2298.

Price-Evans, D.A. (1993) *Genetic Factors in Drug Therapy*, Cambridge University Press, Cambridge.

Prough, R.A., Linder, M.W., Pinaire, J.A., Xiao, G.-H. and Falkner, K.C. (1996) Hormonal regulation of hepatic enzymes involved in foreign compound metabolism, *FASEB Journal*, **10**, 1369–1377.

Prueksaritanont, T., Ma, B., Tang, C., Meng, Y., Assang, C., Lu, P., Reider, P.J., Lin, J.M. and Baillie, T.A. (1999) Metabolic interactions between mibefradil and HMG-CoA reductase inhibitors: an *in vitro* investigation with human liver preparations, *British Journal of Clinical Pharmacology*, **47**, 291–298.

Puga, A., Nebert, D.W., McKinnon, R.A. and Menon, A.G. (1997) Genetic polymorphisms in human drug-metabolizing enzymes: potential uses of reverse genetics to identify genes of toxicological significance, *Critical Reviews in Toxicology*, **27**, 199–222.

Raag, R. and Poulos, T.L. (1989a) The structural basis for substrate-induced changes in redox potential and spin equilibrium in cytochrome $P450_{cam}$, *Biochemistry*, **28**, 917–922.

Raag, R. and Poulos, T.L. (1989b) Crystal structure of the carbon monoxide-substrate-cytochrome $P450_{cam}$ ternary complex, *Biochemistry*, **28**, 7586–7592.

Raag, R. and Poulos, T.L. (1992) X-ray crystallographic structural studies of cytochrome $P450_{cam}$: factors controlling substrate metabolism, *Frontiers in Biotransformation*, **7**, 1–43.

Raag, R., Swanson, B.A., Poulos, T.L. and Ortiz de Montellano, P.R. (1990) Formation, crystal structure and rearrangement of a cytochrome $P450_{cam}$ iron-phenyl complex, *Biochemistry*, **29**, 8119–8126.

Raag, R., Martinis, S.A., Sligar, S.G. and Poulos, T.L. (1991) Crystal structure of the cytochrome $P450_{cam}$ active site mutant Thr252Ala, *Biochemistry*, **30**, 11420–11429.

Rao, S., Aoyama, R., Schrag, M., Trager, W.F., Rettie, A. and Jones, J.R. (2000) A refined

3-dimensional QSAR of cytochrome P450 2C9: computational predictions of drug interactions, *Journal of Medicinal Chemistry*, **43**, 2789–2796.

Ravichandran, K.G., Boddupalli, S.S., Hasemann, C.A., Peterson, J.A. and Deisenhofer, J. (1993) Crystal structure of hemoprotein domain of $P450_{BM3}$, a prototype for microsomal P450s, *Science*, **261,** 731–736.

Rein, H., Jung, C., Ristau, O. and Friedrich, J. (1984) Biophysical properties of cytochrome P450, analysis of the reaction mechanism – thermodynamic aspects, in: *Cytochrome P450* (K. Ruckpaul and H. Rein, eds.) Akademie-Verlag, Berlin, Chapter 4, 163–249.

Renaud, J.P., Boucher, J.L., Vadon, S., Delaforge, M. and Mansuy, D. (1993) Particular ability of liver P450s 3A to catalyse the oxidation of Nω-hydroxyarginine to citrulline and nitrogen oxides and occurrence in NO synthases of a sequence very similar to the heme-binding sequence in P450s, *Biochemical and Biophysical Research Communications*, **192**, 53–60.

Renaud, J.-P., Rochel, N., Ruff, M., Vivat, V., Chambon, P., Gronemeyer, H, and Moras, D. (1995) Crystal structure of the RARγ ligand-binding domain bound to all-trans retinoic acid, *Nature*, **278**, 681–689.

Rendic, S. and DiCarlo, F.J. (1997) Human cytochrome P450 enzymes: a status report summarizing their reactions, substrates, inducers and inhibitors, *Drug Metabolism Reviews*, **29**, 413–580.

Reyes, H., Reisz-Porszasz, S. and Hankinson, O. (1992) Identification of the Ah receptor nuclear translocator protein (Arnt) as a component of the DNA binding form of the Ah receptor, *Science*, **256,** 1193–1195.

Richardson, T.H. and Johnson, E.F. (1994) Alterations of the regiospecificity of progesterone metabolism by the mutagenesis of two key amino acid residues in rabbit cytochrome P450 2C3v, *Journal of Biological Chemistry*, **269**, 23927–23943.

Richardson, T.H. and Johnson, E.F. (1996) The CYP2C Subfamily, in *Cytochromes P450: metabolic and toxological aspects* (C. Ioannides, ed.) CRC Press, Boca Raton, FL, Chapter 7, 161–181.

Ridderstrom, M., Masimirembwa, C., Trump-Kallmeyer, S., Ahlefelt, M., Otter, C. and Andersson, T.B. (2000) Arginines 97 and 108 in CYP2C9 are important determinants of catalytic function, *Biochemical and Biophysical Research Communications*, **270**, 983–987.

Ridley, M. (1996) *Evolution*, 2nd edition, Blackwell, London.

Roberts, G.C.K (1996) The other kind of biological NMR – studies of enzyme-substrate interactions, *Neurochemical Research*, **21**, 1117–1124.

Rodrigues, A.D. (1999) Integrated cytochrome P450 reaction phenotyping: attempting to bridge the gap between cDNA-expressed cytochromes P450 and native human liver microsomes, *Biochemical Pharmacology*, **57**, 465–480.

Ronis, M.J.J., Lindros, K.O. and Ingelman-Sundberg, M. (1996) The CYP2E subfamily, in: *Cytochromes P450 – metabolic and toxicological aspects* (C. Ioannides, ed.) CRC Press, Boca Raton, FL, Chapter 9, 211–239.

Rossi, M., Markovitz, S. and Callahan, T. (1987) Defining the active site of cytochrome P450: the crystal and molecular structure of an inhibitor, SKF 525A, *Carcinogenesis*, **8**, 881–887.

Rowland, K., Ellis, S.W., Lennard, M.S. and Tucker, G.T. (1993) Variation of human CYP2D6 (debrisoquine 4-hydroxylase) expressed in yeast: enzyme kinetics and inhibition by antidepressant drugs, *British Journal of Clinical Pharmacology*, **36**, 157–158.

Rozman, D. and Waterman, M.R. (1998) Lanosterol 14α-demethylase (CYP51) and spermatogenesis, *Drug Metabolism and Disposition*, **26**, 1199–1201.

Ruckpaul, K. (1978) 20 years of investigations on cytochrome P450, *Die Pharmazie*, **33**, 308–309.

Ruckpaul, K. and Rein, H. (1984) *Cytochrome P450*, Akademie-Verlag, Berlin.

Ruckpaul, K. and Rein, H. (1991) *Cytochrome P450 dependent biotransformation of endogenous substrates*, Akademie-Verlag, Berlin.

Ruckpaul, K., Rein, H. and Blanck, J. (1989) Regulation mechanisms of the activity of the hepatic endoplasmic cytochrome P450, *Frontiers in Biotransformation*, **1**, 1–65.

Russell. D.W. (1999) Nuclear orphan receptors control cholesterol catabolism, *Cell*, **97**, 539–542.

Ryan, D.E. and Levin, W. (1990) Purification and characterization of hepatic microsomal cytochrome P450, *Pharmacology and Therapeutics*, **45**, 153–239.

Sachse, C., Brockmoller, J., Bauer, S. and Roots, I. (1997) Cytochrome P450 2D6 variants in a Caucasian population: allele frequencies and phenotypic consequences, *American Journal of Human Genetics*, **60**, 284–295.

Safe, S., Bandiera, S., Sawyer, T., Zmudzka, B., Mason, G., Romkes, M., Denomme, M.A., Sparling, J., Okey, A.B. and Fujita, T. (1985) Effects of structure on binding to the 2,3,7,8-TCDD receptor protein and AHH induction: halogenated biphenyls, *Environmental Health Perspectives*, **61**, 21–33.

Sandhu, P., Guo, Z., Baba, T., Martin, M.V., Tukey, R.H. and Guengerich, F.P. (1994) Expression of modified human cytochrome P4501A2 in *Escherichia coli*: stabilization, purification, spectral characterization and catalytic activities of the enzyme, *Archives of Biochemistry and Biophysics*, **309**, 168–177.

Sariaslani, F.S. (1991) Microbial cytochromes P450 and xenobiotic metabolism, *Advances in Applied Microbiology*, **36**, 133–178.

Sata, F., Sapone, A., Elizondo, G., Stocker, P., Miller, V.P., Zheng, W., Raunio, H., Crespi, C.L. and Gonzalez, F.J. (2000) CYP3A4 allelic variants with amino acid substitutions in exons 7 and 12: evidence of an allelic variant with altered catalytic activity, *Clinical Pharmacology and Therapeutics*, **67**, 48–56.

Sato, H. and Guengerich, F.P. (2000) Oxidation of 1,2,4,5-tetramethoxybenzene to a cation radical by cytochrome P450, *Journal of the American Chemical Society*, **122**, 8099–8100.

Sato, R. and Omura, T. (1978) *Cytochrome P450*, Academic Press, New York.

Sawyer, D.T. (1987) The nature of the bonding and valency for oxygen in its metal compounds, *Comments on Inorganic Chemistry*, **6**, 103–121.

Schenkman, J.B. (1970) Studies on the nature of the type I and type II spectral changes in liver microsomes, *Biochemistry*, **9**, 2081–2091.

Schenkman, J.B, (1992) Steroid metabolism by constitutive cytochromes P450, *Journal of Steroid Biochemistry and Molecular Biology*, **43**, 1023–1030.

Schenkman, J.B. (1993) Protein-protein interactions, in: *Cytochrome P450* (J.B. Schenkman and H. Griem, eds.) Springer-Verlag, Berlin, 527–545.

Schenkman, J.B. and Griem, H. (1993) *Cytochrome P450*, Springer-Verlag, Berlin.

Schenkman, J.B. and Kupfer, D. (1982) *Hepatic Cytochrome P450 Mono-oxygenase System*, Pergamon, Oxford.

Schenkman, J.B., Frey, I., Remmer, H. and Estabrook, R.W. (1967) Sex differences in drug metabolism by rat liver microsomes, *Molecular Pharmacology*, **3**, 516–525.

Schenkman, J.B., Sligar, S.G. and Cinti, D.L. (1981) Substrate interaction with cytochrome P450, *Pharmacology and Therapeutics*, **12**, 43–71.

Schenkman, J.B., Thummel, K.E. and Favreau, L.V. (1989) Physiological and pathophysiological alterations in rat hepatic cytochromes P450, *Drug Metabolism Reviews*, **20**, 557–584.

Scheutz, E.G., Brimer, C. and Sheutz, J.D. (1998) Environmental xenobiotics and the antihormones cyproterone acetate and spironolactone use the nuclear hormone pregnenolone X receptor to activate the *CYP3A23* hormone response element, *Molecular Pharmacology*, **54**, 1113–1117.

Schlichting, I., Berendzen, J., Chu, K., Stock, A.M., Maves, S.A., Benson, D.E., Sweet, R.M., Ringe, D., Petsko, G.A. and Sligar, S.G. (2000) The catalytic pathway of cytochrome $P450_{cam}$ at atomic resolution, *Science*, **287**, 1615–1622.

Schmidt, A., Vogel, R., Holloway, M.K., Rutledge, S.J., Friedman, O., Yang, Z., Rodan, G.A. and Friedman, E. (1999) Transcriptional control and neuronal differentiation by agents that activate the LXR nuclear receptor family, *Molecular and Cellular Endocrinology*, **155**, 57–60.

Schneider, G., Coassolo, P. and Lave, T. (1999) Combining in vitro and in vivo pharmacokinetic data for prediction of hepatic drug clearance in humans by artificial neural networks and multivariate statistical techniques, *Journal of Medicinal Chemistry*, **42**, 5072–5076.

Schulze, J., Tschop, K., Lehnerer, M. and Hlavica, P. (2000) Residue 285 in cytochrome P450 2B4 lacking the NH_2-terminal hydrophobic sequence has a role in the functional association of NADPH-cytochrome P450 reductase, *Biochemical and Biophysical Research Communications*, **270**, 777–781.

Schwarz, D. (1991) Rotational motion and membrane topology of the microsomal cytochrome P450 system as analyzed by saturation transfer EPR, *Frontiers in Biotransformation*, **5**, 94–137.

Schwarze, W., Blanck, J., Ristau, O, Jänig, G.R., Pommerening, K., Rein, H. and Ruckpaul, K. (1985) Spin state control of cytochrome P450 reduction and catalytic activity in a reconstituted P450 LM2 system as induced by a series of benzphetamine analogues, *Chemico-Biological Interactions*, **54**, 127–141.

Scott, J.G., Liu, N. and Wen, Z. (1998) Insect cytchromes P450: diversity, insecticide resistance and tolerance to plant toxins, *Comparative Biochemistry and Physiology*, Part C, **121**, 147–155.

Segall, M.D., Payne, M.C., Ellis, S.W., Tucker, G.T. and Boyes, R.N. (1998) An *ab initio* approach to the understanding of cytochrome P450-ligand interactions, *Xenobiotica*, **28**, 15–20.

Seng, J.E., Gandy, J., Turturro, A., Lipman, R., Bronson, R.T., Parkinson, A., Johnson, W., Hart, R.W. and Leakey, J.A. (1996) Effects of caloric restriction on expression of testicular cytochrome P450 enzymes associated with the metabolic activation of chemical carcinogens, *Archives of Biochemistry and Biophysics*, **335**, 42–52.

Sevrioukova, I., Truan, G. and Peterson, J.A. (1996) The flavoprotein domain of $P450_{BM3}$: expression, purification and properties of the flavin adenine dinucleotide- and flavin mononucleotide-binding subdomains, *Biochemistry*, **35**, 7528–7535.

Sevrioukova, I.F., Li, H., Zhang, H., Peterson, J.A. and Poulos, T.L. (1999) Structure of a cytochrome P450-redox partner electron-transfer complex, *Proceedings of the National Academy of Sciences USA*, **96**, 1863–1868.

Shafiee, A. and Hutchinson, C.R. (1987) Macrolide antibiotic biosynthesis: isolation and properties of two forms of 6-deoxyerythronolide B hydroxylase from *Saccharopolyspora erythraea* (*Streptomyces erythreus*), *Biochemistry*, **26**, 6204–6210.

Shannon, R.D. and Prewitt, C.T. (1970) Revised values of effective ionic radii, *Acta Crystallographica*, **B26**, 1046–1048.

Sharrock, M., Debrunner, P.G., Schulz, C., Lipscomb, J.D., Marshall, V. and Gunsalus, I.C. (1976) Cytochrome $P450_{cam}$ and its complexes. Mössbauer parameters of the heme iron, *Biochimica et Biophysica Acta*, **420**, 8–26.

Shen, S. and Strobel, H.W. (1992) The role of cytochrome P450 lysine residues in the interaction between cytochrome P4501A1 and NADPH-cytochrome P450 reductase, *Archives of Biochemistry and Biophysics*, **294**, 83–93.

Shimada, T., Yamazaki, H., Mimura, M., Inui, Y. and Guengerich, F.P. (1994) Interindividual variations in human liver cytochrome P450 enzymes involved in the oxidation of drugs, carcinogens and toxic chemicals: studies with liver microsomes of 30 Japanese and 30 Caucasians, *Journal of Pharmacology and Experimental Therapeutics*, **270**, 414–423.

Shimizu, T. (1997) Diverse role of conserved aromatic amino acids in the electron transfer of cytochrome P450 catalytic functions: site-directed mutagenesis studies, *Recent Research Developments in Pure and Applied Chemistry*, **1**, 169–175.

Shimura, Y. (1988) A quantitative scale of the spectrochemical series for the mixed ligand complexes of d^6 metals, *Bulletin of the Chemical Society of Japan*, **61**, 693–698.

Shou, M., Grogan, J., Mancewicz, J.A., Kraucz, K.W., Gonzalez, F.J., Gelboin, H.V. and Korzekwa, K.R. (1994) Activation of CYP3A4: evidence for the simultaneous binding of two substrates in a cytochrome P450 active site, *Biochemistry*, **33**, 6450–6455.

Shou, M., Mei, Q., Ettore, M.W., Dai, R., Baillie, T.A. and Rushmore, T.H. (1999) Sigmoid kinetic model for two co-operative substrate-binding sites in a cytochrome P450 3A4 active site: an example of the metabolism of diazepam and its derivatives, *Biochemical Journal*, **340**, 845–853.

Simpson, E.R., Michael, M.D., Agarwal, V.R., Hinshelwood, M.M., Bulun, S.E. and Zhao, Y. (1997) Expression of the CYP19 (aromatase) gene: an unusual case of alternative promoter usage, *FASEB Journal*, **11**, 29–36.

Sligar, S.G. (1976) Coupling of spin, substrate and redox equilibria in cytochrome P450, *Biochemistry*, **15**, 5399–5406.

Sligar, S.G. and Gunsalus, I.C. (1979) Proton coupling in the cytochrome P450 spin and redox equilibria, *Biochemistry*, **18**, 2290–2295.

Sligar, S.G. and Murray, R.I. (1986) Cytochrome $P450_{cam}$ and other bacterial P450 enzymes, in: *Cytochrome P450* (P.R. Ortiz de Montellano, ed.) Plenum, New York, Chapter 12, 429–503.

Sligar, S.G., Debrunner, P.G., Lipscomb, J.D., Namtvedt, M.J. and Gunsalus, I.C. (1974) A role of the putidaredoxin COOH-terminus in $P450_{cam}$ (cytochrome m*) hydroxylations, *Proceedings of the National Academy of Sciences USA*, **71**, 3906–3910.

Sligar, S.G., Cinti, D.L., Gibson, G.G. and Schenkman, J.B. (1979) Spin state control of the hepatic cytochrome P450 redox potential, *Biochemical and Biophysical Research Communications*, **90**, 925–932.

Sligar, S.G., Gelb, M.H. and Heimbrook, D.C. (1984) Bio-organic chemistry and cytochrome P450-dependent catalysis, *Xenobiotica*, **14**, 63–86.

Smith, D.A. (1991) Species differences in metabolism and pharmacokinetics: are we close to an understanding?, *Drug Metabolism Reviews*, **23**, 355–373.

Smith, D.A. (1994) Chemistry and enzymology: their use in the prediction of human drug metabolism, *European Journal of Pharmaceutical Sciences*, **2**, 69–71.

Smith, D.A. and Jones, B.C. (1992) Speculations on the substrate structure-activity relationship (SSAR) of cytochrome P450 enzymes, *Biochemical Pharmacology*, **44**, 2089–2098.

Smith, D.A., Ackland, M.J. and Jones, B.C. (1997a) Properties of cytochrome P450 isoenzymes and their substrates. Part 1: Active site characteristics, *Drug Discovery Today*, **2**, 406–414.

Smith, D.A., Ackland, M.J. and Jones, B.C. (1997b) Properties of cytochrome P450 isoenzymes and their substrates. Part 2: Properties of cytochrome P450 substrates, *Drug Discovery Today*, **2**, 479–486

Smith, D.A., Abel, S.M., Hyland, R. and Jones, B.C. (1998) Human cytochrome P450s: selectivity and measurement *in vivo*, *Xenobiotica*, **28**, 1095–1128.

Smith, G., Stubbins, M.J., Harries, L.W. and Wolf, C.R. (1998a) Molecular genetics of the human cytochrome P450 monooxygenase superfamily, *Xenobiotica*, **28**, 1129–1165.

Smith, G., Modi, S., Pillai, I., Lian, L.Y., Sutcliffe, M.J., Pritchard, M.P., Friedberg, T., Roberts, G.C.K. and Wolf, C.R. (1998b) Determinants of the substrate specificity of human cytochrome P450 CYP2D6: design and construction of a mutant with testosterone hydroxylase activity, *Biochemical Journal*, **331**, 783–792.

Smith, H.J. (1988) *Introduction to the Principles of Drug Design*, Butterworth, London.

Song, B.-J., Gelboin, H.V., Park, S.-S., Yang, C.S. and Gonzalez, F.J. (1986) Complementary DNA and protein sequences of ethanol-inducible rat and human cytochrome P450s, *Journal of Biological Chemistry*, **261**, 16689–16697.

Song, W.V., Funk, C.D. and Brash, A.R. (1993) Molecular cloning of an allene oxide synthase: a cytochrome P450 specialized for the metabolism of fatty acid hydroperoxides, *Proceedings of the National Academy of Sciences USA*, **90**, 8519–8523.

Soucek, P. and Gut, I. (1992) Cytochromes P450 in rats: structures, functions, properties and relevant human forms, *Xenobiotica*, **22**, 83–103.

Spatzenegger, M. and Jaeger, W. (1995) Clinical importance of hepatic cytochrome P450 in drug metabolism, *Drug Metabolism Reviews*, **27**, 397–417.

Spink, D.C., Spink, B.C., Cao, J.Q., DePasquale, J.A., Pentecost, B.T., Fasco, M.J., Li, Y. and Sutter, T.R. (1998) Differential expression of CYP1A1 and CYP1B1 in human breast epithelial cells and breast tumor cells, *Carcinogenesis*, **19**, 291–298.

Squires, E.J. and Negishi, M. (1988) Reciprocal regulation of sex-dependent expression of testosterone 15α-hydroxylase ($P450_{15\alpha}$) in liver and kidney of male mice by androgen, *Journal of Biological Chemistry*, **263**, 4166–4171.

Stayton, P.S. and Sligar, S.G. (1990) The cytochrome $P450_{cam}$ binding surface as defined by site-directed mutagenesis and electrostatic modelling, *Biochemistry*, **29**, 7381–7386.

Stayton, P.S., Poulos, T.L. and Sligar, S.G. (1989) Putidaredoxin competitively inhibits cytochrome b_5 – cytochrome $P450_{cam}$ association: a proposed molecular model for a cytochrome $P450_{cam}$ electron transfer complex, *Biochemistry*, **28**, 8201–8205.

Stegeman, J.J. and Livingstone, D.R. (1998) Forms and functions of cytochrome P450, *Comparative Biochemistry and Physiology*, Part C, **121**, 1–3.

Stegeman, J.J., Seppa, P.L., Knipe, T., Suter, S., Smolowitz, R.M. and Hestermann, E. (1996) Cytochrome P4501A expression in teleost chondroid cells: a possible site of endogenous function of the Ah-receptor CYP1A loop, *Marine Environmental Research*, **42**, 306–307.

Stellwagen, E. (1978) Haem exposure as the determinate of oxidation-reduction potential of haem proteins, *Nature*, **275**, 73–74.

Stier, A. (1976) Lipid structures and drug metabolizing enzymes, *Biochemical Pharmacology*, **25**, 109–113.

Stresser, D.M. and Kupfer, D. (1999) Monospecific antipeptide antibody to cytochrome P450 2B6, *Drug Metabolism and Disposition*, **27**, 517–525.

Strobel, H.W, Nadler, S.G. and Nelson, D.R. (1989) Cytochrome P450: cytochrome P450 reductase interactions, *Drug Metabolism Reviews*, **20**, 519–533.

Strobel, H.W., Stralka, D.J., Hammond, D.K. and White, T. (1993) Extrahepatic microsomal forms: gastrointestinal cytochromes P450, in: *Cytochromes P450* (J.B. Schenkman and H. Griem, eds.) Springer-Verlag, Berlin, 363–371.

Strobel, H.W., Geng, J., Kawashima, H. and Wang, H. (1997) Cytochrome P450-dependent biotransformation of drugs and other xenobiotic substrates in neural tissue, *Drug Metabolism Reviews*, **29**, 1079–1105.

Strobl, G.R., von Kruendener, S., Stöckigt, J., Guengerich, F.P. and Wolff, T. (1993) Development of a pharmacophore for inhibition of human liver cytochrome P450 2D6: molecular modelling and inhibition studies, *Journal of Medicinal Chemistry*, **36**, 1136–1145.

Swanson, B.A., Dutton, D.R., Lunetta, J.M., Yang, C.S. and Ortiz de Montellano, P.R. (1991) The active sites of cytochromes P450IA1, IIB1, IIB2 and IIE1, *Journal of Biological Chemistry*, **266**, 19258–19264.

Swanson, B.A., Halpert, J.R., Bornheim, L.M. and Ortiz de Montellano, P.R. (1992) Topological analysis of the active sites of cytochrome P450IIB4 (rabbit), P450IIB10 (mouse) and P450IIB11 (dog) by *in situ* rearrangement of phenyl-iron complexes, *Archives of Biochemistry and Biophysics*, **292**, 42–46.

Szklarz, G.D. and Halpert, J.R. (1997) Molecular modeling of P450 3A4, *Journal of Computer-Aided Molecular Design*, **11**, 265–272.

Szklarz, G.D. and Halpert, J.R. (1998) Molecular basis of P450 inhibition and activation, *Drug Metabolism and Disposition*, **26**, 1179–1184.

Szklarz, G.D., He, Y.A. and Halpert, J.R. (1995) Site-directed mutagenesis as a tool for molecular modelling of cytochrome P450 2B1, *Biochemistry*, **34**, 14312–14322.

Szklarz, G.D., Graham, S.E. and Paulsen, M.D. (2000) Molecular modeling of mammalian cytochromes P450: application to study enzyme function, *Vitamins and Hormones*, **58**, 53–87.

Tajima, F. and Nei, M. (1984) Estimation of evolutionary distance between nucleotide sequences, *Molecular Biology of Evolution*, **1**, 269–285.

Takemori, S. and Kominami, S. (1984) The role of cytochromes P450 in adrenal steroidogenesis, *Trends in Biological Sciences*, **9**, 393–396.

Tan, Y., White, S.P., Paranawithana, S.R. and Yang, C.S. (1997) A hypothetical model for the active site of human cytochrome P450 2E1, *Xenobiotica*, **27**, 287–299.

Taton, M., Salmon, F. and Rahier, A. (1994) Cytochrome P450 obtusifoliol 14α-methyl demethylase from *zea mays*: enzymology and inhibition, in: *Cytochrome P450 Biochemistry, Biophysics and Molecular Biology* (M.C. Lechner, ed.) Libbey, Paris, 725–728.

Testa, B. (1990) Mechanisms of inhibition of xenobiotic-metabolizing enzymes, *Xenobiotica*, **20**, 1129–1137.

Testa, B. and Jenner, P. (1981) Inhibitors of cytochrome P450s and their mechanism of action, *Drug Metabolism Reviews*, **12**, 1–117.

Thomas, H., Timms, C.W. and Oesch, F. (1990) Epoxide hydrolases: molecular properties, induction, polymorphisms and function, *Frontiers in Biotransformation*, **2**, 278–337.

Tijet, N., Helvig, C., Pinot, F., Le Bouquin, R., Lesot, A., Durst, F., Salaün, J.-P. and Benveniste, I. (1998) Functional expression in yeast and characterization of a clofibrate-inducible plant cytochrome P450 (CYP94A1) involved in cutin monomers synthesis, *Biochemical Journal*, **332**, 583–589.

Tomlinson, E.S., Lewis, D.F.V., Maggs, J.L., Park, B.K. and Back, D.J. (1997) *In vitro* metabolism of side-chain cleaved dexamethasone (9αF-A) is CYP3A4 mediated: rationalization of CYP3A4 and CYP17 (17,20 lyase) involvement in dexamethasone metabolism *in vitro* based on molecular modelling studies, *Biochemical Pharmacology*, **54**, 605–611.

Traylor, T.G. and Xu, F. (1988) Model reactions related to cytochrome P450. Effects of alkene structure on the rates of epoxide formation, *Journal of the American Chemical Society*, **110**, 1953–1958.

Tuck, S.F., Peterson, J.A. and Ortiz de Montellano, P.R. (1992) Active site topologies of bacterial cytochromes P450101 ($P450_{cam}$), P450108 ($P450_{terp}$) and P450102 ($P450_{BM3}$): *in situ* rearrangement of their phenyl-iron complexes, *Journal of Biological Chemistry*, **267**, 5614–5620.

Tuck, S.F., Graham-Lorence, S., Peterson, J.A. and Ortiz de Montellano, P.R. (1993) Active sites of cytochrome $P450_{cam}$ (CYP101) F87W and F87A mutants, *Journal of Biological Chemistry*, **268**, 269–275.

Tucker, G.T., Rostami-Hodjegan, A. and Jackson, P.R. (1998) Determination of drug-metabolizing enzyme activity *in vivo*: pharmacokinetic and statistical issues, *Xenobiotica*, **28**, 1255–1273.

Tukey, R.H. and Johnson, E.F. (1990) Molecular aspects of regulation and structure of the drug-metabolism enzymes, in: *Principles of Drug Action* (W.B. Pratt and P.Taylor, eds.) Churchill-Livingstone, New York, Chapter 6, 423–467.

Uauy, R., Mena, P. and Rojas, C. (2000) Essential fatty acids in early life: structural and functional role, *Proceedings of the Nutrition Society*, **59**, 3–15.

Ueng, Y.-F., Kuwabara, T., Chun, Y.-J. and Guengerich, F.P. (1997) Cooperativity in oxidations catalyzed by cytochrome P450 3A4, *Biochemistry*, **36**, 370–381.

Unno, M., Shimada, H., Toba, Y., Makino, R. and Ishimura, Y. (1996) Role of arg^{112} of cytochrome $P450_{cam}$ in the electron transfer from reduced putidaredoxin, *Journal of Biological Chemistry*, **271**, 17869–17874.

Uvarov, V.Y., Sotnichenko, A.I., Vodovozova, E.L., Molotkovsky, J.G., Kolesanova, E.F., Kyulkin, Y.A., Stier, A., Krueger, V. and Archakov, A.I. (1994) Determination of membrane-bound fragments of cytochrome P450 2B4, *European Journal of Biochemistry*, **222**, 483–489.

Vamecq, J. and Latruffe, N. (1999) Medical significance of peroxisome proliferator-activated receptors, *Lancet*, **354**, 141–148.

Vanden Bossche, H., Lauwers, W., Willemsens, G., Marichal, P., Cornelissen, F. and Cools, W. (1984) Molecular basis for the antimycotic and antibacterial activity of N-substituted imidazoles and triazoles: the inhibition of isoprenoid biosynthesis, *Pesticide Science*, **15**, 188–198.

Vaz, A.D.N. and Coon, M.J. (1994) On the mechanism of action of cytochrome P450: evaluation of hydrogen abstraction in oxygen-dependent alcohol oxidation, *Biochemistry*, **33**, 6442–6449.

Vaz, A.D.N., Roberts, E.S. and Coon, M.J. (1991) Olefin formation in the oxidative deformylation of aldehydes by cytochrome P450. Mechanistic implications for catalysis by oxygen-derived peroxide, *Journal of the American Chemical Society*, **113**, 5886–5887.

Vaz, A.D.N., Pernecky, S.J., Raner, G.M. and Coon, M.J. (1996) Peroxo-iron and oxenoid-iron species as alternative oxygenating agents in cytochrome P450-catalyzed reactions: switching by threonine-302 to alanine mutagenesis of cytochrome P450 2B4, *Proceedings of the National Academy of Sciences USA*, **93**, 4644–4648.

Veitch, N.C. and Williams, R.J.P. (1992) The molecular basis of electron transfer in redox enzyme systems, *Frontiers in Biotransformation*, **7**, 279–320.

Vorhees, C.V., Reed, T.M., Schilling, M.A., Fisher, J.E., Moran, M.S., Cappon, G.D. and Nebert, D.W. (1997) CYP2D1 polymorphism in methamphetamine-treated rats: genetic differences in neonatal mortality and effects on spatial learning and acoustic startle, *Neurotoxicology and Teratology*, **20**, 265–273.

Voznesensky, A.I. and Schenkman, J.B. (1994) Quantitative analyses of electrostatic interactions between NADPH – cytochrome P450 reductase and cytochrome P450 enzymes, *Journal of Biological Chemistry*, **269**, 15724–15731.

Wackett, L.P., Sadowsky, M.J., Newman, L.M., Hur, H.-G. and Li, S. (1994) Metabolism of polyhalogenated compounds by a genetically engineered bacterium, *Nature*, **368**, 627–629.

Wade, R.C. (1990) Solvation of the active site of cytochrome $P450_{cam}$, *Journal of Computer-Aided Molecular Design*, **4**, 199–204.

Walker, C.H. (1998) Avian forms of cytochrome P450, *Comparative Biochemistry and Physiology*, Part C, **121**, 65–72.

Waller, C.L, Evans, M.V. and McKinney, J.D. (1996) Modeling the cytochrome P450-mediated metabolism of chlorinated volatile organic compounds, *Drug Metabolism and Disposition*, **24**, 203–210.

Waller, S.C., He, Y.A., Harlow, G.R., He, Y.Q., Mash, E.A. and Halpert, J.R. (1999) 2,2′,3,3′,6,6′- Hexachlorobiphenyl hydroxylation by active site mutants of cytochrome P450 2B1 and 2B11, *Chemical Research in Toxicology*, **12**, 690–699.

Wang, H., Chen, J., Hollister, K., Sowers, L.C. and Forman, B.M. (1999) Endogenous bile acids are ligands for the nuclear receptor FXR/BAR, *Molecular Cell*, **3**, 543–553.

Wang, M., Roberts, D.L., Paschke, R., Shea, T.M., Masters, B.S.S. and Kim, J.-J.P. (1997) Three-dimensional structure of NADPH-cytochrome P450 reductase: prototype for FMN- and FAD-containing enzymes, *Proceedings of the National Academy of Sciences USA*, **84**, 8411–8416.

Warner, M. and Gustafsson, J.A. (1993) Extrahepatic microsomal forms: brain cytochromes P450, in: *Cytochromes P450* (J.B. Schenkman and H. Griem, eds.) Springer-Verlag, Berlin, 387–397.

Watanabe, I., Nara, F. and Serizawa, N. (1995) Cloning, characterization and expression of the gene encoding cytochrome $P450_{sca\text{-}2}$ from *Streptomyces carbophilus* involved in production of pravastatin, a specific HMG-CoA reductase inhibitor, *Gene*, **163**, 81–85.

Waterman, M.R. (1992) Cytochrome P450: cellular distribution and structural considerations, *Current Opinion in Structural Biology*, **2**, 384–387.

Watkins, P.B. (1990) Role of cytochromes P450 in drug metabolism and hepatotoxicity, *Seminars in Liver Disease*, **10**, 235–250.

Waxman, D.J. (1988) Interactions of hepatic cytochromes P450 with steroid hormones, *Biochemical Pharmacology*, **37**, 71–84.

Waxman, D.J. (1999) P450 gene induction by structurally diverse xenochemicals: central role of nuclear receptors CAR, PXR and PPAR, *Archives of Biochemistry and Biophysics*, **369**, 11–23.

Waxman, D.J. and Azaroff, L. (1992) Phenobarbital induction of cytochrome P450 gene expression, *Biochemical Journal*, **281**, 577–592.

Waxman, D.J. and Chang, T.K.H. (1995) Hormonal regulation of liver cytochrome P450 enzymes, in: *Cytochrome P450* (P.R. Ortiz de Montellano, ed.) Plenum, New York, Chapter 11, 391–417.

Waxman, D.J., Dannan, G.A. and Guengerich, F.P. (1985) Regulation of rat hepatic cytochrome P450: age-dependent expression, hormonal imprinting and xenobiotic inducibility of sex-specific isoenzymes, *Biochemistry*, **24**, 4409–4417.

Waxman, D.J., Lapenson, D.P., Aoyama, T., Gelboin, H.V., Gonzalez, F.J. and Korzekwa, K. (1991) Steroid hormone hydroxylase specificities of eleven cDNA-expressed human cytochrome P450s, *Archives of Biochemistry and Biophysics*, **290**, 160–166.

Weigel, N.L. (1996) Steroid hormone receptors and their regulation by phosphorylation, *Biochemical Journal*, **319**, 657–667.

Weiner, L.V. (1986) Magnetic resonance study of the structure and functions of cytochrome P450, *Critical Reviews in Biochemistry*, **20**, 139–200.

Wheelis, M.L., Kandler, O. and Woese, C.R. (1992) On the nature of global classification, *Proceedings of the National Academy of Sciences USA*, **89**, 2930–2934.

White, R.E. (1991) The involvement of free radicals in the mechanisms of monooxygenases, *Pharmacology and Therapeutics*, **49**, 21–42.

White, R.E. (1998) Short- and long-term projections about the use of drug metabolism in drug discovery and development, *Drug Metabolism and Disposition*, **26**, 1213–1216.

White, R.E. and Coon, M.J. (1980) Oxygen activation by cytochrome P450, *Annual Review of Biochemistry*, **49**, 315–356.

White, R.E. and McCarthy, M. (1986) Active site mechanics of liver microsomal cytochrome P450, *Archives of Biochemistry and Biophysics*, **246**, 19–32.

Whitlock, J.P. (1989) The control of cytochrome P450 gene expression by dioxin, *Trends in Pharmaceutical Sciences*, **10**, 285–288.

Whitlock, J.P. (1999) Induction of cytochrome P4501A1, *Annual Review of Pharmacology and Toxicology*, **39**, 103–125.

Whitlock, J.P. and Denison, M.S. (1995) Induction of cytochrome P450 enzymes that metabolize xenobiotics, in: *Cytochrome P450* (P.R. Ortiz de Montellano, ed.) Plenum, New York, Chapter 10, 367–390.

Whitlock, J.P., Okino, S.T., Dong, L., Ko, H.P., Clarke-Katzenberg, R., Ma, Q. and Li, H. (1996) Induction of cytochrome P4501A1: a model for analyzing mammalian gene transcription, *FASEB Journal*, **10**, 809–818.

Whitlock, J.P., Chichester, C.H., Bedgood, R.M., Okino, S.T., Ko, H.P., Ma, K., Dong, L., Li, H. and Clarke-Katzenberg, R. (1997) Induction of drug metabolizing enzymes by dioxin, *Drug Metabolism Reviews*, **29**, 1107–1127.

Wickramasinghe, R.H. and Villee, C.A. (1975) Early role during chemical evolution for cytochrome P450 in oxygen detoxification, *Nature*, **256**, 509–511.

Williams, J.A., Martin, F.L., Muir, G.H., Hewer, A., Grover, P.L. and Phillips, D.H. (2000a) Metabolic activation of carcinogens and expression of various cytochromes P450 in human prostate tissue, *Carcinogenesis*, **21**, 1683–1689.

Williams, P.A., Cosme, J., Sridhar, V., Johnson, E.F. and McRee, D.E. (2000b) Mammalian cytochrome P450 monooxygenase: structural adaptations for membrane binding and functional diversity, *Molecular Cell*, **5**, 121–131.

Willson, T.M., Brown, P.J., Sternbach, D.D. and Henke, B.R. (2000) The PPARs: from orphan receptors to drug discovery, *Journal of Medicinal Chemistry*, **43**, 527–550.

Winkler, M. and Wiseman, A. (1992) Recombinant cytochrome P450 production in yeast, *Biotechnology and Genetic Engineering Reviews*, **10**, 185–208.

Wiseman, A. (1993) Genetically-engineered mammalian cytochromes P450 from yeasts – potential applications, *Trends in Biotechnology*, **11**, 131–135.

Wiseman, A. (1994) Better by design: biocatalysts for the future, *Chemistry in Britain*, 571–573.

Wiseman, A. (1996a) Therapeutic proteins and enzymes from genetically engineered yeasts, *Endeavour*, **20**, 130–132.

Wiseman, A. (1996b) Genetically-engineered mammalian cytochromes P450 from yeasts-potential applications, *Trends in Biotechnology*, **11**, 131–136.

Wiseman, A. (1996c) Novel biocatalysts will work even better for industry, *Journal of Chemical Education*, **73**, 55–59.

Wiseman, A. (1996d) Biocatalyst replacement by mimetics in bioconversions: can the time-scale be predicted?, *Journal of Chemical Technology and Biotechnology*, **65**, 3–4.

Wiseman, A., Lewis, D.F.V., Ridgway, T. and Wiseman, H. (2000) Cytochromes P450 enzymes for clean food processing: limitations of imitations, *Journal of Chemical Technology and Biotechnology*, **75**, 3–5.

Wiseman, H. and Lewis, D.F.V. (1996) The metabolism of tamoxifen by human cytochromes P450 is rationalized by molecular modelling of the enzyme-substrate interactions: potential importance to its proposed anti-carcinogenic/carcinogenic actions, *Carcinogenesis*, **17**, 1357–1360.

Woese, C.R. (1987) Bacterial evolution, *Microbiological Reviews*, **51**, 221–271.

Wolff, T. and Strecker, M. (1992) Endogenous and exogenous factors modifying the activity of human liver cytochrome P450 enzymes, *Experimental Toxicology and Pathology*, **44**, 263–271.

Wrighton, S.A. and Stevens, J.C. (1992) The human hepatic cytochromes P450 involved in drug metabolism, *Critical Reviews in Toxicology*, **22**, 1–21.

Wrighton, S.A., Stevens, J.C., Becker, G.W. and Vanden Branden, M. (1993) Isolation and characterization of human liver cytochrome P450 2C19: correlation between 2C19 and S-mephenytoin 4′-hydroxylation, *Archives of Biochemistry and Biophysics*, **306**, 240–245.

Xia, Z.-X. and Mathews, F.S. (1990) Molecular structure of flavocytochrome b_2 at 2.4Å resolution, *Journal of Molecular Biology*, **212**, 837–863.

Yabusaki, Y. and Ohkawa, H. (1991) Genetic engineering of cytochrome P450 mono-oxygenases, *Frontiers in Biotransformation*, **4**, 169–189.

Yamano, S., Tatsuno, J. and Gonzalez, F.J. (1990) The CYP2A3 gene product catalyzes coumarin 7-hydroxylation in human liver microsomes, *Biochemistry*, **29**, 1322–1329.

Yang, C.S., Yoo, J.-S.H., Ishizaki, H. and Hong, J. (1990) Cytochrome P450IIE1: roles in nitrosamine metabolism and mechanisms of regulation, *Drug Metabolism Reviews*, **22**, 147–159.

Yang, D., Oyaizu, Y., Oyaizu, H., Olsen, G.J. and Woese, C.R. (1985) Mitochondrial origins, *Proceedings of the National Academy of Sciences USA*, **82**, 4442–4447.

Yasukochi, T., Okada, O., Hara, T., Sagara, Y., Sekimizu, K. and Horiuchi, T. (1994) Putative functions of phenylalanine-350 of *Pseudomonas putida* cytochrome $P450_{cam}$, *Biochimica et Biophysica Acta*, **1204**, 84–90.

Yeom, H. and Sligar, S.G. (1997) Oxygen activation by cytochrome $P450_{BM3}$: effects of mutating an active site acidic residue, *Archives of Biochemistry and Biophysics*, **337**, 208–216.

Yeom, H., Sligar, S.G., Li, H., Poulos, T.L. and Fulco, A.J. (1995) The role of Thr268 in oxygen activation of cytochrome $P450_{BM3}$, *Biochemistry*, **34**, 14733–14740.

Yin, H., Anders, M.W., Korzekwa, K.R., Higgins, L., Thummel, K.E., Kharasch, E.D. and Jones, J.P. (1995) Designing safer chemicals: predicting the rates of metabolism of halogenated alkanes, *Proceedings of the National Academy of Sciences USA*, **92**, 11076–11080.

Yoshida, Y. and Aoyama, Y. (1991) Cytochromes P450 in the ergosterol biosynthesis, *Frontiers in Biotransformation*, **4**, 127–148.

Yoshikawa, K.-I., Noguti, T., Tsujimura, M., Koga, H., Yasukochi, T., Horiuchi, T. and Go, M. (1992) Hydrogen bond network of cytochrome $P450_{cam}$: a network connecting the heme group with helix K, *Biochimica et Biophysica Acta*, **1122**, 41–44.

Zakharieva, O., Grodzicki, M., Trautwein, A.X., Veeger, C. and Rietjens, I.M.C.M. (1998) Molecular orbital study of porphyrin-substrate interactions in cytochrome P450 catalysed aromatic hydroxylation of substituted anilines, *Biophysical Chemistry*, **73**, 189–203.

Zhai, S., Dai, R., Friedman, F.K. and Vestal, R.E. (1998) Comparative inhibition of human cytochromes P450 1A1 and 1A2 by flavonoids, *Drug Metabolism and Disposition*, **26**, 989–992.

Zhukov, A. and Ingelman-Sundberg, M. (1999) Relationship between cytochrome P450 catalytic cycling and stability: fast degradation of ethanol-inducible cytochrome P450 2E1 (CYP2E1) in hepatoma cells is abolished by inactivation of its electron donor NADPH-cytochrome P450 reductase, *Biochemical Journal*, **340**, 453–458.

Zimniak, P. and Waxman, D.J. (1993) Liver cytochrome P450 metabolism of endogenous steroid hormones, bile acids and fatty acids, in: *Cytochrome P450* (J.B. Schenkman and H. Griem, eds.) Springer-Verlag, Berlin, 123–144.

Index

acetone 96
active site 160
adrenal gland 8
Ah (aryl hydrocarbon) receptor 29, 118, 121, 122, 125–6; ligand binding affinity 129–32
Ah receptor nuclear translocator (ARNT) 118, 126
aldehyde oxygenation 57
alkane hydroxylation 80
allelic variants 15, 160
allene oxide synthase (AOS) 3
amphibians 26
androstenedione 85
aniline 96
animals: evolution 25–8; land colonization 21, *see also specific groups*
antioxidants 110
arachidonic acid 80, 100
archaebacteria 20, 23
aromatase *see* CYP19
atmospheric oxygen levels 20–1, 28

Bacillus megaterium 25, 32, 125
bacteria 18; archaebacteria 20, 23; cyanobacteria 20, 38; endosymbiosis 31; P450 families 43, 48–9; P450 functions 80; P450 substrates 82
banded-iron formations (BIFs) 20–1
basal transcription element (BTE) 123
benzene 96, 97
benzodiazepines 72
benzphetamine 62, 90
binding spectra 11–12
biotechnological applications 16–17
birds 27
black swallowtail butterfly 28, 30
blue-green algae 21

Caenorhabditis elegans 25, 31, 49
caffeine 87, 134
Cambrian period 21, 26
camphor 54–5, 59, 63, 80
carbon monoxide adduct 1, 2, 10–11, 12, 133
carcinogens 89, 126; activation of 89, 97, 98, 110–11, 127
catalase 109
catalytic cycle *see* reaction cycle
chlorzoxazone 96, 97
cholesterol 78, 80, 85
cholic acid 85
circular dichroism (CD) 146
clearance rates 70–5
co-evolutionary processes 27, 28–9, 38
COMPACT method 126–9
constitutive androstane receptor (CAR) 92, 121, 125, 128
corticosterone 85
cortisol 82
coumaric acid 80
coumarin 89–90
cyanobacteria 20, 38
cyclic AMP 123, 125
cyclohexane oxidation 60
cyclophosphamide 90
CYP1 family 29, 41, 129; evolution 26, 27; functions 82, 83, 85, 121; inhibitors 138; regulation 118–22, 125–6, 128, 130–2; structural studies 148; substrate recognition sites 34; substrates 86–9, 104; toxic activation 110, 127
CYP1A subfamily: evolution 27; functions 83; structural studies 150, 157; substrates 87, 89; toxic activation 110

CYP1A1 86, 125; carcinogenesis and 89; evolution 27, 28; polymorphism 15; regulation 121, 125–6; structural studies 149; substrates 86–7
CYP1A2 86, 122, 134; biotechnological applications 16; evolution 27, 28; inhibitor 136; polymorphism 15; regulation 125, 126; structural studies 156–7; substrate interactions 113–14; substrates 86–7, 90, 105, 107, 109, 113
CYP1B subfamily 27
CYP1B1 86, 87, 125
CYP2 family 41–2, 129; evolution 26, 27; functions 82, 83, 85, 121; regulation 121, 132; structural studies 148, 151, 157; substrate recognition sites 15–16, 34–7; substrates 87, 104
CYP2A subfamily 83; evolution 27, 28; functions 85, 89; regulation 121, 122, 125; substrates 89–90, 91
CYP2A5 89, 157
CYP2A6 28, 89, 157; polymorphism 15; substrates 89, 90, 92, 105, 107, 109
CYP2B subfamily 74–5, 83, 128; evolution 27; functions 85; inhibitors 73, 134, 138; regulation 121, 125, 128; substrates 89, 90–3
CYP2B4: site-directed mutagenesis 17; substrate binding 73
CYP2B6 90; inhibitor 134; substrates 90–1, 93, 105, 107
CYP2C 145
CYP2C subfamily 83; evolution 27; functions 85; regulation 121, 125; substrates 93–4, 121
CYP2C8 94, 97, 121
CYP2C9 112, 157; inhibitor 134; polymorphism 15; substrate interactions 114; substrates 73, 93–5, 105, 107, 109, 115, 121
CYP2C19 157; inhibitor 135; polymorphism 15, 28; substrates 94, 96, 102, 106, 107, 109
CYP2D subfamily 29; evolution 27; functions 83; inhibitors 135; regulation 121, 122; substrates 94–5, 98
CYP2D1 95, 135
CYP2D6 28, 157; inhibitors 72–3, 135, 138; polymorphism 15; substrate interactions 114; substrates 94–5, 99, 106, 107, 109, 115, 158
CYP2D9 95
CYP2E subfamily: evolution 27; functions 83; inhibitors 73, 110, 138; regulation 121, 122, 123; substrates 89, 96–8, 100, 127; toxic activation 110, 127–8
CYP2E1 89, 112, 128, 157; inhibitors 74, 112, 136; substrates 98, 101, 106, 107
CYP3 family 42, 128, 129; evolution 25–6; functions 82, 83, 85, 121; regulation 122, 128, 132; structural studies 148; substrate recognition sites 34; substrates 87, 104
CYP3A 74–5; functions 82–3; regulation 125, 128, 132, 134; substrates 98–9, 102
CYP3A1 136
CYP3A4: inhibitors 134, 136; substrate interactions 114, 152; substrates 72, 98, 99, 102, 103, 106, 107, 109
CYP3A5 98
CYP4 family 39, 43, 128, 129; evolution 25–6, 80; regulation 122, 132; substrates 80, 84
CYP4A subfamily 100–2
CYP4A1 112; biotechnological applications 16
CYP4A11 112
CYP5 family 44
CYP6 family 26, 27, 39, 44
CYP6B1 28, 80
CYP7 family 44; functions 85; regulation 122, 123
CYP8 family 44
CYP9 family 39, 44
CYP10 family 39, 44
CYP11 family 25, 45; functions 78, 85; regulation 123; structural studies 148
CYP11A1 85
CYP11B1 44
CYP11B2 85
CYP12 family 39
CYP14 family 31
CYP15 family 39
CYP17 family 44, 45; evolution 26; functions 85; regulation 123; structural studies 17, 148
CYP18 family 39
CYP19 (aromatase) family 29–30, 45; functions 78, 85; inhibitors 134, 138; regulation 123; structural studies 148
CYP21 family 45; evolution 26; functions 78, 85; regulation 123; structural studies 148
CYP23 family 31
CYP24 family 45, 85
CYP26 family 45, 123

CYP27 family 45, 85, 123
CYP33 family 31
CYP34 family 31
CYP35 family 31
CYP36 family 31
CYP51 family 44, 46; functions 77–8, 80; inhibitors 134, 138; structural studies 148
CYP52 family 46, 80
CYP53 family 46
CYP54 family 46
CYP55 family 46, 80; functions 80; structure 141, 152
CYP101 48; biotechnological applications 16; inhibitors 133; reaction cycle 54, 68–9; structural studies 133, 141, 148, 152, 157
CYP102 25, 33, 48, 70, 80, 144–5; functions 80; regulation 125; structural studies 141–3, 149, 150, 152–6
CYP107 141, 152
CYP108 141, 143, 144, 152
cytochromes 17–19, *see also* P450 superfamily; *specific cytochromes*

Daisyworld hypothesis 29
detoxification 109–10; plant toxins 27, 28–9, 30; reactive oxygen species 25, 38
Devonian period 21, 26–7
dichlorodibenzo-*p*-dioxins (DCDDs) 129
diclofenac 93, 94
dietary preferences 28
dioxygen reduction 54, 57, 60–2
diphenylmethane degradation 16
drug metabolism 1, 7, 16, 38; polymorphism 15, 28, 30, *see also specific drugs*; substrates
drug metabolizing enzymes (DMEs) 83, 88
drug–drug interactions 16, 112

eburicol 77
eicosanoids 26, 80
electron spin resonance (ESR) 146–7
endoplasmic reticulum (ER) 4–5, 7–8, 31
endosymbiosis 31
epoxide hydrase (EH) 7
ergosterol 78, 80
erythromycin 80, 99
estradiol 83, 85, 87, 125
estrogen 29, 121–2
estrogen receptor (ER) 121, 122, 125
estrone 85
estrus cycle 122, 125; modulation 27, 29
ethanol 96
7-ethoxyresorufin 86–7
evolution 21–2; co-evolutionary processes 27, 28–9, 38; extinction events 21–2; P450 sequences 31–8; phylogeny 22–30
EXAFS 147
extinction events 21–2

farnesoid X receptor (FXR) 122, 123
fatty acids 80, 100, 112
fetal development 83, 122
fish 26
flavin adenine dinucleotide (FAD) 32–3
flavin mononucleotide (FMN) 32–3, 68, 144
fluvoxamine 16
free oxygen radicals *see* reactive oxygen species
fungi: P450 families 46; P450 functions 80; P450 substrates 82
Fusarium oxysporum 80

G-protein coupled receptors (GPCRs) 29
gene regulation 118, 123–6
gene sequences 9, 31–8
genetic variation 15–16
genotype 160
gestodene 136
globin proteins 37, 152
glucocorticoid receptor (GR) 99, 121, 122, 128, 132
gluconeogenesis 83

heme prosthetic group 2, 143; inhibition mechanisms 132–3
heme-thiolate enzymes 2, 17–19
hepatic nuclear factor (HNF-4) 121, 122, 125
hepatocytes 4–5
human evolution 28
hydrogen peroxide 51–2, 57, 58
hydroperoxide 57, 59

immunoquantification 12–15
induction 160
infrared (IR) spectroscopy 146
insects 26, 27, 39; insect–plant co-evolution 27, 28–9, 38; P450s 39, 43, 44, 49
ifosfamide 90
isolation 9
isozymes 160

ketoconazole 134
kinetics 60–2, 65–75, *see also* thermodynamics of P450 reaction

land colonization: by animals 21, 26; by plants 26
lanosterol 44, 77–8, 80
lauric acid 100, 112
liver 4–5, 85–6; P450 content 8, 13, 14, 118, 119
liver X receptor (LXR) 122, 123
lung cancer susceptibility 15

magnetic circular dichroism (MCD) 146
magnetic resonance spectroscopy 146–7
mammals 39; evolution 27–8; P450 families 41–3, 44, 50; P450 functions 76–9, 80–6
mephenytoin 91, 94
metabolic intermediate (MI) 133–4
metal toxicity 136–7
7-methoxyresorufin 87, 109
mibefradil 136
mitochondria 31
mixed-function oxidases 2, 110, *see also* P450 superfamily; *specific cytochromes*
molecular dynamics (MD) simulations 155
molecular modelling studies 150–7
molecular orbital calculations 140, 158
monkeys 28
mono(2-ethylhexyl)phthalate (MEHP) 100
mono-oxygenases *see* P450 superfamily; *specific cytochromes*
Mössbauer (MB) spectroscopy 147
myristic acid 100

naphthalene degradation 16
neurotransmitter metabolism 83
nifedipine 99
nitric oxide synthases (NOS) 39, 80
p-nitrophenol 96
nomenclature 17–19, 38–50
nuclear magnetic resonance (NMR) 146–7
nuclear receptor binding affinity 129–32

obtusifoliol 77
omeprazole 102
optical rotatory dispersion (ORD) 146
orphenadrine 133–4
oxygen levels, atmospheric 20–1, 28
oxygen radicals *see* reactive oxygen species

P450 superfamily 17–19, 160; biotechnological applications 16–17; characterization/detection 9–15; discovery of 1–4; evolution 2–3; functions 76–7; genetic variation 15–16; inhibition of 132–7; isolation/purification 9; microsomal P450s 25–6, 32, 37, 76; mitochondrial P450s 25, 32, 76; nomenclature 17–19, 38–50; occurrence/distribution 4–8; phylogeny 22–30; reaction *see* reaction cycle; regulation 118–26; sequences 9, 31–8; substrate selectivity *see* substrates; toxicity 109–12, 126–9, *see also specific cytochromes*
palmitic acid 100
palmitoleic acid 80
Papilio polyxenes (black swallowtail butterfly) 28, 80
paracetamol 97
PB-responsive enhancer module (PBREM) 125, *see also* phenobarbital
Pelomyxa 31
pentadecanoic acid 80
7-pentoxyresorufin 90
Per-ARNT-Sim (PAS) gene family 118
peroxidases 19
peroxide 56–7, 59, 109
peroxisome proliferator-activated receptor (PPAR) 100, 122, 125, 132
phenobarbital (PB) 90, 92, 121–3, 125
phenotype 161
phenyldiazenes 134
photosynthesis, oxygen levels and 20–1
phylogeny 22–30
phytoestrogens 27, 29
planar polyaromatic hydrocarbons (PAHs) 86, 127
plant toxins 28–9; detoxification 27, 28–9, 30; xanthotoxin 28, 80, 101
plants: evolution 25, 27, 28–9, 38; land colonization 26; P450 families 47–8; P450
functions 80, 82; P450 substrates 81
polyaromatic hydrocarbons (PAHs) 129–32
polychlorobiphenyls (PCBs) 129, 130
polyhalogenated hydrocarbons 16
polymorphism 15–16, 28, 30, 161

porphyrin 2, 5, 133
pregnane-X receptor (PXR) 99, 122, 125, 128
pregnenolone 85
primates 28
progesterone 85
prostacyclin synthase (PGIS) 3
prostaglandin 76, 80
protein kinase A (PKA) 123
protein kinase C (PKC) 123
protein sequence alignments 15–16, 37
purification 9
putidaredoxin 55, 63
pyrene degradation 16

quantitative structure–activity relationship (QSAR) analyses 70–5, 107–8, 112–15, 157, 161; P450 inducers 129–32; P450 inhibitors 138–9

reaction cycle 51–75; coupling of spin 62–5; kinetics 60–2, 65–75; redox equilibria 62–5; stages in 54–60; thermodynamics 53–4, 56, 60–2, 67
reactive oxygen species 59, 97, 109–10, 127–8; detoxification 25, 38, 109; formation of 2, 54, 127–8, 136
redox potential 32–3, 63–5
reptiles 27
residue modification 149–51
resonance Raman (RR) spectroscopy 56, 146
retinoic acid receptor (RAR) 94, 121, 123
retinoid X receptor (RXR) 121, 123, 125
reverse Type I binding spectrum 12

selectivity 102–9, 114–15, 161
SF-1 nuclear factor 123, 125
site-directed mutagenesis 17, 147–50, 151, 152
SKF-525A 133–4
sodium channel blockers 72
specificity 161, *see also* selectivity
spectroscopy 145–7
spin-redox coupling 62–5
SRC-1 protein 125
STAT5b transcription factor 125
steroid biosynthesis 25, 26, 29–30, 43–4, 76–9, 85, 123
steroid hormone receptors 29–30, 121
steroid hormone regulation 26, 27, 38, 83, 121–2
structure 33–8; molecular modelling studies 150–7; molecular orbital calculations 140, 158; site-directed mutagenesis/residue modification 147–50; spectroscopy 145–7; X-ray crystallography 140, 141–5, 151–2
substrate recognition sites (SRSs) 15–16, 34–7; structural studies 147, 148–9, 151
substrates 161; endogenous 76–85; exogenous 85–102; quantitative structure–activity relationships (QSARs) 112–15, 157; selectivity 102–9, 114–15, 161; substrate structure–activity relationships (SSARs) 158; turnover rates 65–75, *see also specific cytochromes*
superoxide 55–6, 59, 109
superoxide dismutase (SOD) 109

tacrine 16
tamoxifen 102
tertiary structure *see* structure
testosterone 82–5, 89, 121, 122
2,3,7,8-tetrachlorodibenzo-*p*-dioxin (TCDD) 86, 118–21, 122, 126, 129
thermodynamics of P450 reaction 53–4, 56, 60–2, 67
thromboxane synthase (TXAS) 3
toxic activation 109–12, 126–9, *see also* carcinogens
toxins, plants 28–9; detoxification 27, 28–9, 30; xanthotoxin 28, 80, 101
triacetyloleandomycin (TAO) 134
4-trifluoromethyl-7-ethoxycoumarin 91
trimethoprim 102
Type I binding spectrum 11
Type II binding spectrum 11–12

UV absorption spectrum 1–3, 10–12, 14, 55, 133, 145–6
UV spectrophotometry 145–6

vitamin D 76, 80, 85
vitamin D receptor (VDR) 123

warfarin, bleeding complications 15

X-ray crystallography 140, 141–5, 151–2
xanthotoxin 28, 80, 101